Project Directors

Angela Hall	Emma Palmer
Robin Millar	Mary Whitehouse

Editors

Mike Kalvis	Emma Palmer
Carol Levick	Jean Scrase
Mary Whitehouse	

Authors

Michael Brimicombe	Mike Kalvis	Jean Scrase
Dariel Burdass	Carol Levick	Mary Whitehouse
Andrew Hunt	Emma Palmer	Bryan Williams

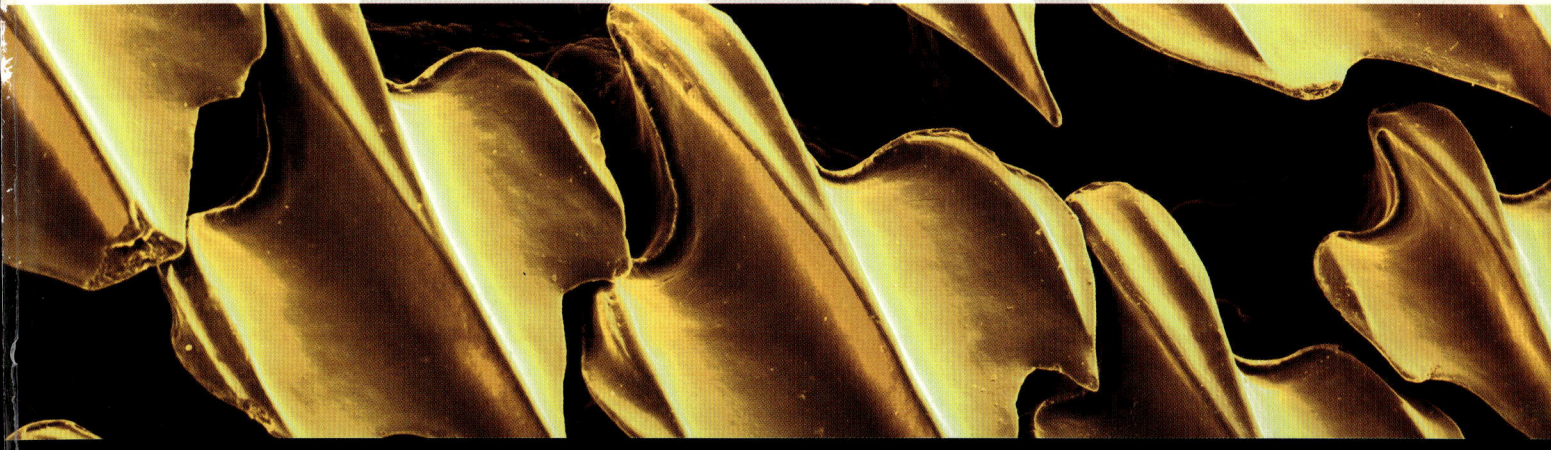

THE UNIVERSITY *of York*

THE SALTERS' INSTITUTE

Nuffield Foundation

OCR
RECOGNISING ACHIEVEMENT

OXFORD
UNIVERSITY PRESS

Official Publisher Partnership

Contents

How to use this book 4
Structure of assessment 6
Command words 7
Work-related portfolio 8
Making sense of graphs 11

A1 Sport and fitness 14

A winning combination 16
Injury treatment 19
Sport and fitness in health 20
Lifestyle, health, and fitness 22
Selling shoes 23
Healthy-lifestyle guidelines 24
The respiratory system 26
The cardiovascular system 29
Assessing health and fitness 32
Performance-enhancing drugs 33
Joints, muscles, and movement 34
Control of body temperature 36
New world records 38
Measuring body temperature 41
Measuring pulse rate 42
Measuring blood pressure 43
Measuring body mass index 44
Performing a step test 45
Preparing a blood slide 46
Using a microscope 47
Summary 48

A2 Health care 52

Road-traffic accident 54
Emergency: cardiac arrest 56
Diagnose and treat 58
Antenatal care 60
The National Health Service 62
Postnatal care 66
The female reproductive system 68
Pregnancy 69
IVF 72
Clinical tests 74
Ultrasound scanning 76
Birth 77
Measuring blood pressure 78
Testing urine 79
Summary 80

A3 Monitoring and protecting the environment 84

Cape Farewell 86
Working in environmental protection 88
The Environment Agency 90
Good laboratory practice 92
Setting standards 94
Detectives on the case 96
Evidence in colour 97
Carrying out tests 98
Analysing and evaluating results 100
Collecting samples 103
Recording visual information 105
Measuring colour 107
Turbidity 109
Good laboratory practice: weighing a sample 110
Good laboratory practice: using a pipette 111
Reading a linear scale 112
Kick sampling 113
Testing pH by colour matching 114
Measuring turbidity with a turbidity tube 115
Summary 116

A4 Scientists protecting the public 120

Solving crime 122
Dangerous dyes 124
Consumer protection 126
Law enforcement 128
Colourimetry 130
Chromatography 132
Microscopic examination 136
Electron microscopy 138
Electrophoresis 140
Bioinformation 142
Preparing a temporary slide 144
Estimating concentrations by colour matching 145
Preparing a calibration graph for a colourimeter 146
Paper chromatography 148
Summary 150

B1 Sports equipment — 154

Sports performance	156
Faster materials	158
Heat management	160
Setting standards	162
The calibration chain	164
Choosing materials	166
Mechanical properties	168
Classes of material	170
Making rigid structures	172
Matching the material to its purpose	174
Standard procedures – 1	176
Standard procedures – 2	178
Standard procedures – 3	180
Summary	182

B2 Stage and screen — 186

The lighting director	188
Safe performance	190
Managing performance	192
Making it sound right	194
Light sources	196
Filters	197
Room lighting	198
Mirror illusions	200
Recording images	202
Stage lighting	204
Refractive index	206
Amplifying sound	207
Acoustic properties	208
Sound control	210
Focal length of a converging lens	212
Measuring sound levels	213
Summary	214

B3 Agriculture, biotechnology, and food — 218

Brewing	220
Baking bread	222
Outbreak	223
Food industries	224
The dairy industry	226
Harnessing microorganisms	228
Food poisoning	229
Growing wheat for bread production	230
Rearing dairy cows	232
Yoghurt production	234
Choosing the right organisms	235
The fermentation process	236
Biotechnology in food production	238
Controlling bioreactors	240
Genetic modification	242
Keeping food safe	244
Testing the freezing point of milk	246
Testing milk quality	247
Growing yeast	248
Preserving food	250
Flaming a wire loop	251
Culturing microorganisms on agar plates	252
Making a hard cheese	254
Testing germination rates	256
Measuring crop yields	257
Summary	258

B4 Making chemical products — 262

What's in beauty products?	264
Taking your medicine	266
Designer colours	267
The chemical industry	268
Developing a new medicine	270
The work of laboratory technicians	271
Making and testing formulations	272
Safe practice	274
Acids	276
Alkalis	277
Salts	278
Chemical quantities	280
Yields	282
Controlling reaction rates	283
Choosing a synthetic route	285
Scaling up	286
A chemical language	288
Chemical formulae	290
Concentrations	291
Emulsions and suspensions	292
Making a soluble salt	294
Making an insoluble salt	296
Measuring reaction rates	297
Preparing a solution with a known concentration	298
Making an emulsion	299
Summary	300

Glossary	304
Index	310
Appendices	313

How to use this book

Welcome to Twenty First Century Science. This book has been specially written by a partnership between OCR, the University of York Science Education Group, the Nuffield Foundation Curriculum Programme, and Oxford University Press.

On these two pages you can see the types of page you will find in this book, and the features on them. Everything in the book is designed to provide you with the support you need to help you prepare for your examinations and achieve your best.

Module Openers

Why study?: This explains why what you are about to learn is useful to people who apply science in their work.

Find out about: Every module starts with a short list of the things you'll be covering.

The Science: This box summarises the science behind the module you're about to study.

What you already know: This list is a summary of the things you've already learnt that will come up again in this module. Check through them in advance and see if there is anything that you need to recap on before you get started.

Main Pages

Tabs: The coloured tabs help you to identify different sections and navigate through the book.

Questions: Use these questions to see if you've understood the topic.

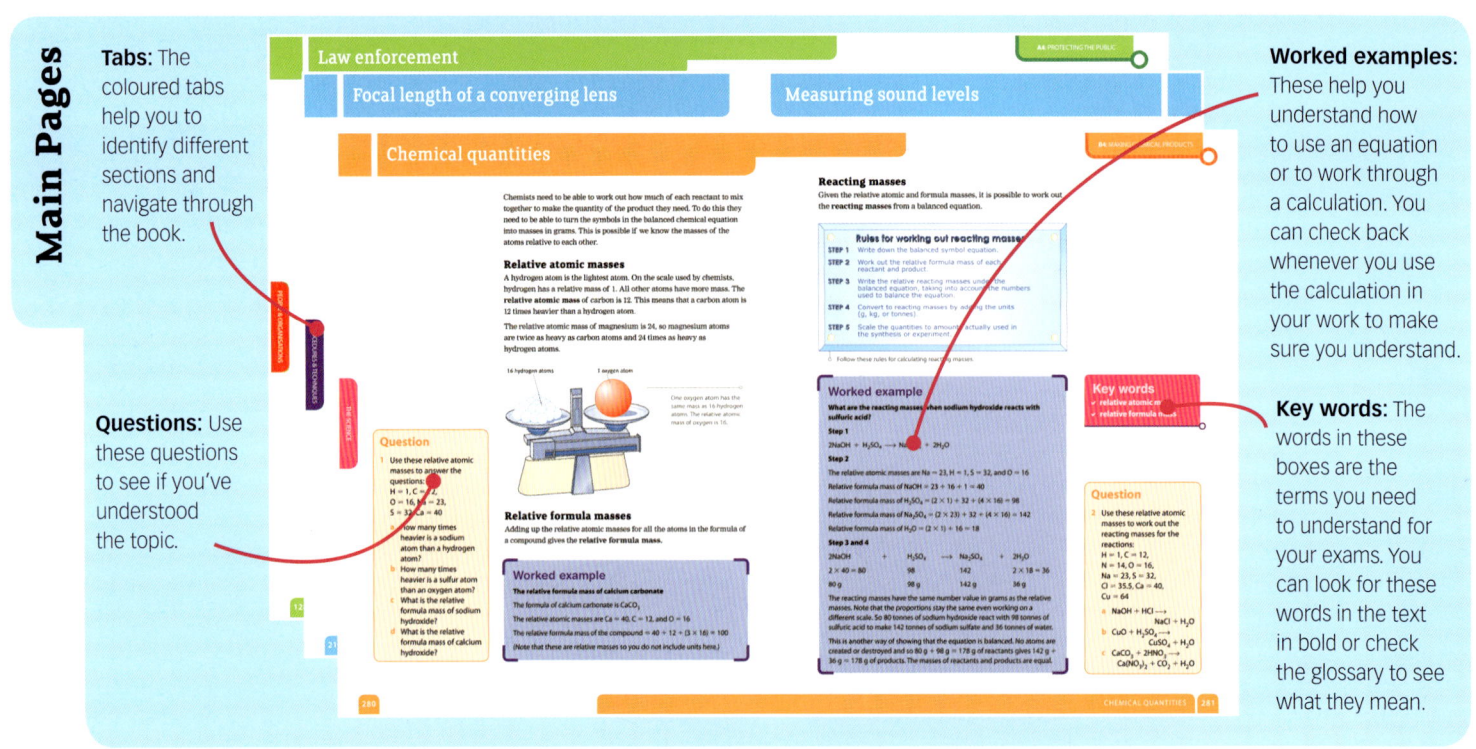

Worked examples: These help you understand how to use an equation or to work through a calculation. You can check back whenever you use the calculation in your work to make sure you understand.

Key words: The words in these boxes are the terms you need to understand for your exams. You can look for these words in the text in bold or check the glossary to see what they mean.

You should know: This is a summary of the main ideas in the unit. You can use it as a starting point for revision, to check that you know about the big ideas covered.

Review questions: You can begin to prepare for your exams by using these questions to test how well you know the topic.

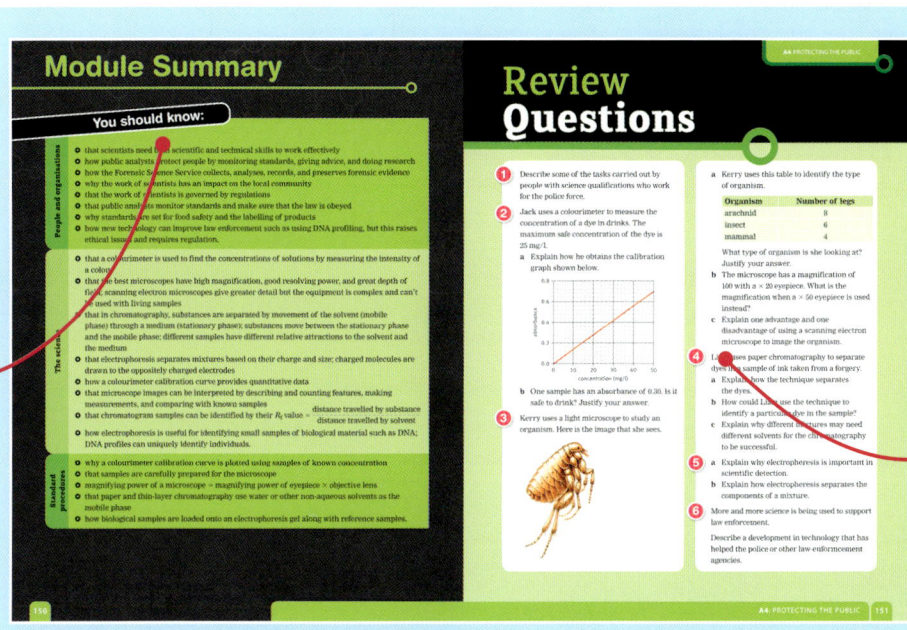

Visual summary: Another way to start revision is to use a visual summary, linking ideas together in groups so that you can see how one topic relates to another. You can use this page as a starting point for your own summary.

Structure of assessment

GCSE Additional Applied Science Assessment

The content of the topics in this book matches the topics in the specification.

The diagram below shows you which topics are in each exam paper. It also shows you how much of your final mark you will be working towards in each paper.

Unit Code	Unit Title	Topics Tested	Percentage	Type	Time	Marks Available
A191	Science in Society	A1 Sport and fitness A2 Health care A3 Monitoring and protecting the environment A4 Scientists protecting the public	20%	Written Exam	1 h	50
A192	Science of Materials and Production	B1 Sports equipment B2 Stage and screen B3 Agriculture, biotechnology, and food B4 Making chemical products	20%	Written Exam	1 h	50
A193	Science Work-Related Portfolio	Controlled Assessment • Standard procedures • Suitability test • Work-related report	60%	Portfolio	Approximately 38 h	120

Command words

The list below explains some of the common words you will see used in exam questions.

Calculate
Work out a number. You can use your calculator to help you. You may need to use an equation. The question will say if your working must be shown. (Hint: don't confuse with 'Estimate' or 'Predict'.)

Compare
Write about the similarities and differences between two things.

Describe
Write a detailed answer that covers what happens, when it happens, and where it happens. Talk about facts and characteristics. (Hint: don't confuse with 'Explain'.)

Discuss
Write about the issues related to a topic. You may need to talk about the opposing sides of a debate, and you may need to show the difference between ideas, opinions, and facts.

Estimate
Suggest an approximate (rough) value, without performing a full calculation or an accurate measurement. Don't just guess – use your knowledge of science to suggest a realistic value. (Hint: don't confuse with 'Calculate' and 'Predict'.)

Explain
Write a detailed answer that covers how and why a thing happens. Talk about mechanisms and reasons. (Hint: don't confuse with 'Describe'.)

Evaluate
You will be given some facts, data, or other kind of information. Write about the data or facts and provide your own conclusion or opinion on them.

Justify
Give some evidence or write down an explanation to tell the examiner why you gave an answer.

Outline
Give only the key facts of the topic. You may need to set out the steps of a procedure or process – make sure you write down the steps in the correct order.

Predict
Look at some data and suggest a realistic value or outcome. You may use a calculation to help. Don't guess – look at trends in the data and use your knowledge of science. (Hint: don't confuse with 'Calculate' or 'Estimate'.)

Show
Write down the details, steps, or calculations needed to prove an answer that you have given.

Suggest
Think about what you've learnt and apply it to a new situation or context. Use what you have learnt to suggest sensible answers to the question.

Write down
Give a short answer, without a supporting argument.

Top Tips

Always read exam questions carefully, even if you recognise the word used. Look at the information in the question and the number of answer lines to see how much detail the examiner is looking for.

You can use bullet points or a diagram if it helps your answer.

If a number needs units you should include them, unless the units are already given on the answer line.

GCSE Additional Applied Science Assessment

This Additional Applied Science course aims to help you:

- carry out specific scientific procedures and understand the results that matter
- apply science knowledge and techniques to solve problems
- learn about a variety of science-based workplaces
- select, organise, and communicate information clearly and logically.

You will show your progress with these skills through your work-related portfolio.

In GCSE Additional Applied Science your internal assessment counts for 60% of your total mark. Your school or college may give you details of the mark scheme for each part of it. This will help you understand how to get the most credit for your work.

Marks are given for a work-related portfolio, which includes:

- four standard procedures (12%)
- one suitability test (24%)
- one work-related report (24%).

Standard procedures

A standard procedure is a series of practical steps, often including scientific techniques, that will achieve the same result no matter who carries it out. It involves following instructions, working safely, making and recording observations, and processing the results.

You will carry out four standard procedures, each accounting for 3% of the marks. Each task will take about two hours to complete.

You will be awarded marks for:

- collecting and recording data accurately and in an appropriate format
- processing the results, using graphs or calculations specified by the procedure
- evaluating how the risks are managed
- writing a clear report using specialist terms.

Suitability test

Suitability tests are another example of how science is used in the workplace. Three types of test you might carry out are:

- finding the most suitable material for a particular purpose
- finding the most suitable procedure for a particular purpose
- testing the suitability of a device for a particular purpose.

You will carry out one suitability test, accounting for 24% of the marks. The task will take about 12 hours.

You will be awarded marks for:

Researching the purpose of the test

- Use secondary data to support your description of the purpose of the material, procedure, or device in the workplace.
- Use secondary data to support your description of the desirable properties of the material, procedure, or device.

Planning and risk assessment

- Carry out and explain a full risk assessment.
- Describe your plan, linking the procedures to the properties you are testing.
- Manage the risks successfully.
- Evaluate your risk assessment after carrying out the work.

Collecting data

- Devise and use suitable tables to record data.
- Collect data carefully across an appropriate range.
- Carry out appropriate repeats.
- Record data to a sensible degree of precision.

Processing and analysing data

- Present your data to make clear any patterns in the results, including lines of best fit and levels of uncertainty.
- Describe patterns in the data, using scientific language and taking account of uncertainty of the evidence.

Evaluating

- Evaluate the practical methods used and suggest improvements to apparatus or techniques.
- Evaluate the quality of the data, including repeatability and uncertainty.

Justifying a conclusion

- Draw a conclusion justified by the patterns in the results.
- Link the conclusion clearly to the purpose of the test.
- Discuss limitations, such as the range over which it is suitable.

Presenting your report

- Make sure that your report is laid out clearly in a sensible order.
- Make full use of scientific terms.
- Use diagrams, tables, charts, and graphs to present information.
- Take care with your spelling, grammar, and punctuation.

Work-related report

This task gives you the opportunity to find out about some science-related activity carried out in a real workplace, such as a hospital, laboratory, or factory. You present your findings in a report. Your report should focus on specific aspects of workplace practice, and make links to relevant scientific knowledge and skills from one of the eight topics.

You will produce one work-related report, accounting for 24% of the marks. The task will take about 18 hours, including, if possible, a visit to a practitioner or workplace.

You will be awarded marks for:

Collecting primary and secondary data

- Collect information from a range of sources including during a practitioner or workplace visit.
- Identify your sources of information clearly using references that are accurate, fully detailed, and dated.

Analysing the work carried out

- Analyse the importance of the roles of the practitioner to the organisation.
- Analyse the purpose of the work and its importance to the wider organisation.
- Analyse the factors influencing the location of the organisation and its impact on society.

Analysing the skills used in the workplace

- Analyse the technical skills used in the workplace.
- Analyse the expertise needed by the practitioner and explain the relevance of vocational qualifications and the personal qualities needed.

Analysing the scientific knowledge applied in the workplace

- Analyse the scientific knowledge needed and explain its importance in the work.
- Analyse the impact of two examples of financial or other regulatory factors.

Presenting your report

- Make sure your report is laid out clearly in a sensible order, including contents listing, references, and page numbers.
- Make full use of scientific terms.
- Use diagrams, tables, charts, and graphs to present information.
- Take care with your spelling, grammar, and punctuation.

Writing a work-related report
Where do I start?

Read the briefing sheet given to you by your teacher.

You might use the Internet, the school library, newspapers and magazines.

You should also gather information from specific people or organisations, interviewing a practitioner or visiting a workplace. To obtain useful information, prepare detailed questions in advance. Explain what you are doing.

Use the information selectively, picking out parts relevant to your topic and writing in your own words. Make sure you record information about the sources used.

Plan your report so it has a clear and sensible structure. Use charts, graphs, and pictures to convey information.

Tip

The best advice is 'plan ahead'. Give your work the time it needs and work steadily and evenly over the time you are given. Your deadlines will come all too quickly, especially if you have coursework to do in other subjects.

Making sense of data

Scientists use graphs and charts to present data clearly and to look for patterns in the data. You will need to plot graphs or draw charts to present data in the standard procedures and suitability test and then describe and explain what the data is showing. Examination questions may also give you a table of data or a graph and ask you to explain what the data is telling you.

Presenting data

This table shows the dates of flood events in the River Derwent area since 1890.

Year	Month	Location
1890	September	Seathwaite
1891	September	Keswick (River Derwent)
1891	August	Keswick (River Derwent)
1895	November	Keswick (River Greta)
1896	Unknown	Cockermouth
1897	December	Bassenthwaite (Derwent)
1898	November	Northern part of the Lake District, including Keswick
1900	December	Cockermouth (River Derwent)
1909	July	Cockermouth
1909	October	Cockermouth
1911	October	Keswick (River Derwent and River Greta)
1914	August	Borrowdale (River Derwent)
1918	September	Cockermouth (River Derwent)
1932	December	Cockermouth
1938	Unknown	Cockermouth
1955	Unknown	Derwentwater and Bassenthwaite
1966	Unknown	Cockermouth
1995	January	Thirlmere
2005	Unknown	Cockermouth
2009	November	Cockermouth, Keswick, Workington, Ulverston, Kendal and Burneside (River Derwent)

The graph presents the information in a different way.

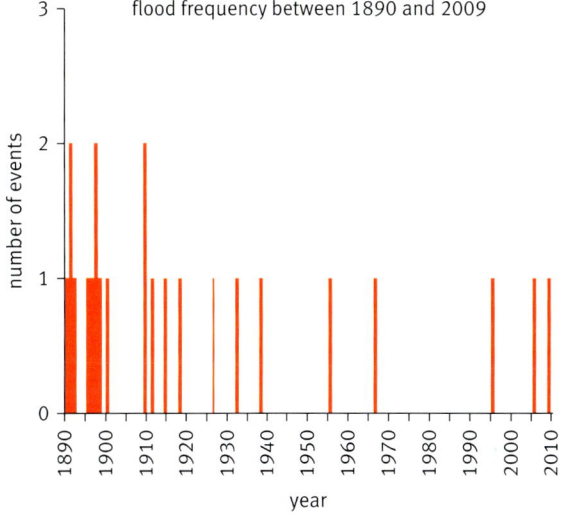

flood frequency between 1890 and 2009

Q What are the advantages of the graph over the table for presenting this information?

Q What information is easier to give in the table?

Reading the axes

Look at these two graphs. They both provide data about a world-record-breaking 100-m sprint by Usain Bolt in 2009.

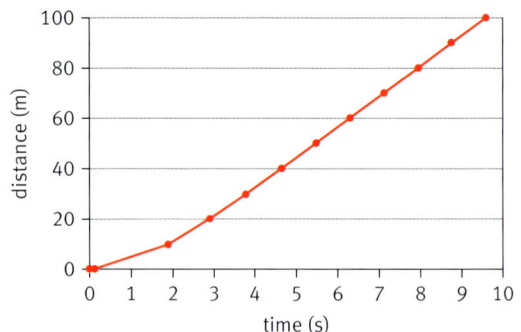

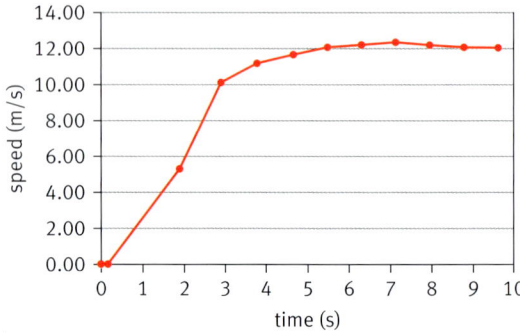

Graphs showing data from a world-record-breaking 100-m sprint by Usain Bolt in 2009.

Why are the graphs so different if they both represent information about the same race?

Look at the labels on the axes.

One shows distance travelled against time, while the other shows speed against time.

The first graph shows that after 6.0 s, Usain Bolt had travelled 56 m, while the second graph shows that at this point in the race he was travelling at 12.1 m/s.

First rule of reading graphs: read the axes and check the units.

Line of best fit

Usain Bolt is very tall. A student wanted to find out whether there is a link between height and how fast people run. She measured the height of eight students in her class and then measured the time it took for them to run 100 m. She calculated the speed of each runner and plotted her results on a graph.

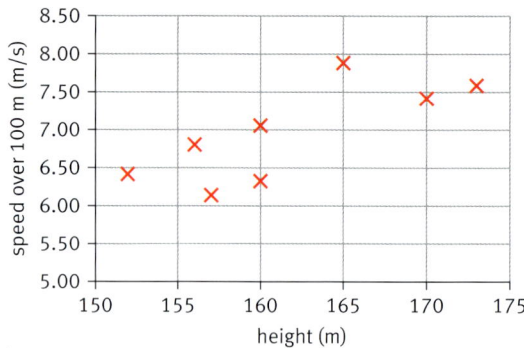

Graph to show height and speed over 100 m for eight school students.

Q Does the graph show a pattern? In other words, is there a correlation?

When there are lots of points that don't all fall on a straight line, a line of best fit helps to show if there are any patterns in the data.

To draw a line of best fit, balance the number of points above the line with the number of points below the line, and keep all the points as close to the line as you can.

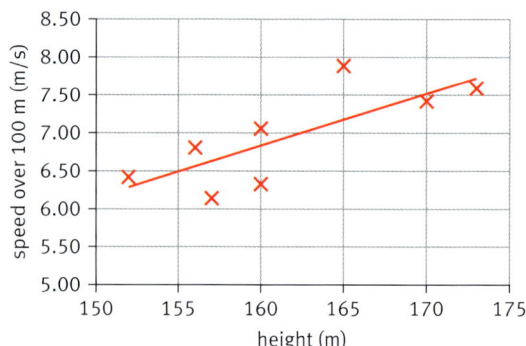

Q Did the student collect enough data to reach a conclusion? How could she improve her investigation?

Mean

The mean is the total of all the values divided by the number of data points.

Q What measurements would you need to take and what calculations you would need to do to find out the mean speed of all the students in your class over 100 m?

Range

Statistics are used to describe data. As well as the mean, a useful statistic to describe data is its range.

Class interval	Tally	Frequency
60–65			1
65–70	||||	4
70–75	ഺ ഺ ||	12
75–80	ഺ |||	8
80–85	ഺ	5
85–90			1
	Total	**31**

A data set of pulse rates from a class of 31 pupils tallied in a frequency table.

Q How would you calculate the mean of this data?

The range of this data set can be expressed as 'between x beats per minute (lowest value) and y beats per minute (highest value)'.

It could also be presented in a chart like this:

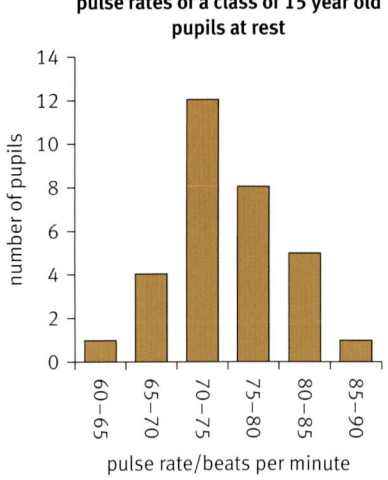

Frequency data can be shown in a bar chart.

Q If you are comparing the pulse rates of two different classes in your school, why would it be useful to have both of these statistics – mean and range?

Looking for relationships

Gradient of the graph

The gradient of the graph describes the way one variable changes relative to the other. Often the x axis is the time axis. The gradient then describes the rate of change.

Look at this graph, which shows how a population of bacteria changes over a period of time.

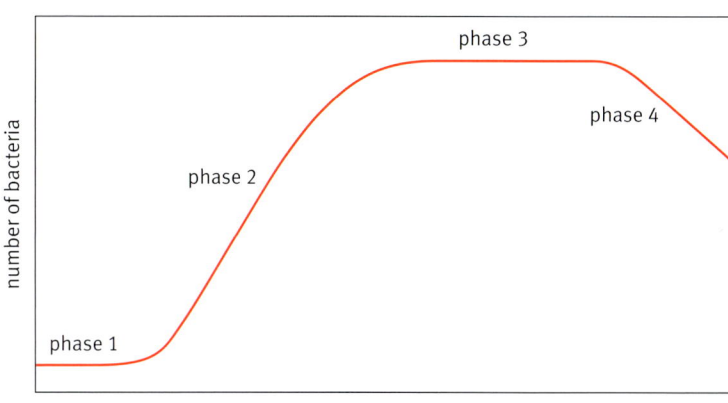

Phases of bacterial growth.

How many different gradients can you see?

There are four parts to the graph, so you should describe each phase:
- Phase 1: the number of bacteria stays constant for a time.
- Phase 2: the number of bacteria increases rapidly to a maximum.
- Phase 3: the number of bacteria again stays constant for a time.
- Phase 4: the number of bacteria decreases.

Explaining graphs

When a graph shows that there is a correlation between two sets of data, scientists try to find out if a change in one factor causes a change in the other. They use science ideas to look for an underlying mechanism to explain why two factors are related.

For example, an explanation of the graph of bacterial population growth links the changing gradient and population to science ideas about how the bacteria reproduce. In phase 1, the bacteria are making enzymes and preparing for growth. In phase 2, the bacteria are reproducing rapidly, but growing and splitting, doubling in number regularly. Find out more about phases 3 and 4 in Topic B3: Agriculture, biotechnology, and food.

A1 Sport and fitness

Why study sport and fitness?

Everyone knows that physical activity is important for health and wellbeing. Taking a walk, going for a swim, playing a game of tennis...these all make you feel good. Science explains how being active keeps you fit and healthy. Competitive sport allows people to measure themselves against one another to find who is the best. A better diet and the right training program can make all the difference between winning and losing. Scientific instruments are used to measure the performance of athletes. Science also detects cheats, for example, people who take performance-enhancing drugs to try to obtain an unfair advantage.

What you already know

- the important parts of a healthy diet
- your body reacts to changes in conditions to maintain a stable internal environment
- glucose and oxygen allow your muscles to do work.

Find out about

- how your lungs and heart get oxygen to your muscles
- how your muscles move your bones
- how your body changes to control your temperature
- tests for assessing fitness
- how to measure time and distance accurately for sporting events.

The Science

Sports scientists use their understanding of the human body to help athletes perform at their best. A healthy lifestyle involves the right amount of exercise to keep your body in shape. It also requires a balanced diet with just enough of each of the different things your body needs. Scientists understand how your lungs and heart help your muscles and bones to do their work. They also know how to help your body to recover quickly from injury.

A winning combination

Anna Bevan storms to victory

Successful athletes need good coaches to develop their strengths and skills. Here, coach Dan Forde describes how he worked with Anna Bevan to improve her speed and fitness, making her a world-class race winner.

Dan, how did your partnership with Anna Bevan start?

I coach the athletics team at the university. Anna was studying sports science. During a physiology practical her aerobic fitness (oxygen consumption during exercise) was measured. She gave excellent results.

Her lecturer asked me to do further tests on Anna. A step test and a bleep test confirmed that she was aerobically very fit. Anna had been keen on sports at school, playing tennis, netball, hockey, and even rugby. But she said her first love was athletics, even though she had not been successful in sprinting. I offered to train her, and Anna accepted.

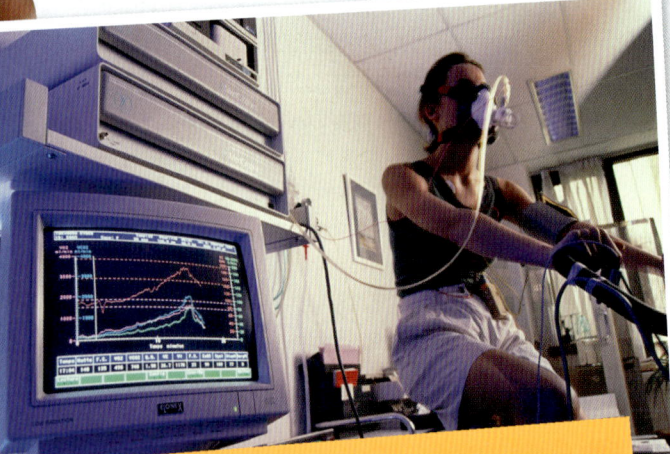

This is how Anna's aerobic fitness was tested. The equipment measures the maximum volume of oxygen she consumes in one minute as she cycles harder and harder. High values of oxygen consumption show high levels of aerobic fitness.

Did you start training with Anna immediately?

No. First we spent some time talking and got to know each other. And I had to check her health before she started any intensive training.

But surely the fitness test showed that Anna was healthy?

Not really. Health and fitness are not the same thing. Anna had performed well in the fitness tests. She had good aerobic fitness, stamina, and strength. However, she could still have a medical condition that would make it dangerous for her to train seriously. As an elite athlete, she would have to put her body under extreme physical stress.

What did the health check involve?

First Anna completed a physical activity readiness questionnaire (PAR-Q). Next, our athletics-team doctor noted down Anna's gender and age and asked her questions about her lifestyle. He then asked about her medical history: whether she was taking any medication, whether she might be pregnant, details of any previous injuries, treatments, or operations. He took blood and urine samples and measured her body temperature, resting heart rate, and blood pressure. Anna also had an **electrocardiogram** (ECG) to make sure her heart was healthy.

The doctor also measured Anna's weight and height to make a calculation of her body mass index (BMI). He took skinfold measurements to estimate her body fat content.

All of these checks helped us to design a training programme that reduced the risk of Anna getting injured.

Were there any problems?

Anna was very healthy. However, her BMI and skinfold measurements showed that her body fat was not ideal for an elite athlete. It was not unhealthy, but a little high. I referred her to a sports nutritionist who knows how foods and drinks affect sporting performance. The nutritionist gave Anna an eating plan that reduced her body fat but gave her enough energy to train.

Did you do anything else to prepare for training?

Yes. I wondered why Anna had not done well at sprints in school athletics competitions, so I asked for a muscle biopsy. The results showed

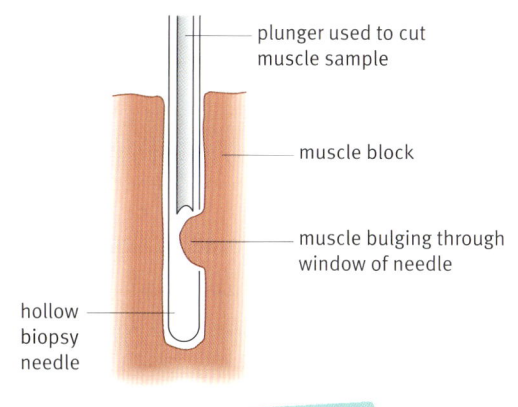

plunger used to cut muscle sample

muscle block

muscle bulging through window of needle

hollow biopsy needle

A muscle biopsy needle.

This photomicrograph of a thin slice of human muscle shows fast-twitch fibres (light) and slow-twitch fibres (dark).

that Anna's proportion of fast-twitch to slow-twitch muscle fibres was more suited to middle-distance events than sprinting. After Anna agreed to try a new distance, we put together a training programme.

What did the training programme involve?

Anna did sessions of several sprints with rest, walking, or light jogging in between. Once a week, she did one long slow run. Before each session, we measured her resting heart rate and body temperature. She also weighed herself to check she was well-hydrated. Water is lost during intensive exercise and has to be replaced before another intensive session.

We made regular videos of Anna running during the training programme. Reviewing the videos meant we could focus on aspects of her running style and link the style to her racing times.

What's been the biggest test of your working relationship?

Late last year, Anna started arriving late for training sessions. I suspected she was losing motivation. When we talked, I found out that her family were planning a trip to Australia during the running season. I asked her to think hard about what she really wanted to do and convinced her that I believed she could be a world-class runner. She talked to her family and explained the importance of this year to her running career. They decided to postpone the trip. Since then, she has always been early for training sessions, focussed on her running, and nothing stops her giving her best.

Any problems?

Halfway through her first season, Anna strained her hamstring, a muscle in her thigh. The start of a race was delayed and she got cold. When she pushed out of the blocks, her muscle tore, which was really painful. The sports physiotherapist was on hand with icepacks and a compression bandage for immediate first aid. The physiotherapist advised Anna to use the 'RICE' treatment: as much rest as possible and to keep her leg elevated most of the time, with regular 20-minute icepack treatments. It was difficult because Anna still had to go to work, which meant standing up a lot of the day, and quite a bit of walking.

The physiotherapist also said that to protect the muscle Anna shouldn't run for at least four weeks – a real blow to the training programme. Anna stayed aerobically fit by swimming (the physiotherapist said this was OK as long as she was careful not to push off on a turn). We worked on other muscles in the gym with light weights to keep her strong. Gradually Anna built up the damaged muscle with gentle stretches and careful exercise – until it was safe to start running again.

Injury treatment

These four steps are the best first aid for a muscle injury and for some joint injuries.

An athlete in training can't spend 24 hours a day resting, with a limb elevated.

But it's really important to protect an injured muscle or joint until it is completely healed. If not, an athlete risks long-term pain and permanent damage.

You can protect a damaged muscle or joint with a splint or firm bandage. You might use crutches to protect a leg injury. An athlete must avoid intensive exercise that might cause more damage before the injury is healed – even if this means planning a new training programme.

After an injury, you might need to restore aerobic fitness or muscle strength.

Exercises for fitness are usually at low or medium intensity for long sessions. This might mean jogging or running slowly, or working with moderate weights and many repetitions of a movement.

Exercises to build muscle are usually high intensity for short sessions, with rests in between. This could mean sprinting hard for 10–15 seconds, or working with heavy weights for a few repetitions.

REST means immobilising the injured part, for example, keeping the weight off a torn muscle.

COMPRESSION usually involves wrapping a bandage around the injured part to reduce swelling. The bandage should be snug but not too tight.

SIMPLE STRETCHING ROUTINES help to regain mobility, but only when swelling goes down.

ICE acts as an anaesthetic, reduces swelling, and slows the flow of blood to the injured area. To avoid damaging the tissue, the ice is applied indirectly, for example, in a tea towel or plastic bag. This is done for up to 20 minutes at a time with 30 minutes between applications.

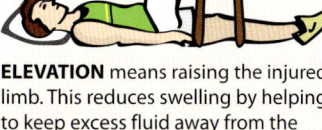

ELEVATION means raising the injured limb. This reduces swelling by helping to keep excess fluid away from the damaged area.

AEROBIC EXERCISING of the injured part is not restarted until it has regained at least 75% of the previous level of strength, and then only moderately. This exercise helps build muscle and return the athlete to peak fitness.

> Recovery from a sports injury often involves RICE followed by stretching and strengthening exercises.
> RICE stands for Rest, Ice, Compression, and Elevation.

Sport and fitness in health

PEOPLE & ORGANISATIONS

Key words
- litigation
- obese

How did you become... a personal trainer?

Personal trainers often have to work at unsocial hours, for low initial pay.

To become a qualified personal trainer you have to hold a current NVQ level 2 gym qualification and a current cardiopulminary resuscitation (CPR) qualification. You can also study for a personal trainer diploma.

You should also join the register of exercise professionals. This requires instructors to work within a code of ethical practice.

Exercise professionals must have the ability to work with clients in a way that demonstrates good practice. They must be able to:

- develop a detached yet personal relationship with the client
- make judgements when clients' statements and evidence conflict
- recognise the importance of team work
- consider the whole person, including family, workplace, and community contexts.

Improving health through sport and exercise

Regular sport and exercise have many health benefits. Sport has a social role in communities. Sport and exercise also play an important part in **rehabilitation** after illness, or injury. Local authorities provide facilities for sport and recreation in the community, such as sports and leisure centres and playing fields. They have to balance the need for sports amenities with the need to control local taxes. Some employers, sports clubs, privately run gyms, and health centres also provide facilities for sport.

Personal trainers

A personal trainer is a fitness professional who teaches people how to exercise correctly, lose weight, develop physical strength, and adopt a healthier lifestyle.

They work with their clients to help them reach their personal goals, by developing nutritional guides and individual exercise programmes. An exercise programme might include aerobic fitness exercises such as jogging or swimming, muscle-building exercises using weights, and flexibility training.

How did you become ... a coach?

A good coach is enthusiastic, cares about people, and can communicate well.

Each sport has a national governing body with its own coaching qualifications. You must have one of these qualifications to coach in that sport. Although there are many jobs available for qualified coaches, most coaches work part-time or as volunteers.

The Fitness Industry Association (FIA) has a Code of Practice that sets standards for its member health clubs and gyms. The code is designed to ensure that customers are provided with a safe environment in which to exercise. Together with the UK Government, the FIA has also developed a list of qualified fitness professionals, along with a list of the clubs that employ them.

Sports and leisure centres must make sure they comply with government regulations to protect the health and safety of employees and clients. For example, the Health and Safety at Work Act 1974 requires employers to:

- use risk assessments to identify hazards to employees and clients, such as gym equipment and swimming-pool chemicals
- take reasonable actions to remove hazards or reduce risks to a safe level
- record accidents and injuries in an accident book – serious injuries or death must be reported to the Health and Safety Executive for investigation.

Risk assessments reduce the risk of clients and employees getting injured. They also protect practitioners and organisations from legal proceedings (**litigation**) should anything go wrong.

A sports coach at work.

Physiotherapy

If you are injured as a result of sport or exercise you might visit a sports physiotherapist. These specialist physiotherapists are trained to return you to full fitness as quickly, but as safely, as possible.

How did you become ... a sports physiotherapist?

Physiotherapists use physical methods to treat injury, and to prevent and relieve pain. The work is quite strenuous.

To become a physiotherapist you need five GCSEs including two sciences, three A levels, a relevant degree, and you need to become a Chartered Physiotherapist.

To specialise as a sports physiotherapist you do a special postgraduate course. You also need practical experience.

Questions

1 What qualifications do you need to be a personal trainer?

2 Give two personal qualities that a good coach should have. For each one, say why you think these qualities are important.

3 Name an independent organisation that cooperates with the UK Government to improve health through sport.

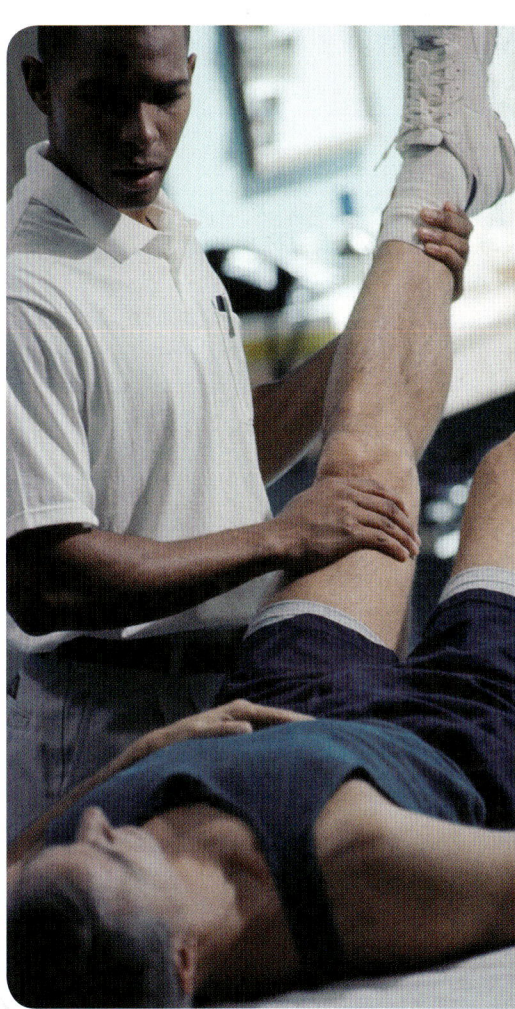
A physiotherapist treating an injury.

People from many organisations work together to achieve common goals, such as better health and longer life expectancy.

In the UK two major causes of premature death are **cancer** and **coronary heart disease (CHD)**. It has been estimated that CHD alone costs the UK economy £9 billion a year. Another growing problem is obesity, particularly in children. Young people today are becoming less active and the problem is made worse by the popularity of high-fat convenience foods. People's chances of suffering from CHD, some cancers, and obesity are affected by their lifestyle. People can improve their lifestyle if they:

- eat a more healthy diet
- exercise more
- drink less alcohol
- stop smoking
- manage their responses to stress.

Many organisations work to encourage people to follow a healthy lifestyle, for example:

- The British Heart Foundation (BHF) provides support and information for heart patients and their families. It funds research into heart disease and educates the public and health professionals about heart issues.
- The Food Advertising Unit (FAU) is a centre for information, communication, and research in the area of food advertising, particularly TV advertising to children.

Questions

1 List three health problems that may be associated with lifestyle choices people make.

2 Why do you think it is important to monitor and research the effect of food advertising on children?

The UK has one of the highest rates of death from CHD in the world. A poor diet, high in saturated fats, may contribute to this. Over 120 000 UK deaths a year are caused by CHD.

Selling shoes

A serious athlete should never choose running shoes on looks alone. However fashionable and comfortable, the wrong choice can result in lower performance on the track. In some cases, the wrong choice can even result in injury.

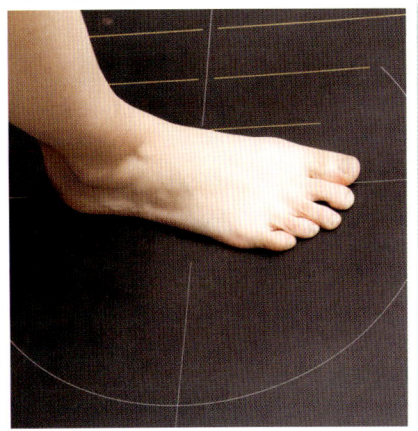
Measuring the shape of a sole with a pressure-sensitive surface.

Trying out trainers on a treadmill.

They may look and feel good, but are they going to improve your performance on the track?

Video analysis

Some shops use the latest technology to make sure their customers choose the correct shoe.

These include:

- pressure-sensitive surfaces that precisely map the shape of the sole
- treadmills for customers to put shoes through their paces
- video cameras to record running action and provide customers with still pictures
- computers that analyse video data for performance likely to lead to injury.

The technology is not cheap, but a satisfied customer is more likely to come back again. Trainers can wear out in months, so shops that invest in the right equipment can make more money.

Questions

1 Describe the technology installed in shops that helps customers choose the correct shoes.

2 Give two reasons why athletes need to choose their shoes carefully.

Healthy-lifestyle guidelines

Regular exercise helps to keep you healthy and can be fun too.

Aerobic exercise

To improve health and fitness, people should take **aerobic exercise** that increases the heart rate. They should exercise for at least 15 minutes, three times a week. Examples include riding a bicycle, walking briskly, jogging, playing football, aerobics, dance classes, skateboarding, and many more. Sport and exercise help to:

- reduce body fat
- strengthen bones
- improve coordination, balance, and flexibility
- improve stamina and concentration
- fight depression and anxiety.

Fruit and vegetables

Fresh fruit and vegetables contain important minerals and vitamins. These improve general health. **Antioxidants** may help to prevent cancer.

Fruit and vegetables are also an important source of fibre in the diet, reducing the risk of diseases of the gut.

Saturated fat

Many 'fast foods' contain high levels of saturated fat and cholesterol (a white waxy substance). These may increase the risk of CHD. Cholesterol can build up in the arteries, making them narrower or even blocking them. If the coronary artery becomes blocked, a heart attack may result.

Salt

High levels of salt in the diet can cause raised blood pressure, which in turn can lead to heart disease.

Sugar

A diet high in sugar can cause tooth decay. Most sugary foods are also energy rich. Consuming more energy than you need leads to an increase in body weight.

People who are 30% overweight are said to be **obese**. Obesity puts an increased strain on the joints and the heart, and increases blood pressure. It is associated with a higher risk of **diabetes**.

Do you eat enough fruit and vegetables?

Smoking

Smoking causes lung diseases such as lung cancer, bronchitis, and **emphysema**. Smoking also causes narrowing of the arteries. This leads to high blood pressure and an increased risk of CHD.

Drug use

Many substances, both legal and illegal, can be bad for the body. **Drugs** have to be broken down by the liver or passed out in the urine by the kidneys, so drug abuse can damage the liver and kidneys.

Drug abuse can also cause alterations in the brain, leading to dependency and depression, and changes in behaviour.

Excess alcohol

Drinking too much alcohol can lead to:

- alcoholism
- liver damage
- increased aggression and violence
- cancers of the liver, oesophagus, and mouth
- increased risk of **stroke** (a blood clot forming in a blood vessel in the brain).

For an adult female, drinking more than 14 units per week is bad for her health. Adult males should drink no more than 21 units per week.

Stress

Stress can cause:

- physical symptoms, such as high blood pressure, chest pain, and weight loss
- psychological symptoms, such as moodiness and depression.

You can reduce how stress affects your health by:

- exercising
- reducing stimulant intake, such as cola, coffee, and tea
- talking to someone about how you feel
- making time to have some fun
- rewarding yourself – recognise your achievements and feel good about them
- getting regular sleep.

A unit of alcohol is $10\,cm^3$ of pure alcohol. Counting units of alcohol can help you keep track of the amount you're drinking.

2 units — pint of ordinary-strength lager

3 units — pint of strong lager

2 units — pint of bitter

2 units — pint of ordinary-strength cider

2 units — $175\,cm^3$ glass of red or white wine

1 unit — pub measure of spirits

1.5 units — alcopop

Key words

✔ **aerobic exercise**
✔ **obese**

Questions

1 Explain in your own words why it is sensible to:
   a eat five portions of fruit and vegetables a day
   b have a low intake of salt
   c drink only moderate amounts of alcohol
   d not smoke tobacco
   e take regular aerobic exercise.

2 Explain how stress can adversely affect health.

The respiratory system

Key words

- breathing system
- respiratory system
- lungs
- trachea
- bronchus (bronchi)
- bronchioles
- alveolus (alveoli)

All human cells carry out respiration to provide them with energy. Respiration uses glucose from your food and oxygen from the air. Carbon dioxide and lactic acid are produced as waste products. The **respiratory system** or human **breathing system** supplies oxygen to the body, and removes carbon dioxide. This exchange of oxygen for carbon dioxide is called gas exchange. Gas exchange happens in the **lungs** where the gases can move from air to blood or blood to air.

The airways

The **airways** are a system of tubes carrying air deep into the lungs.

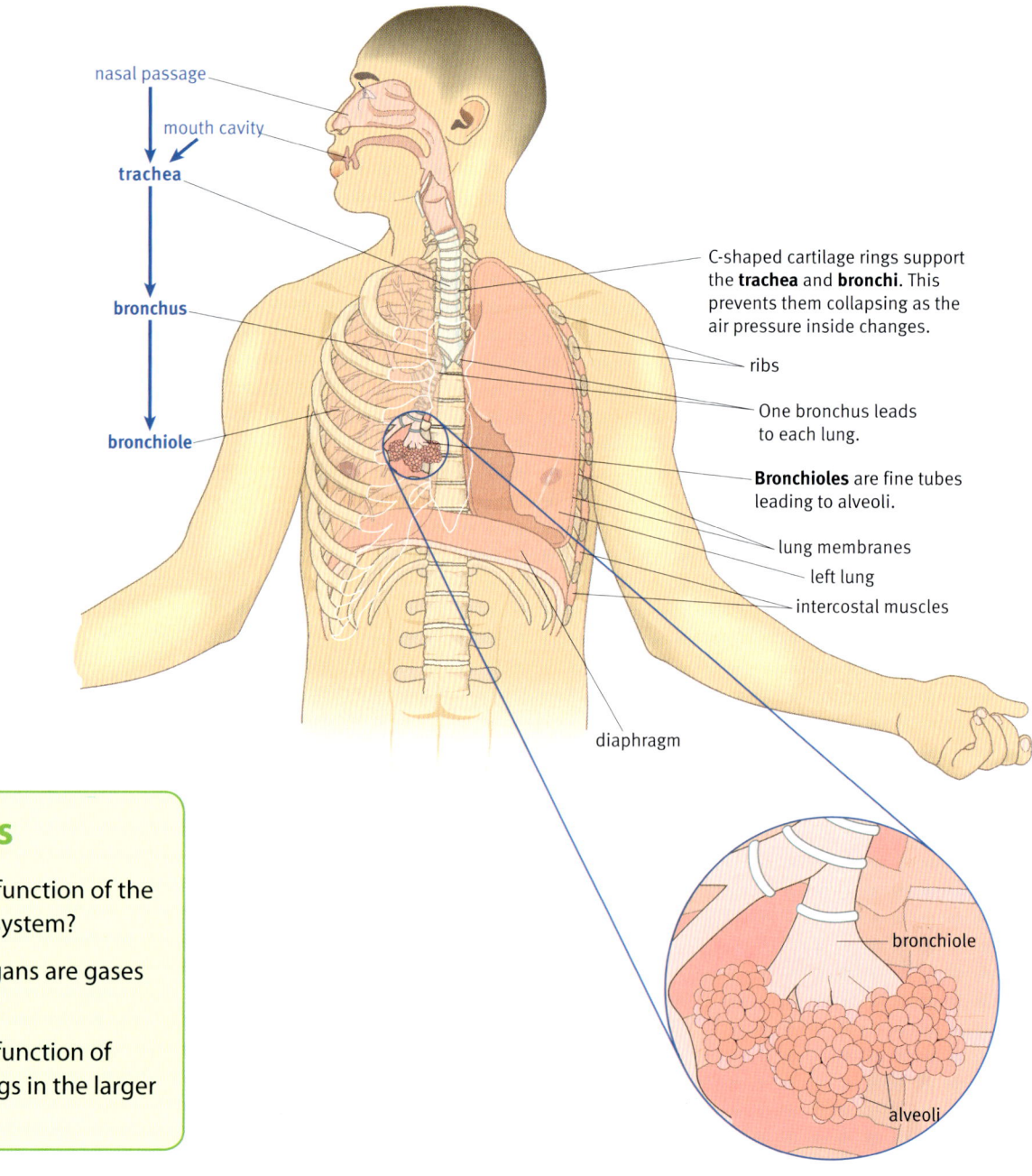

nasal passage

mouth cavity

trachea

bronchus

bronchiole

C-shaped cartilage rings support the **trachea** and **bronchi**. This prevents them collapsing as the air pressure inside changes.

ribs

One bronchus leads to each lung.

Bronchioles are fine tubes leading to alveoli.

lung membranes

left lung

intercostal muscles

diaphragm

bronchiole

alveoli

THE SCIENCE

Questions

1 What is the function of the respiratory system?

2 In which organs are gases exchanged?

3 Explain the function of cartilage rings in the larger airways.

Gas exchange in the alveoli

Gas exchange takes place in the **alveoli**. These are tiny air sacs surrounded by fine blood capillaries.

- Oxygen passes from the air in the alveoli into the blood.
- Carbon dioxide passes from the blood into the air in the alveoli.

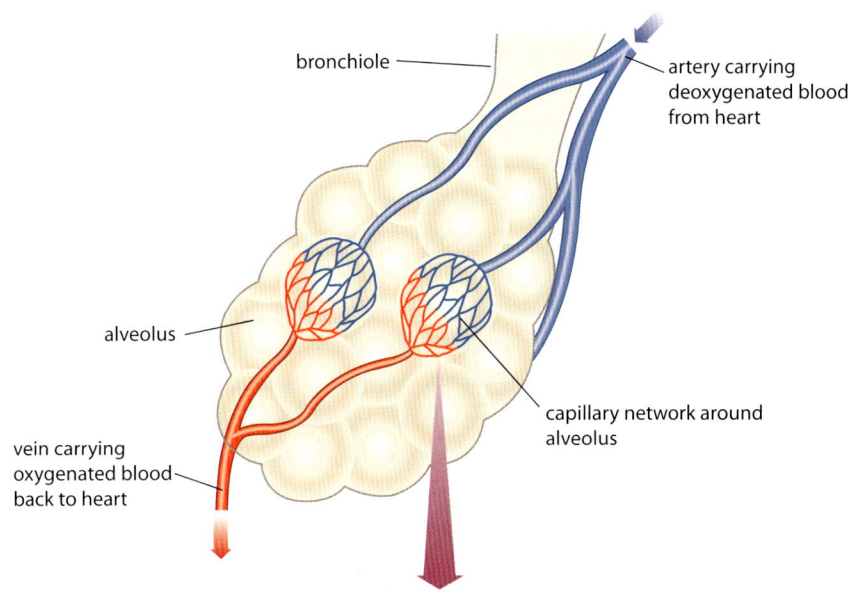

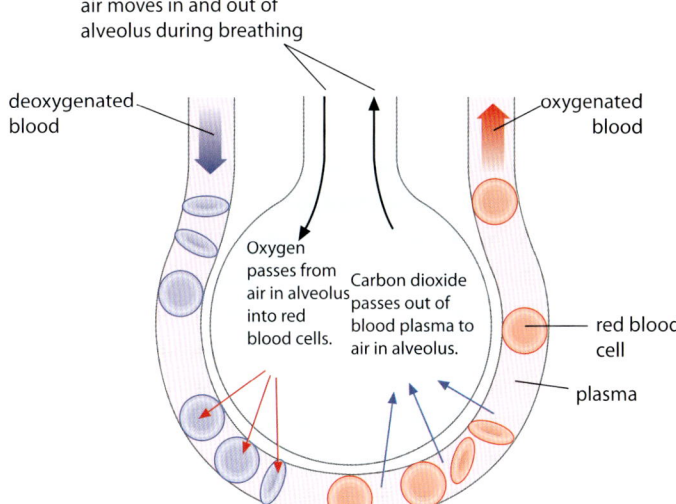

The circulatory system then carries the oxygen to all the body cells for respiration. It also collects their carbon dioxide. The alveoli and capillaries are well suited to their function of gas exchange:

- There are many millions of alveoli, together making a huge surface area for gas exchange.
- The walls of the alveoli and the blood capillaries are very thin, and so gases can pass through.
- The distance between the air in the lungs and the blood is very small.

Questions

4. In which direction do the following gases move in the alveoli, air to blood or blood to air?
   a carbon dioxide
   b oxygen

5. Describe the path of a molecule of oxygen from the air entering the nose, moving through the airways, and finishing in the blood. List the structures through which it passes.

6. List three ways in which the gas-exchange system is adapted to its function.

Ventilating the lungs

Blood continually flows through the capillaries surrounding the alveoli, carrying away oxygen and bringing carbon dioxide to the lungs. The air in the alveoli is also constantly replaced, supplying more oxygen and removing carbon dioxide.

The air is replaced by breathing or **ventilation**. This is the rhythmic movement of the lungs, taking air in (inhalation) and pushing it out (exhalation).

Breathing in – inhalation

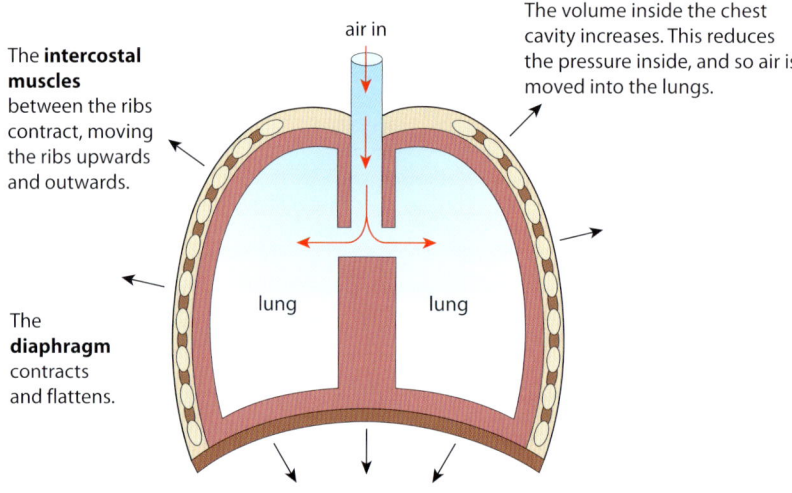

The **intercostal muscles** between the ribs contract, moving the ribs upwards and outwards.

air in

The volume inside the chest cavity increases. This reduces the pressure inside, and so air is moved into the lungs.

lung   lung

The **diaphragm** contracts and flattens.

Breathing out – exhalation

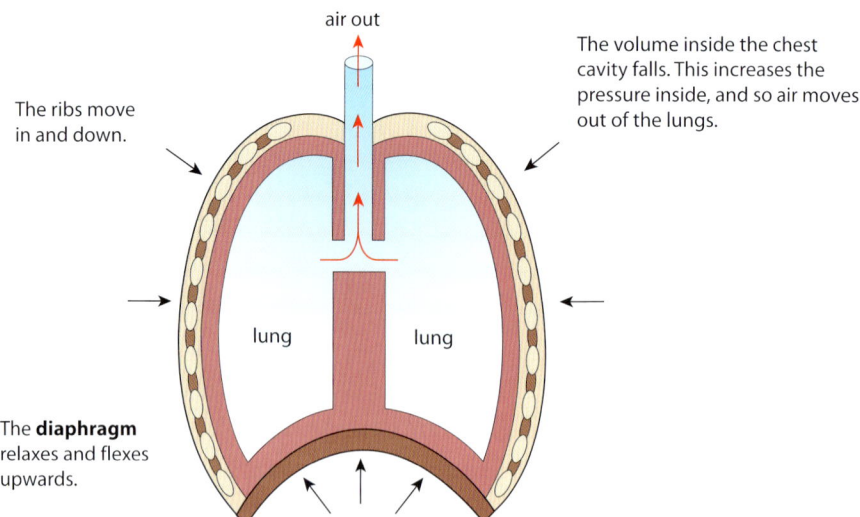

air out

The ribs move in and down.

The volume inside the chest cavity falls. This increases the pressure inside, and so air moves out of the lungs.

lung   lung

The **diaphragm** relaxes and flexes upwards.

One complete breath is an inhalation, an exhalation, and a pause. The **breathing rate** is the number of complete breaths in one minute.

At rest, most people breathe 16–18 times a minute. During exercise breathing becomes deeper and faster.

Questions

7 Explain in your own words how the lungs are inflated and deflated.

8 Describe the effect on gas exchange if ventilation stopped. How would the air in the alveoli be different?

9 Why does the breathing rate need to increase during exercise?

The cardiovascular system

The **cardiovascular system** carries substances around the body. It has three parts:

- blood
- blood vessels
- the heart.

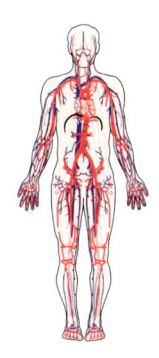

Blood

You have 5–7 litres of blood circulating in your body.

Blood:

- carries oxygen and nutrients such as glucose to all your tissues including muscles for respiration
- removes the waste products of respiration such as carbon dioxide and lactic acid, and other wastes such as urea
- transports hormones
- helps to regulate the body's temperature and water content.

Looking at blood under the microscope shows some of its structures.

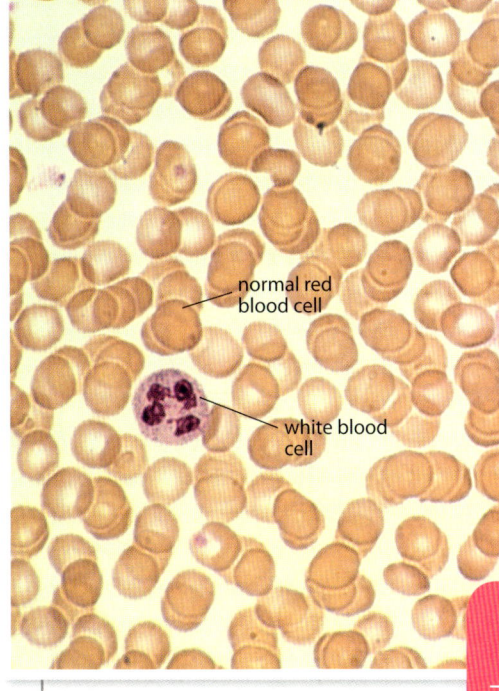

normal red blood cell

white blood cell

Photomicrograph of red and white blood cells from a healthy person. (Magnification × 1000.)

Plasma is a pale-yellow watery fluid containing dissolved chemicals including nutrients such as glucose, and waste products such as carbon dioxide, lactic acid, and urea.

Red blood cells contain the red pigment **haemoglobin**, which carries oxygen. Their biconcave shape increases their surface area, making gas exchange more efficient. They have no nucleus. Red blood cells also carry the waste product carbon dioxide.

White blood cells vary in shapes and have a nucleus. They defend the body against infection. You have more white blood cells in your blood when you have an infection.

Platelets are small fragments of cells involved in blood clotting. Clotting is important in stopping the flow of blood after an injury.

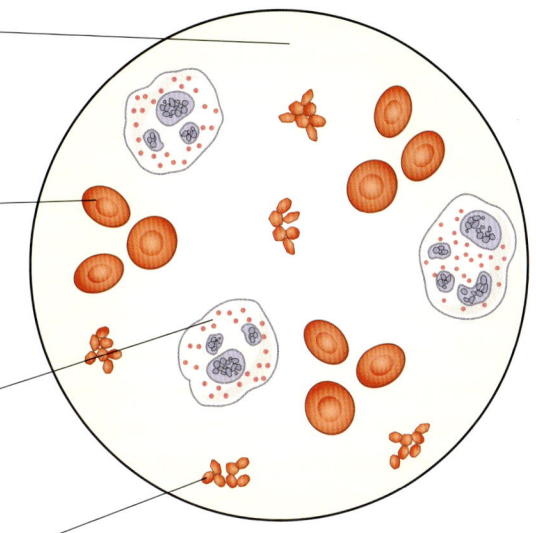

Questions

1 Name two substances that are transported around the body in blood.

2 How do red blood cells and white blood cells differ in:
   a   their appearance?
   b   their function?

Key words

- ✔ **red blood cells**
- ✔ **white blood cells**
- ✔ **platelets**
- ✔ **cardiovascular system**

Blood vessels

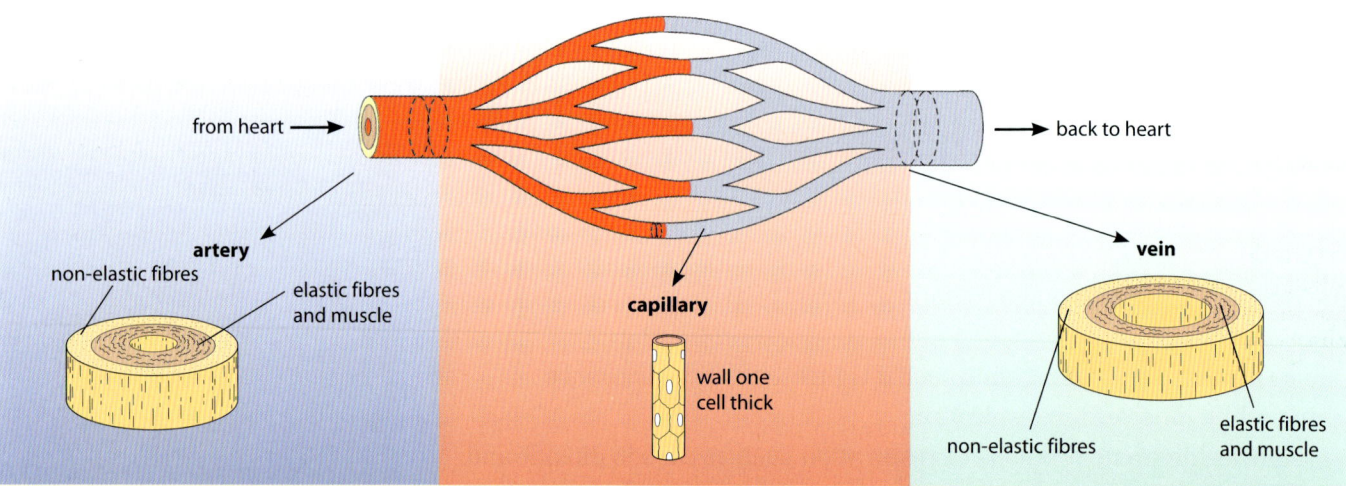

Arteries carry blood away from the heart. They have thick muscular walls that cope with the high pressure of blood as it leaves the heart. Arteries divide into capillaries.

Capillaries branch and spread through the body taking blood to every cell. The walls are very thin and leaky. Plasma carrying oxygen, nutrients, carbon dioxide, and water can pass out to the cells. Capillaries join up to form veins.

Veins carry blood back to the heart. They have thinner walls and are wider inside than arteries. Blood pressure is low. When you move around, your body muscles help push the blood along veins. Veins have valves to stop the blood flowing backwards.

THE SCIENCE

Key words
- ✔ **artery**
- ✔ **capillary**
- ✔ **vein**

The blood vessels form a continuous system around the body. They carry blood through the heart twice on each complete trip around the body. Blood goes:
- to the lungs to pick up oxygen
- back to the heart for a pressure boost
- around the body to deliver oxygen
- back to the heart, and then the whole cycle repeats.

Questions

3   Make a table to compare the three types of blood vessel.

4   Arteries have thicker walls than veins. Why is this important?

5   The human circulation is called a double circulation. Why do you think this is?

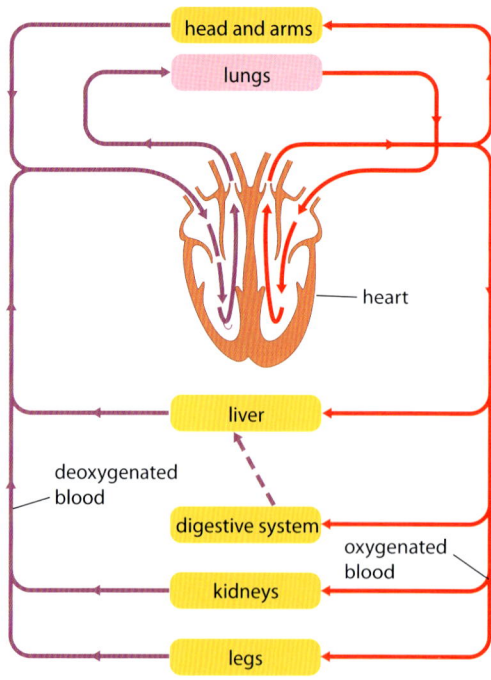

The total length of the blood vessels is about 150 000 km. End to end they would stretch $3\frac{1}{2}$ times around the world.

The heart

The heart is the pump that makes blood circulate. The heart beats automatically but the rate varies with the body's level of stress and activity.

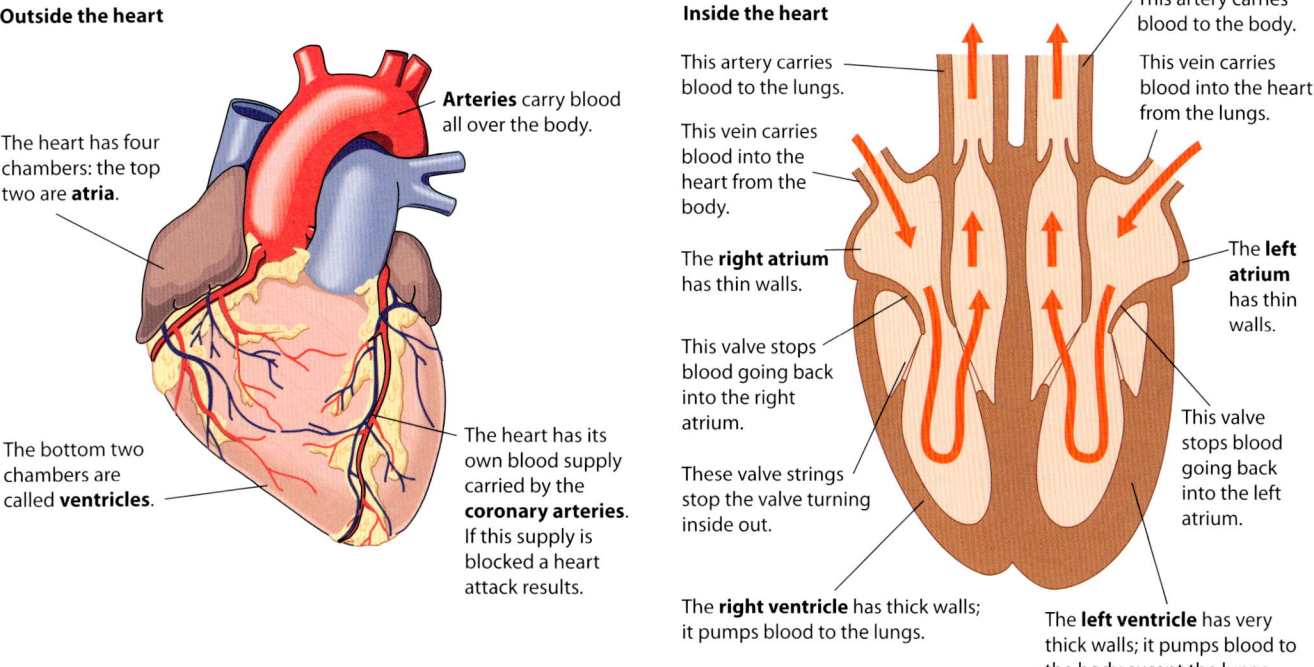

Outside the heart

Arteries carry blood all over the body.

The heart has four chambers: the top two are **atria**.

The bottom two chambers are called **ventricles**.

The heart has its own blood supply carried by the **coronary arteries**. If this supply is blocked a heart attack results.

Inside the heart

This artery carries blood to the lungs.

This vein carries blood into the heart from the body.

The **right atrium** has thin walls.

This valve stops blood going back into the right atrium.

These valve strings stop the valve turning inside out.

The **right ventricle** has thick walls; it pumps blood to the lungs.

This artery carries blood to the body.

This vein carries blood into the heart from the lungs.

The **left atrium** has thin walls.

This valve stops blood going back into the left atrium.

The **left ventricle** has very thick walls; it pumps blood to the body except the lungs.

Blood pressure and pulse

The **blood pressure** is the force of blood per unit area as it flows through the blood vessels.

- The **systolic blood pressure** is the highest pressure, when the left ventricle contracts to pump blood into the arteries.
- The **diastolic blood pressure** is lower. It is the pressure in the arteries when the heart is relaxed and filling with blood.

As the left ventricle contracts, surges in blood pressure cause the arteries to expand and contract. You can feel this as a **pulse** in the major arteries.

Questions

6 Look at the cross section of the heart here and the circulation diagram on page 30. Describe the path of the blood into and out of:
   a the right side of the heart
   b the left side of the heart.

7 What are the two readings quoted for blood pressure? Explain the difference between them.

Key words

✓ atrium (atria)
✓ ventricle
✓ blood pressure
✓ pulse

Assessing health and fitness

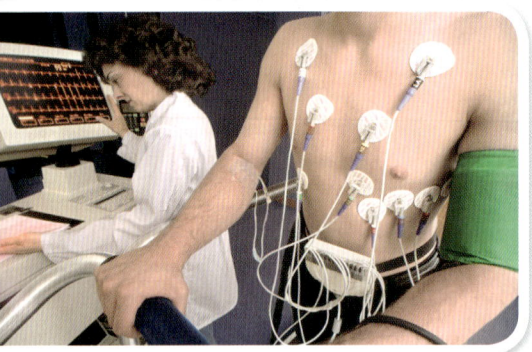

To take an ECG, electrodes are placed on the chest and limbs. The screen shows how the heart responds as the subject exercises.

Blood pressure and pulse rate

A normal value for blood pressure is about 120/80 mm Hg. A raised blood pressure can be a sign of kidney, heart, or circulatory disease.

A normal resting pulse rate is about 60–100 beats per minute. The pulse rate may be raised during exercise, or because of a health problem. A low pulse rate may be a sign of low blood pressure, or a heart disorder.

Regular aerobic exercise lowers the resting pulse rate, and also helps lower the blood pressure.

An **electrocardiogram** (ECG) shows the electrical changes in the heart as it beats. ECGs also show any abnormalities in the heartbeat.

Measuring body fat: BMI and skinfold measurements

The **body mass index (BMI)** is a calculation to show whether someone is underweight, overweight, or at a desirable weight. BMI can also be read from a chart. BMIs are used to produce health guidelines. However, BMI tables assume that everyone with a high BMI has a high level of body fat and is therefore unhealthy. This may not always be true, for example, body-builders with very large muscles may have a high BMI but a low body-fat content.

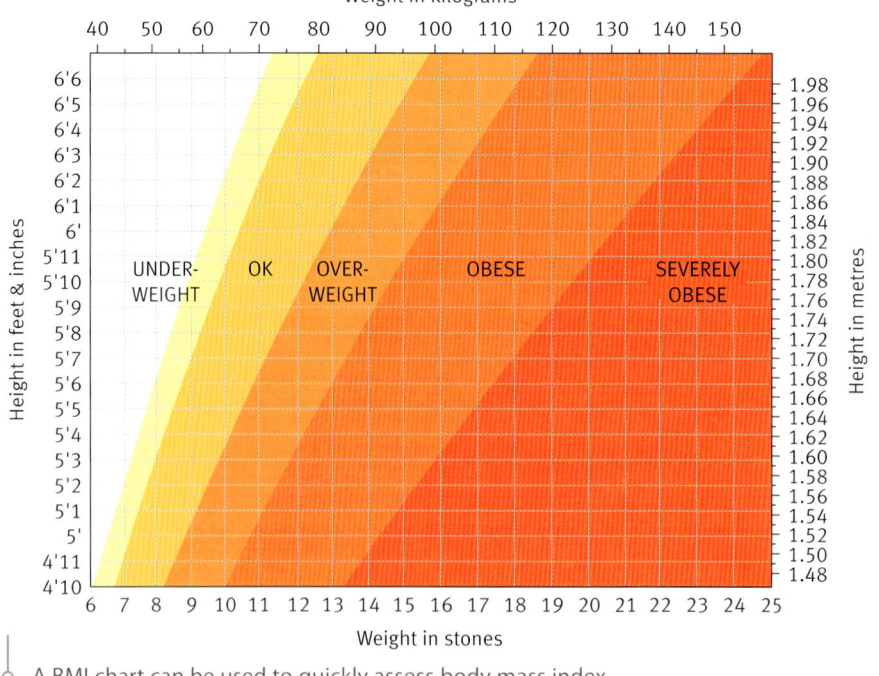

A BMI chart can be used to quickly assess body mass index.

Special calipers are used to take skinfold measurements.

Folds in the skin are thicker when there is more fat in the body. **Skinfold measurements** are another method used to estimate the level of body fat.

Performance-enhancing drugs

A small number of athletes cheat by taking drugs to improve their performance.

Drug	Benefit to athlete	Danger to athlete
anabolic steroids	promotes growth of muscle, making them bigger and stronger	causes liver damage, mood swings, and depression
amphetamines	stimulates the athlete, making them more alert and aggressive, increasing heart and pulse rates	raises blood pressure and can result in shaking
alcohol	relaxes the athlete, reducing their stress levels before competing	impairs mental function, affecting their judgement
cortisone	reduces muscle pain, so masks the effect of an injury	can weaken bones and muscles

All drugs leave traces in the body. Tests on blood, urine, and sometimes hair samples reveal these tell-tale signs. Athletes give samples before and after important races. Over the years, many athletes have been suspended or disqualified after testing positive for performance-enhancing drugs.

Making urine

The **kidneys** have three main functions.

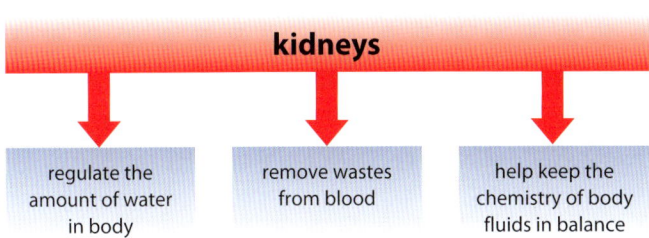

kidneys
- regulate the amount of water in body
- remove wastes from blood
- help keep the chemistry of body fluids in balance

Blood passes through the kidneys and is filtered. Useful substances such as protein and glucose are retained in the blood. Excess water and waste leave the body in **urine**.

Marion Jones winning the 200-m sprint in the Sydney 2000 Olympics. She won five medals altogether, three of them gold. All were taken away from her when she tested positive for steroids.

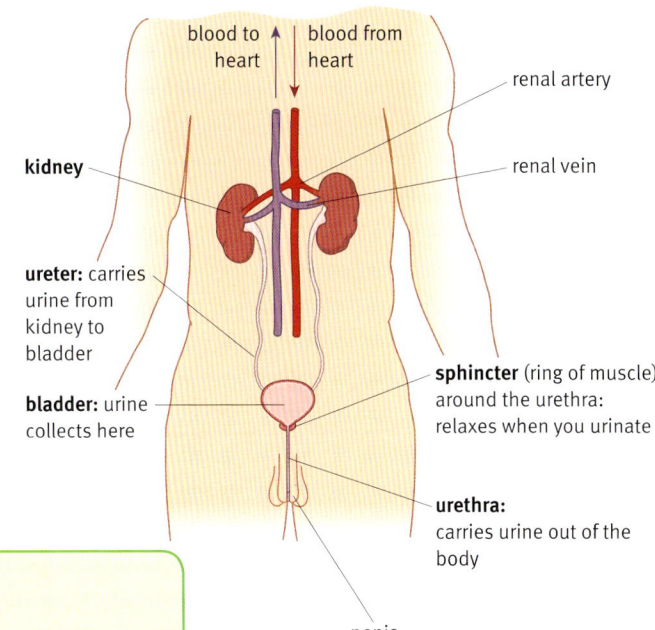

blood to heart
blood from heart
renal artery
kidney
renal vein
ureter: carries urine from kidney to bladder
sphincter (ring of muscle) around the urethra: relaxes when you urinate
bladder: urine collects here
urethra: carries urine out of the body
penis

Questions

1 Name two performance-enhancing drugs. Describe why an athlete might be tempted to use them and the problems they might cause.

2 How are athletes tested to find out if they have taken performance-enhancing drugs?

3 What are the functions of the kidneys?

Key words
- ✓ kidneys
- ✓ urine

Where two or more **bones** meet you have a **joint**. Some joints are wrapped in a layer of special tissue that makes and holds fluid around and between the bones. This synovial fluid protects and lubricates the joint.

Diagram of the knee.

muscle

tendon: a tough band of inelastic tissue attaching muscle to bone

femur

patella

joint capsule

ligaments: bands of tough elastic tissue holding bones to each other

synovial fluid: this lubricates and nourishes the tissues in the joint capsule

cartilage: a smooth protective surface covers the bone ends, providing easy movement

synovial membrane: this tissue lines the joint capsule and secretes synovial fluid

Different types of joint allow different sorts of movement. Joints such as the knee and elbow work like a hinge and can only move at certain angles. In joints like the hip and shoulder, a ball fits into a socket. This means the joint can move into a wider range of positions.

How muscles move bones

Skeletal **muscles** can be moved at will. A muscle contracts to pull on a bone and move it at a joint.

Muscles can only pull. They pull by contracting and getting shorter. Once they relax, they can be stretched again. This happens when another muscle contracts and pulls them. So at least two muscles are working at every joint:

- One contracts to bend the joint.
- The other contracts to straighten the joint.

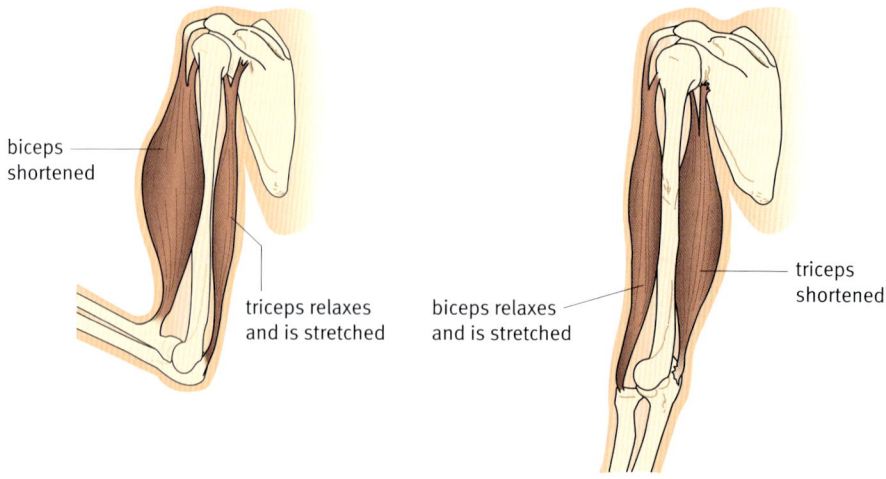

biceps shortened

triceps relaxes and is stretched

biceps relaxes and is stretched

triceps shortened

Muscle forces

Jim's personal trainer has developed a training programme for him that includes muscle-building exercises. He trains by lifting 20 kg dumbbells. Each dumbbell weighs 200 N so it pulls the end of Jim's arm down with a force of 200 N. Jim's biceps muscle has to balance this with an upwards force on the arm. However, the force from the biceps muscle is applied close to the elbow joint, so it has to be 2000 N. This is 10 times bigger at this angle.

This shows the forces on Jim's arm while he is training.

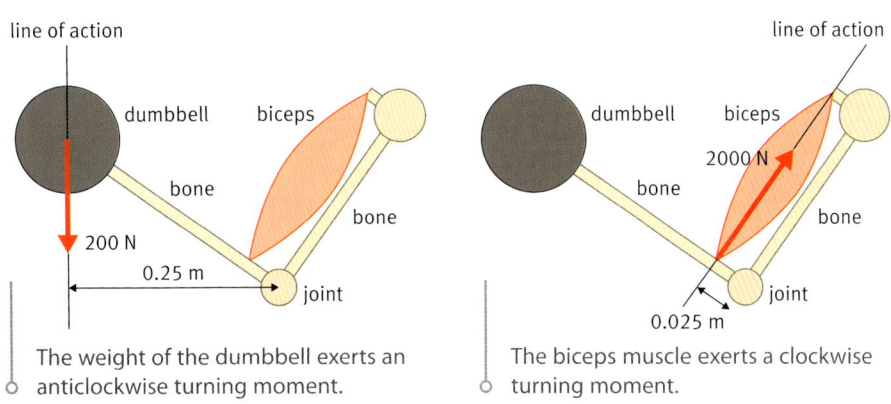

The weight of the dumbbell exerts an anticlockwise turning moment.

The biceps muscle exerts a clockwise turning moment.

The diagrams above model what is going on. The weight of the dumbbell is turning the forearm bone anticlockwise. The line of action of the downwards 200 N force is 0.25 m away from the joint, so the anticlockwise turning moment is 200 N × 0.25 m = 50 Nm.

The biceps muscle balances this out by trying to turn the forearm bone clockwise. Its line of action is only 0.025 m from the centre of the joint.

To hold the dumbbell in this position, the clockwise and anticlockwise moments must cancel each other out. This happens when the muscle force is 2000 N. The clockwise turning moment is then 2000 N × 0.025 m = 50 Nm.

Questions

1. What is the difference between a tendon and a ligament?
2. Name the parts of a synovial joint and explain their functions.
3. Explain why skeletal muscles always occur in pairs.
4. What is the rule for calculating a turning moment?

Key words
- ✔ bone
- ✔ joint
- ✔ muscle
- ✔ tendon
- ✔ ligament
- ✔ cartilage

Control of body temperature

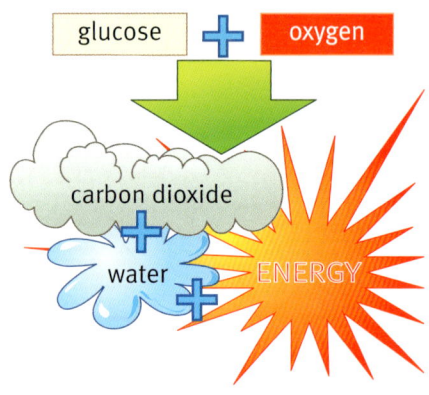

glucose + oxygen

carbon dioxide + water + ENERGY

Key words
- ✓ **core body temperature**
- ✓ **respiration**
- ✓ **sweating**
- ✓ **shivering**

A constant body temperature

Heat energy flows from warmer objects to cooler ones. The bigger the temperature difference the greater the rate of heat-energy transfer.

The **core body temperature** is the temperature of the internal vital organs. Humans keep their core body temperature constant at around 37 °C no matter how warm or cold their surroundings.

The temperature regulatory centre in the brain continually monitors the body temperature. It sends messages to the skin and other systems that react to keep the temperature within a narrow range.

Gaining energy	Losing energy
• **Respiration** warms your body. • On a hot sunny day the Sun warms your body.	• If your surroundings are cooler than your body, your body will lose energy to the surroundings. • Energy leaves your body in your warm urine and faeces. • **Evaporation** causes energy loss. You lose energy when sweat evaporates and in the moist air you breathe out.

Respiration releases energy. Exercise makes you hot because of respiration in muscle cells.

The skin and temperature control

Cross section of the skin. The layer of fatty tissue insulates the body and limits energy loss to the surroundings.

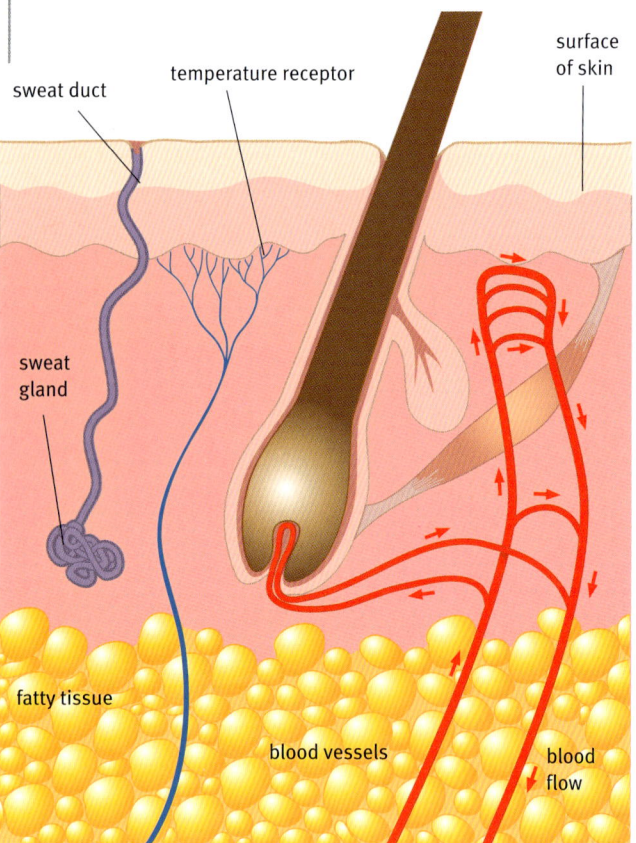

sweat duct

temperature receptor

surface of skin

sweat gland

fatty tissue

blood vessels

blood flow

Sweating...
When you are hot you **sweat**. Water evaporates from your skin, transferring energy away from you.

... and shivering
When you are cold you **shiver**. Your muscles contract, and this generates heat to help you warm up. Exercise such as running around or swinging your arms has the same effect.

The body's response to changes in air temperature

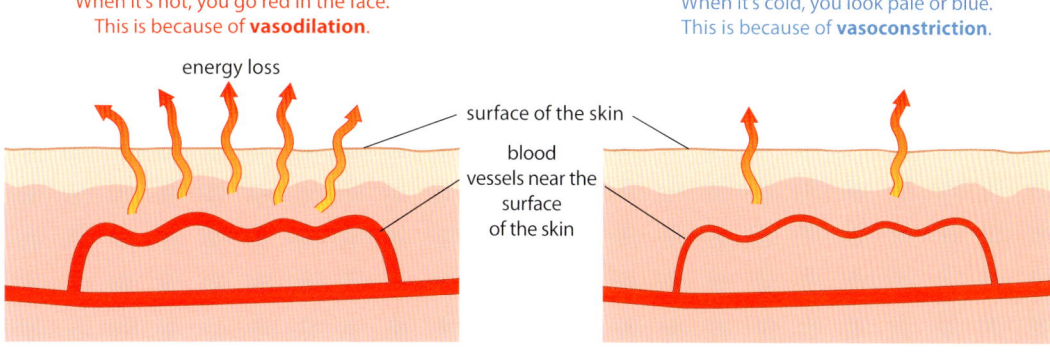

When it's hot, you go red in the face. This is because of **vasodilation**.

energy loss

surface of the skin

blood vessels near the surface of the skin

Vasodilation helps you cool down.

The blood vessels near the skin surface get wider (dilate). This diverts blood closer to the surface. Energy is transferred through the skin to the cooler air.

When it's cold, you look pale or blue. This is because of **vasoconstriction**.

Vasoconstriction helps you stay warm.

The blood vessels near the skin surface get narrower (constrict), so less blood flows through them. Less energy is transferred through the skin to the cooler air.

The control of body temperature

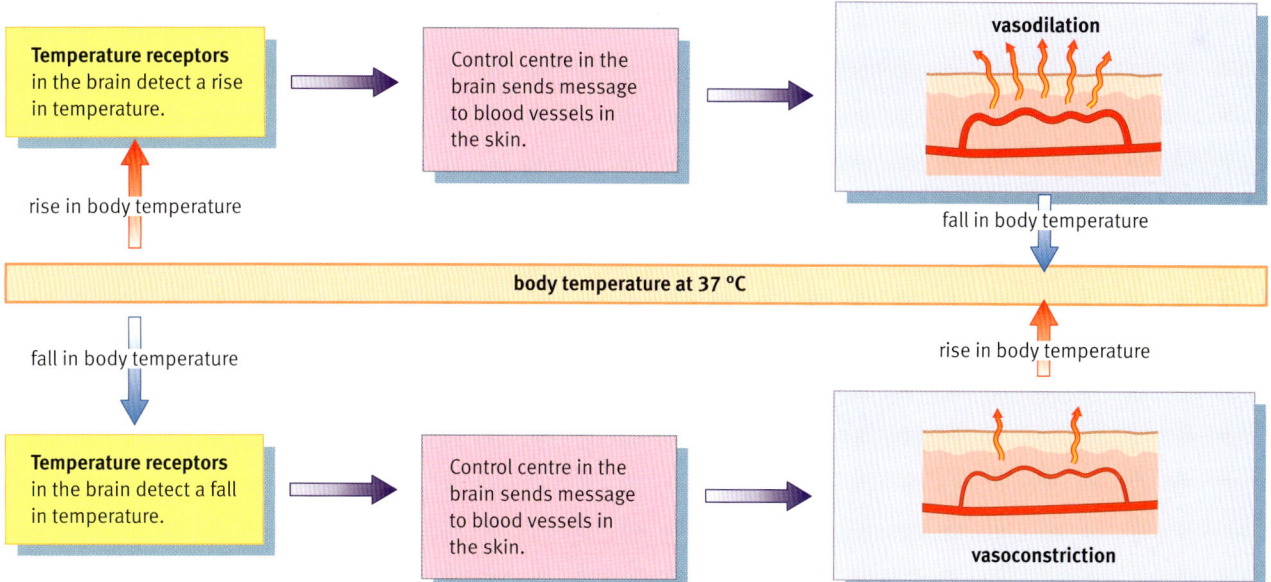

Temperature receptors in the brain detect a rise in temperature.

Control centre in the brain sends message to blood vessels in the skin.

vasodilation

rise in body temperature

fall in body temperature

body temperature at 37 °C

fall in body temperature

rise in body temperature

Temperature receptors in the brain detect a fall in temperature.

Control centre in the brain sends message to blood vessels in the skin.

vasoconstriction

Questions

1 Give two ways in which energy is transferred:
   a to your body, warming you
   b from your body, cooling you.

2 Why does your skin go red when you are hot, and pale when you are cold?

3 Explain the role of temperature receptors in the brain.

4 Your skin may feel cool to someone else when you come in from the cold. Does this mean your body core temperature has fallen? Explain your answer.

Key word
✓ **temperature receptors**

Usain Bolt breaking the world speed record for 100 m in 2008.

In 2008 Usain Bolt became the fastest runner on the planet. He ran a distance of 100 m in a time of only 9.72 s, setting a new world record.

Athletes in training need to know how fast they are running. A race winner might not run as fast as possible all the way. Successful athletes learn how to pace themselves – this means running fast enough in the early stages but still having enough energy for a good final lap.

Calculating speed

You calculate **speed** with the formula:

$$\text{speed (in metres per second)} = \frac{\text{distance (in metres)}}{\text{time (in seconds)}}$$

To calculate speed in metres per second (m/s) you need to measure the distance in metres and the time in seconds. To calculate speed in kilometres per hour (km/h) you need to measure the distance in kilometres and the time in hours.

So how fast did Usain Bolt go?

$$\text{speed} = \frac{\text{distance (in metres)}}{\text{time (in seconds)}}$$

$$\text{speed} = \frac{100 \text{ m}}{9.72 \text{ s}} = 10.3 \text{ m/s}$$

This is about 23 mph, which is faster than the road speed limit outside most schools.

Measuring speed

For accurate speed calculations you need accurate measures of both distance and time. Accurate measures of distance are made with lasers or infrared sensors. Accurate measures of time for short, fast races need accurate clocks. They also need systems to start and stop the clock. When you use a stopwatch, there is a slight delay in starting the timer and stopping it. This means you are likely to overestimate the time. This will mean your calculation of speed is an underestimate – if only slightly.

Try measuring a regular event with a stopwatch and see how accurately you can measure time.

Farther but slower

Long-distance runners are slower than sprint racers. This is because they cannot sustain sprinting speeds over long distances.

The longest race of the Olympics is the marathon, a distance of 42.195 km. The world record for the fastest time is 2 hours 3 minutes 59 seconds, set by Haile Gebrselassie in 2008.

So how fast did he go?

$$\text{speed} = \frac{42195 \text{ m}}{2 \times 60 \times 60 \text{ s} + 3 \times 60 \text{ s} + 59 \text{ s}} = \frac{42195 \text{ m}}{7439 \text{ s}} = 5.7 \text{ m/s}$$

Notice that the **distance** and **time** values have to be converted to metres and seconds before calculating the speed.

The right distance

The International Association of Athletics Federations (IAAF) has very strict rules about world records for races. For example, they require that the distance from the start line to the finish line for the 100-m race must be correct to within 1 cm. How can you do that?

A tape measure isn't accurate enough because it stretches too much. So a laser rangefinder is used. This fires a pulse of laser light from the start line to a mirror on the finish line and times how long it takes to come back. Then it uses the speed of light (299 792 km/s) to calculate the distance. A precision of ±2 mm is easily possible.

Accurate timing

An electronic clock is used to time record-breaking attempts. It needs an accuracy of 0.01 s.

The clock starts when the trigger on the starting gun is pulled. As the winner goes through the finishing line they break a laser beam pointing across the track. This stops the clock.

Long-distance runners go at about half the speed of sprint runners.

A laser rangefinder.

Pressure sensors are built into the starting blocks to detect false starts.

Faster and faster

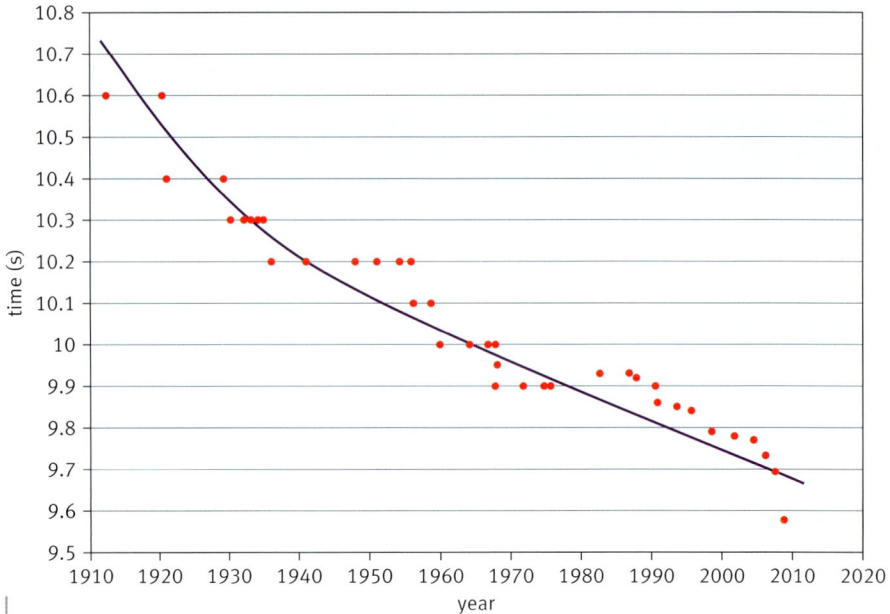

World-record times for the 100-m sprint over the past 100 years. Up to 1977, the timing was done manually. Since then, electronic timing has been used, making the world-record times accurate to 0.01 s.

World-record times for running are still falling, mainly thanks to sport scientists. Here are some of the things they are working on:

*   improving track surfaces so that athletes waste less energy
*   making better running shoes
*   devising better training regimes to get athletes to peak performance
*   working out the right diet.

Questions

1   The fastest time for running 1 km is 2 minutes 11.96 seconds.
    a   What is this distance in metres?
    b   What is this time in seconds?
    c   Calculate the speed of the runner.

2   Suggest reasons why running speed records continue to be broken.

Measuring body temperature

Equipment

Several devices are used to measure core body temperature. Two are shown here.

In a clinical thermometer mercury in the bulb expands as the temperature goes up, and rises up a narrow glass tube. The tube is marked to show the temperature.

A digital thermometer has a probe connected to an electronic display.

A liquid-crystal thermometer measures skin temperature. It changes colour to show the temperature.

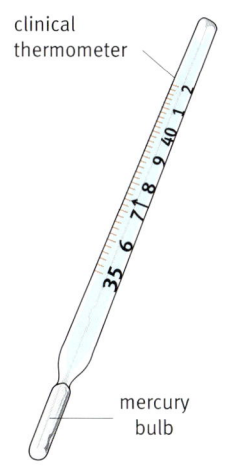

clinical thermometer

mercury bulb

Procedure

Always use the same body site to monitor somone's body temperature, as temperature varies over the body.

Approximate core body temperatures can be taken by placing the mercury bulb or temperature probe:
- in the mouth, under the tongue
- under the arm in the armpit
- with a special probe in the ear.

More accurate core body temperatures would be measured with probes placed in the **oesophagus** or the **rectum**.

Shake a clinical thermometer before using to return the mercury to the bulb.

Record the body temperature in **degrees Celsius** (°C).

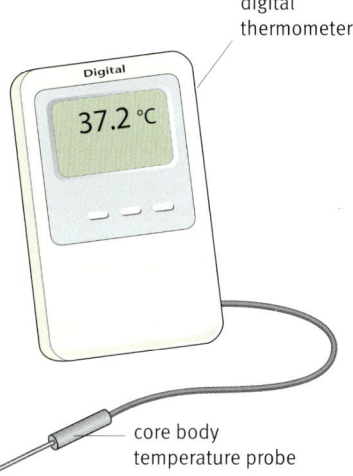

digital thermometer

Digital

37.2 °C

core body temperature probe

Interpreting the reading

The normal range of core body temperature is between 36 °C and 37.5 °C. Children have slightly higher temperatures due to their higher **metabolic rate**.

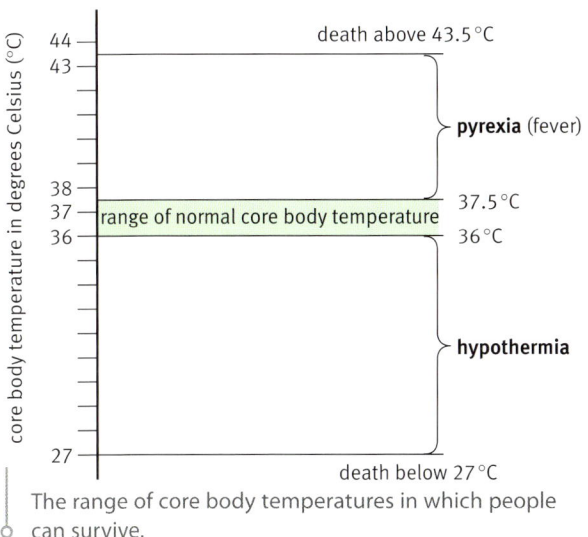

death above 43.5 °C

pyrexia (fever)

37.5 °C

range of normal core body temperature

36 °C

hypothermia

death below 27 °C

core body temperature in degrees Celsius (°C)

44
43
38
37
36
27

The range of core body temperatures in which people can survive.

PROCEDURES & TECHNIQUES

Measuring pulse rate

Equipment
- wristwatch with a second hand

Procedure

The pulse can be **palpated** (felt) where you can press an artery near the body surface against a firm structure such as bone. The diagram opposite shows some places where you can feel the pulse. It is usually taken in the radial artery.

The person should be relaxed.

1 Place your first and second fingers along the radial artery and press gently against the bone.

2 Apply enough pressure to feel the pulse but not so much that the artery is blocked.

3 Count the number of pulses in 60 seconds. This gives the pulse rate in beats per minute.

An electronic pulse monitor may be used in place of this manual procedure.

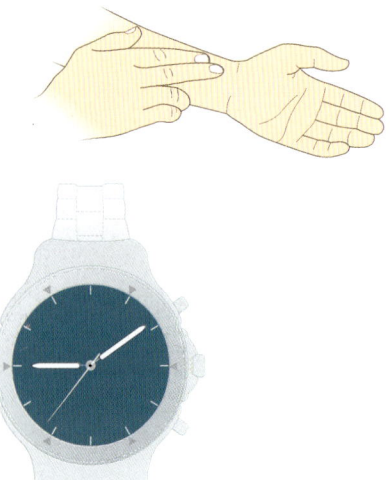

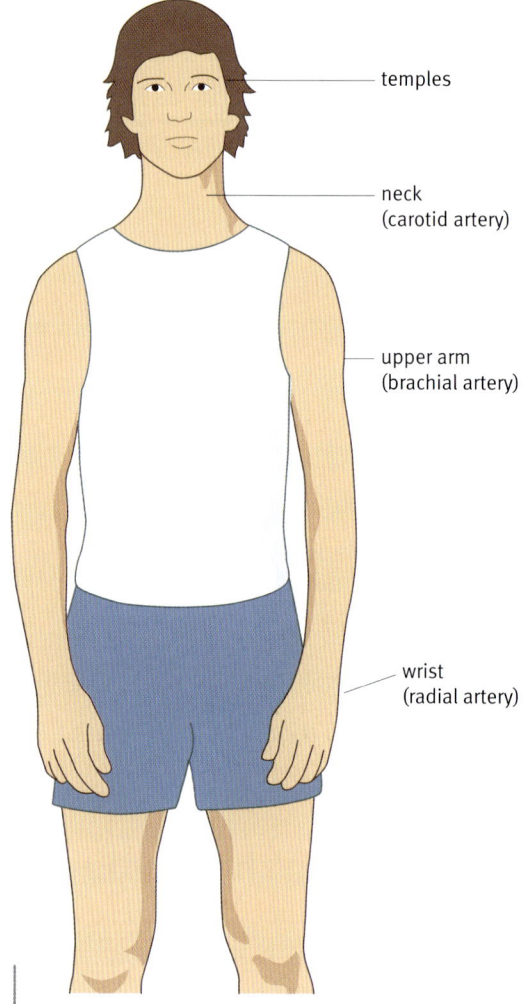

temples

neck (carotid artery)

upper arm (brachial artery)

wrist (radial artery)

Major pulse points of the upper body.

Interpreting the reading
- Healthy adults have a pulse rate of 60–100 beats per minute.
- Infants and children have much higher pulse rates.
- The pulse rate increases during exercise, panic attack, anxiety, or if the person has a heart disorder.
- A weak pulse may be a sign of low blood pressure, shock, or a heart disorder.
- A low pulse rate can result from illness. Fit athletes also have low pulse rates, as their heart and lungs work efficiently.

Measuring blood pressure

Equipment

Blood pressure is measured using a manual or electronic **sphygmomanometer**. This consists of:

- an inflatable sleeve or cuff
- a means of inflating the cuff (by squeezing a rubber bulb, with compressed air, or with an air pump)
- a device for measuring the pressure in the cuff.

As well as the sphygmomanometer, you need something that can detect the blood flow through the vessels in the arm. This could be a stethoscope that someone uses to listen to the pulse, or a sensor in the electronic device.

The reading

- Blood pressure is measured in **millimetres of mercury** (mm Hg).
- There are two numbers, for example, 130/80 mm Hg ("130 over 80").
- The first figure is always higher.

Interpreting the reading

Blood pressure varies with age, fitness, and stress levels. A healthy blood pressure is usually lower than 140/90 mm Hg. A 'normal' reading is about 120/80 mm Hg.

The higher reading is the **systolic blood pressure**. This is the maximum pressure in the arteries when the ventricles of the heart are contracting and forcing blood to flow.

The lower reading is the **diastolic blood pressure**. This is the lowest pressure in the arteries, at the point where the ventricles are relaxed and filling with blood from the atria.

A high blood pressure is unhealthy because it means you are at risk of blood vessels bursting. If this happened to capillaries in your brain you would have a stroke.

A low blood pressure is unhealthy because it means that some parts of your body will not get enough oxygen and nutrients.

Procedure

An electronic sphygmomanometer takes readings automatically. You place the cuff on your wrist or arm and the monitor on your finger. When you press 'start', the cuff inflates then deflates. An electronic readout shows you the blood pressure.

The manual procedure is shown on the right.

1 Place the cuff around the upper arm. Hold a stethoscope to the brachial artery just below the cuff. Listen for the pulse.

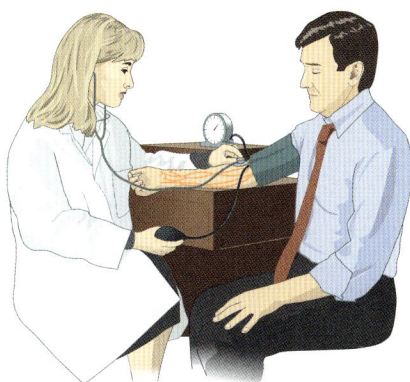

2 Inflate the cuff until you can no longer hear the pulse.

no pulse heard

3 Slowly deflate the cuff until you hear high-pressure blood swishing through. Record the reading as the systolic pressure.

pulse heard

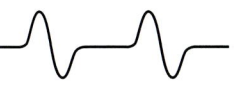 

4 Deflate the cuff further. The surging pulse sound will suddenly die away. When it starts to die away, record the reading as the diastolic pressure.

no pulse heard

Measuring body mass index

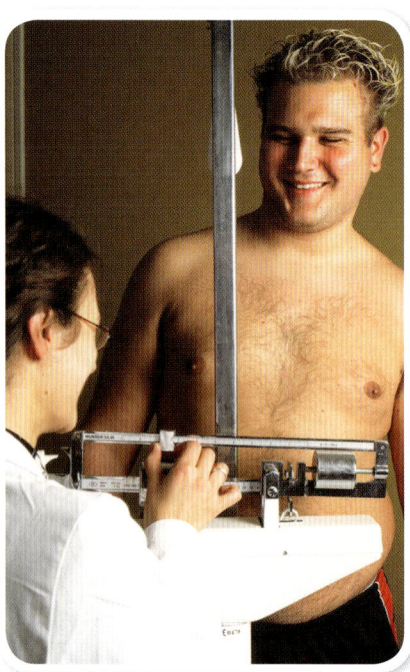

This man is being weighed and measured so that his body mass index can be calculated.

Equipment

- stadiometer or accurate weight scales and height rule

The **body mass index** (**BMI)** is found by dividing the body mass in kilograms by the square of the height in metres.

Procedures

Measuring body mass
- Use scales on a hard, flat, uncarpeted floor.
- Wear light clothing and no shoes.
- Record mass in kilograms to one decimal place.

Measuring body height
- Ideally, use a stadiometer. This combines a weight scale with a height measurer.
- The person should stand tall, facing forwards.
- Mark the measurement at the highest point of the head.
- Record the height in metres to one decimal place.

Calculating BMI

Calculate the BMI using the formula:

$$\text{BMI} = \frac{\text{body mass in kg}}{(\text{height in m})^2}$$

This index is used to estimate whether a person's body weight is at an advisable level, too high, or too low.

Interpreting the result

Most authorities use the following guidelines:

BMI	Condition	Advice given
under 20	underweight	may need to gain weight
20.0–24.9	advisable range	
25.0–29.9	overweight	some weight loss may be beneficial to health
30.0–34.9	obese	need to lose weight
35 and over	severely obese	urgent need to lose weight; consult a doctor

Being overweight (BMI over 25) leads to increased health risks such as high blood pressure and diabetes. The risk gradually increases as the BMI rises further. With a BMI over 35, the risk of premature death is doubled.

Performing a step test

Step tests are designed to assess **aerobic fitness**.

Equipment
- a step or bench no more than 50 cm high
- a stopwatch
- a metronome

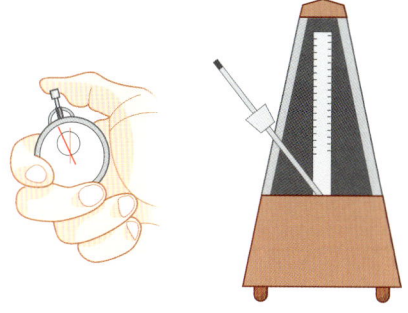

Procedure
First complete a physical activity readiness questionnaire (PAR-Q).
If you have any medical problems do not do a step test without a doctor's approval.

1   Set the metronome at two beats per second.

2   Start the stopwatch and step in time to the beat:
    1   first foot up
    2   second foot up
    3   first foot down
    4   second foot down
    1   first foot up . . .

3   Repeat for five minutes or until you are too tired to continue. Do not stop the stopwatch.

4   Sit down immediately you stop. Note the time.

5   Count your pulse (see page 42) for 30 seconds exactly one, two, and three minutes after you stop.

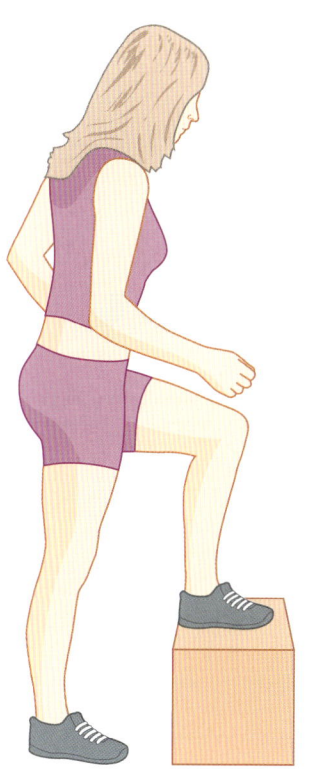

Interpreting the results
Calculate your step test score:

- time you stepped for in seconds × 100 (A)
- add up the three pulse counts, then multiply by 2 (B)
- step-test score $= \dfrac{A}{B}$

The higher the score, the more aerobically fit you are.

Male score	Female score	Guideline for 16-year-olds
over 90	over 86	excellent
80–90	76–86	above average
65–79	61–75	average
55–64	50–60	below average
below 55	below 50	poor

Example
A person stepped for three minutes, with pulse counts of 65, 50, and 35:
- A is $3 \times 60 \times 100 = 18\,000$
- B is $65 + 50 + 35 = 150$
- Step test score
  $= \dfrac{18\,000}{(150 \times 2)} = 60$

Equipment

- two microscope slides
- blood (use ink to practise the technique, or animal blood as supplied) with dropper

Procedure

1 Place a small drop of blood on a clean slide.

2 Hold a second clean slide at an angle to the first one.

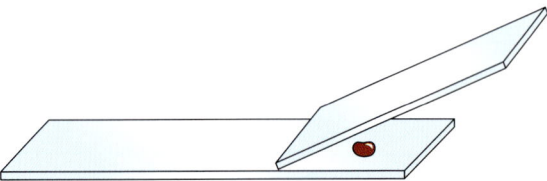

3 Make sure the slides are touching. Pull the top slide back to touch the drop. The drop will start to spread.

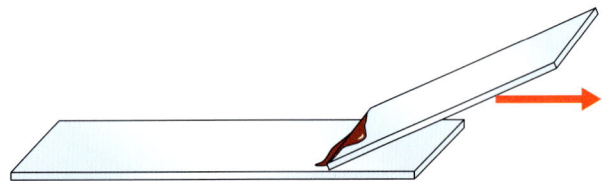

4 Keep the slides firmly touching. Push the top slide away in one smooth motion to produce a smear. It should be thin at one end and thicker near the original drop of blood.

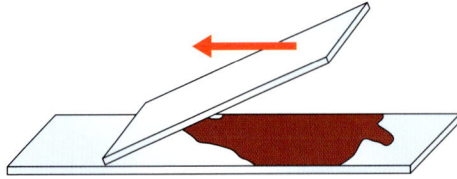

Interpreting the results

Examine the blood slide under a microscope (see page 47).

Blood slides give information such as:

- number of red blood cells (red-cell count)
- proportion of red to white blood cells
- number of reticulocytes (immature red blood cells with nucleus)
- deficiency anaemia (few red blood cells)
- sickle-cell anaemia (sickle-shaped red blood cells)

Review Questions

1 Describe the job of a person employed at a sports centre who could help you to improve your fitness. Describe some of the personal qualities needed in a good fitness coach.

2 A personal trainer collects some basic information about a client before giving them a fitness programme.
  a Make a list of the information they will need to collect.
  b Explain why it is important to collect this information

3 Research shows that people who have a BMI value over 25 have additional health risks. Mick has a mass of 84 kg and a height of 1.8 m. Use the equation $BMI = \dfrac{body\ mass}{height^2}$ to comment on his risk.

4 Describe how to measure a person's blood pressure.

5 Here are some parts of the human breathing system.

diaphragm   intercostal muscles
lungs   ribs

Use these words to explain how we breathe air in and out.

6 Red blood cells in arteries carry oxygen to the muscles.
  a Name two other parts of the blood and state their function.
  b What else does the blood carry to muscles? How do the muscles use this substance?
  c Explain why arteries have thick walls.
  d Name two substances that veins carry away from muscles.
  e Explain why veins have thin walls.

7 Exercise can be used to cure a skeletal–muscle injury.
  a Describe a set of exercises suitable for treating a skeletal–muscle injury.
  b Explain the role of a physiotherapist in treating the injury.

8 Here are the results of three different races.

Runner	Distance (m)	Time (s)
Alfie	100	15
Belinda	200	25
Catherine	500	120

Who was the fastest? Justify your answer.

9 An athlete running a marathon puts great stress on her body.

  a Describe the role of her kidneys in keeping her fluid levels balanced.
  b The runner's body temperature will change during the race. Explain how her body responds to overheating while running and then also tries to stop her getting cold when she stops.

capillaries

arteries

veins

blood vessels

platelets

plasma

white blood cells

red blood cells

blood

circulation system

valves

atria

ventricles

nerve

fluid balance

urine

turning moments

kidneys

energy

respiration

joints

temperature response

human breathing system

test for performance-enhancing drugs

ribs and diaphragm

lungs

airways

regulation

health and safety

fitness industry association

litigation

community

organisations

local authority sports centres

privately owned gyms

WORKING IN SPORTS AND FITNESS

people
- coach
  - communication
  - skills
  - qualifications
- physiotherapist
  - communication
  - qualifications
  - skills
  - role
    - assessment of injury
    - exercises to aid recovery
- personal trainer
  - communication
  - skills
  - qualifications
  - role
    - training programmes
      - monitoring and modelling
      - movement sequences
      - aerobic fitness exercises
      - muscle-building exercises
      - weights
    - baseline assessment
      - health
        - medication
        - personal history
      - fitness
        - stamina
        - strength
      - lifestyle
        - stress
        - tobacco
        - alcohol
        - food

the human body

fitness assessment
- standard procedures
  - aerobic fitness
    - step test
    - stopwatch
  - pulse rate
  - blood pressure
    - sphygmomanometer
  - body mass index (BMI)
    - body mass (kg)
    - height (m)
  - body temperature
    - crystal thermometer
    - liquid-crystal thermometer

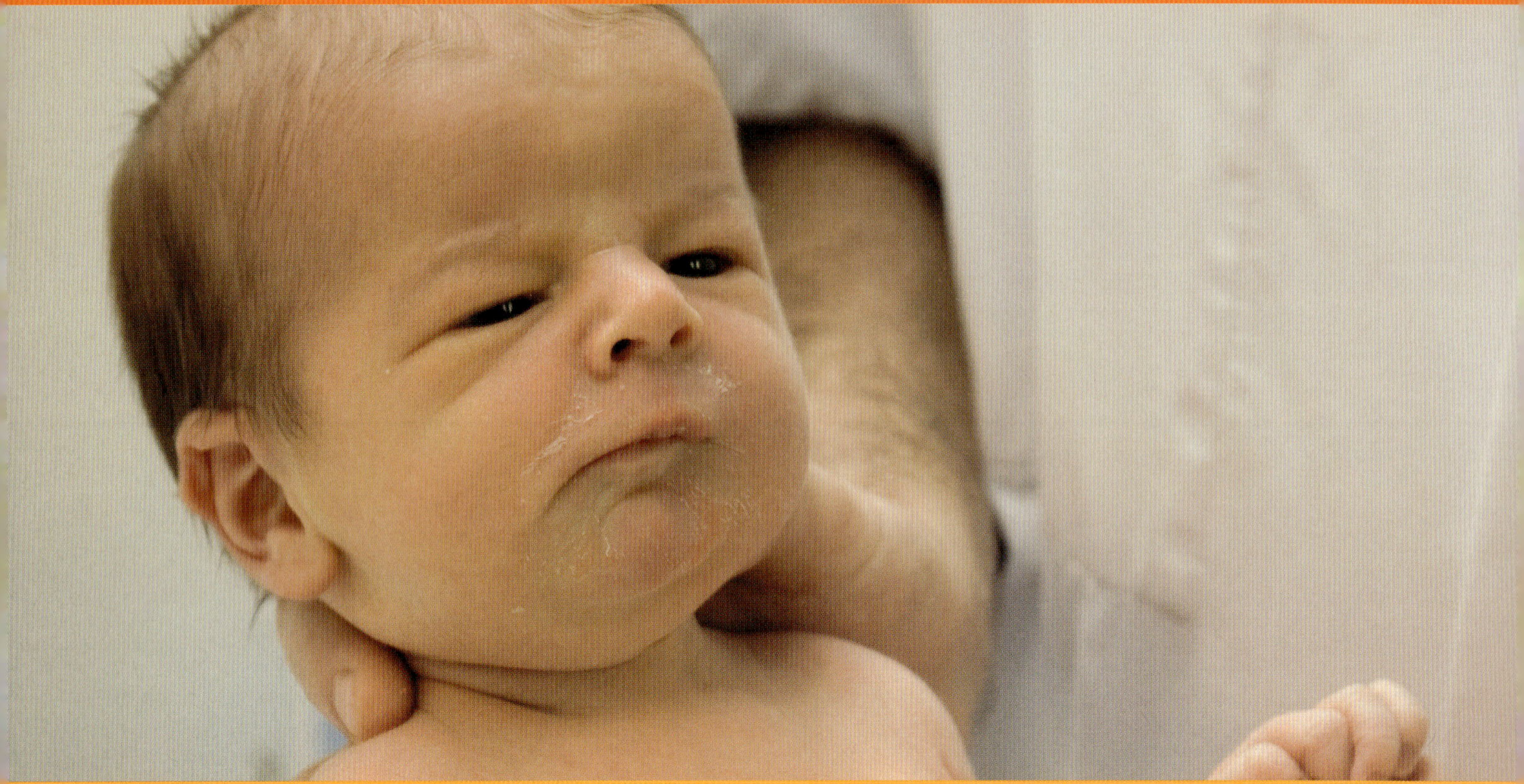

A2 Healthcare

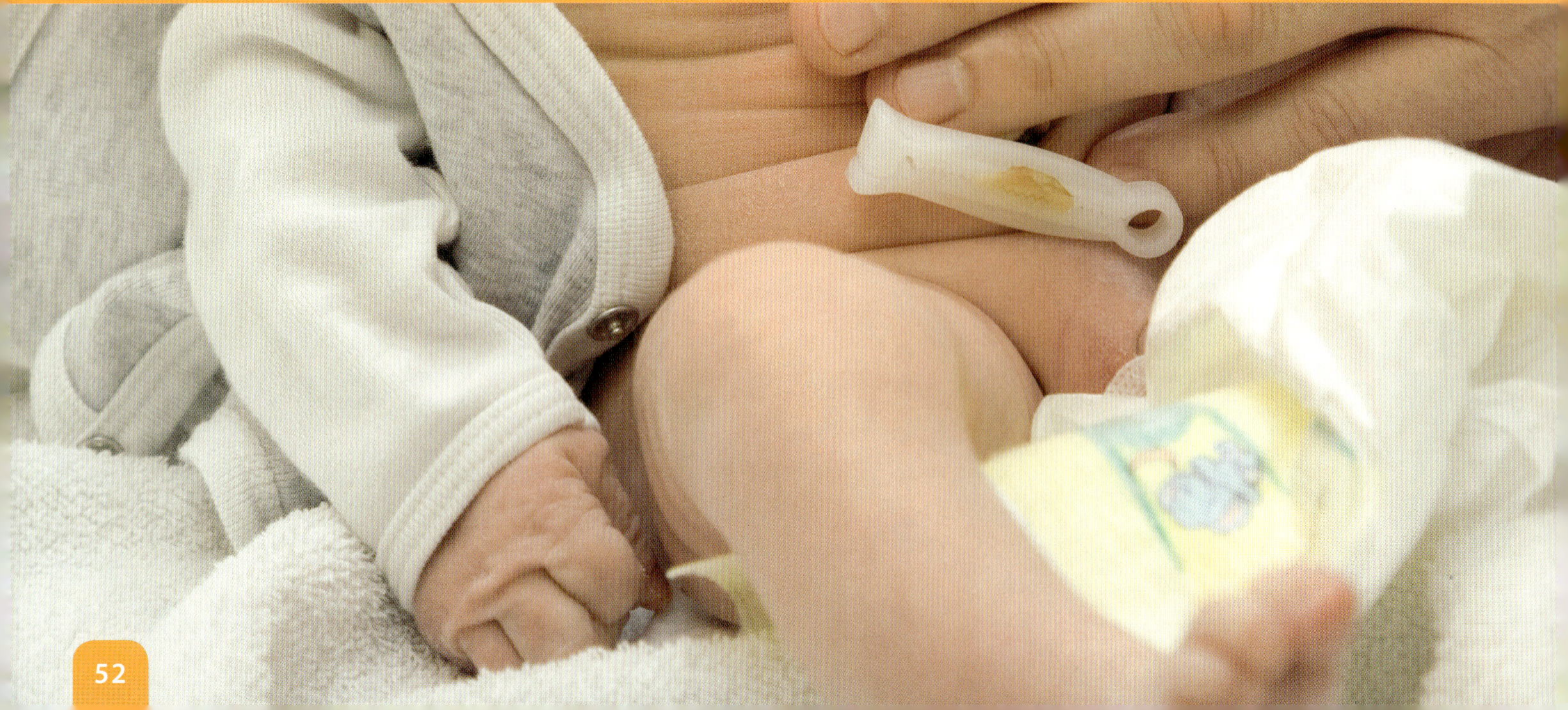

Why study healthcare?

Being born is possibly the most hazardous thing that we have all had to do. Science has determined how babies and their mothers are cared for before, during, and after birth. Following an accident you might need to go to hospital. On arrival, doctors, nurses, and other healthcare professionals will use their expertise to decide what is wrong with you. They will explain the options open to you, and give you the treatment that you choose. You need to know enough to be able to make an informed choice.

What you already know

- how your body resists infection
- how vaccines and antibiotics work to keep you healthy
- the role of your kidneys in keeping your water balance correct.

Find out about

- how healthcare professionals deal with accident victims
- the female reproductive system
- how to help a mother and her baby safely through the birth process
- the tests used to check on the health of mother and baby during pregnancy.

The Science

Doctors gather evidence from what their patients say and from examining them, as well as from laboratory tests and scans. This evidence is used to make a diagnosis and decide on a range of suitable treatments. Chemicals devised and tested by scientists are used as medicines to make people better.
Scientists know about the potential risks of childbirth. Scientific instruments monitor the progress of mother and child throughout pregnancy to minimise these risks and make childbirth as safe as possible.

Road-traffic accident

Life-saving professionals who handle accidents and emergencies must move quickly and confidently. They are trained to cope with a variety of situations.

At the ambulance station: an interview with paramedic Neil Hutchinson

What happens when you arrive at a road-traffic accident?

First we check for hazards, so we can work safely without endangering the patient's life further.

How do you decide who to treat first?

We have a system called triage. It is just like the system used by **triage nurses** in the accident-and-emergency (A&E) department of hospitals. **Triage** means deciding which patients need attention straight away, and which ones can safely wait until later. Patients with breathing or circulatory problems are treated first. We follow the ABC of first aid.

ABC?

A is for airway. We make sure it is open by tilting back the chin and looking for obstructions. Then we check breathing – that's B – giving rescue breathing if necessary. C is for circulation. If a person has no heartbeat, we carry out cardiopulmonary resuscitation, or CPR. If the person can't talk or is unconscious, we ask witnesses to tell us exactly what happened.

And then?

If there is no heartbeat we need to get the patient breathing, restore the heartbeat, give oxygen, and take them to hospital quickly. One paramedic continues with CPR while the other sets up the oxygen mask, the heart monitor, and the defibrillator. Sometimes we have to use a tube to keaep the airways open. If the patient begins breathing we put them in the recovery position.

The police officer watches traffic so the paramedics can treat the casualty in safety. Witnesses can give vital information.

And time is of the essence?

Oh yes, it's critical. The first hour after injury is called the 'golden hour'. For example, I recently treated a man who came off his motorbike. His blood pressure was low, which suggested internal bleeding. Thankfully, we got him to the operating theatre 45 minutes after the accident. So he was alright. The air ambulance speeds things up.

Who is treated next, after the ABC patients?

We assess those with major trauma, such as broken limbs or large wounds. Someone losing blood will be treated quickly to staunch the flow. Someone with a head injury might have dilated pupils or blood in the whites of their eyes, or their walking or speech may be affected.

What else might you have to do?

Well, some broken bones need to be treated straight away. A broken femur can be dangerous. The muscles in the thigh are so strong that they tend to push the bone up into the abdomen. That's when you need traction.

And then it's to the hospital?

Yes. We deliver patients to the A&E department. If we have radioed ahead, a team is waiting to take over. We give the A&E staff details of any treatment we have given, and how the patient is.

The A&E staff are trained to prioritise patients requiring treatment, using the triage system. They have to manage resources carefully to make sure that they are used effectively for patient care.

A&E teams also practise to make sure that they know what to do if there is a major incident, such as a multiple car crash.

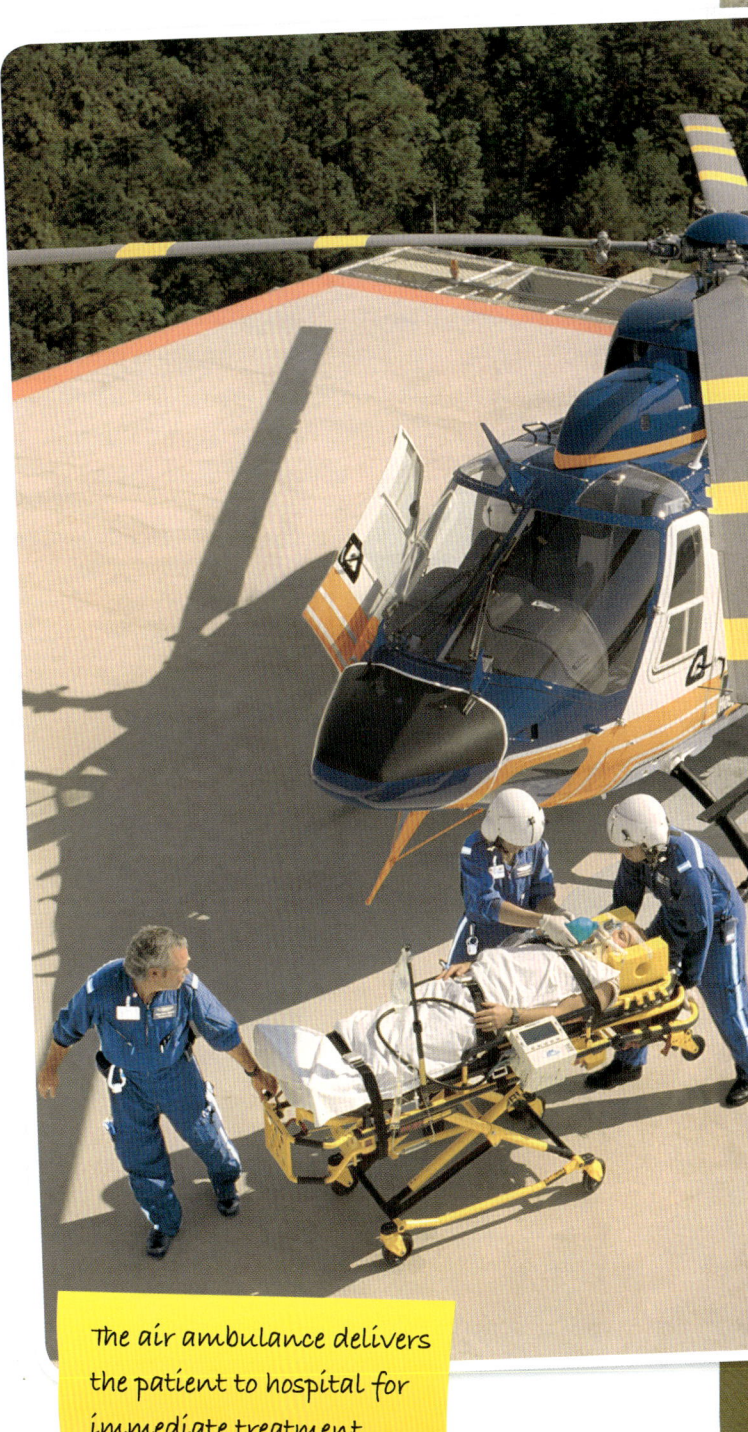

The air ambulance delivers the patient to hospital for immediate treatment.

Emergency: cardiac arrest

At the A&E department: an interview with junior doctor Mike Yousif

What happens if a patient arrives at A&E with a suspected cardiac arrest?

We take details from the paramedics, and follow the ABC standard procedure.

Ah … I know all about that now. Airways, Breathing, and Circulation. But what next?

Once cardiac arrest is confirmed, we put a line into a blood vessel to give fluids or drugs. If the blood pressure is falling, we give fluids to maintain the blood pressure. We then get ready to use the defibrillator.

That gives an electric shock?

Yes, that's right. We check the airways again and put in a tube to keep the airways open. A cycle of shocks and chest compressions is carried out, and we use an airbag to inflate the lungs. We do this three times.

Mike Yousif at work in A&E at St Thomas's Hospital.

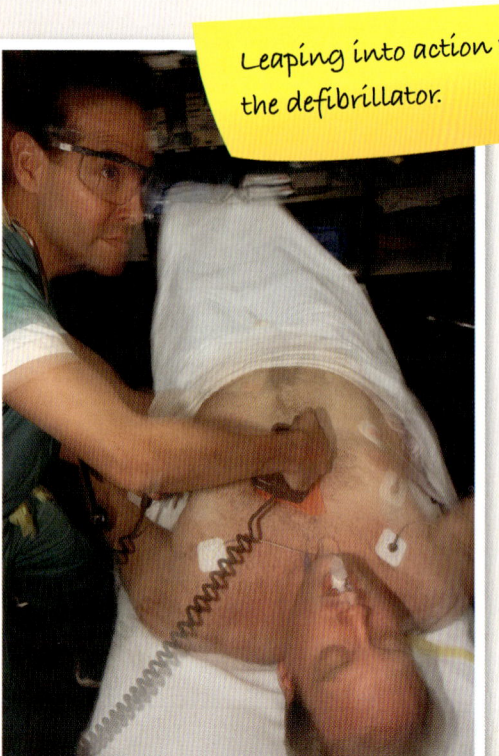

Leaping into action with the defibrillator.

If there's still no response, we give adrenaline. If the patient's heartbeat and breathing pattern are restored, we keep them in A&E until they are stable.

What problems might the patient have?

Heart failure or shock. We watch the heart monitor for signs of heart failure. We also monitor blood pressure closely.

Once the patient is stable, what happens next?

They are transferred to the cardiac care unit. If their condition remains serious, they will go to ITU – the intensive therapy unit.

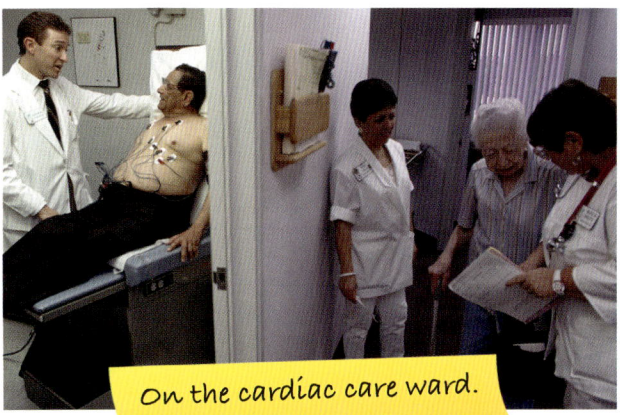

On the cardiac care ward.

On the cardiac ward: an interview with staff nurse Jane Ratliff

What happens when a patient arrives on the ward?

They stay on the heart monitor while we take details such as name, address, age, sex, next of kin, medication, and so on. We monitor vital signs and record them on a patient's chart. Once the doctor has seen the patient, we devise a care plan.

What exactly is a care plan?

It records how often we should take observations, and what medication to give. It also shows details of diet and fluids, and any proposed surgery.

What happens if the patient has another heart attack?

We bring the crash trolley, which has a defibrillator, oxygen, and drugs. We carry out CPR.

And once the patient is recovering?

We continue checking the electrocardiogram (ECG) trace on the heart monitor, along with temperature, pulse, breathing rate, and blood pressure. Once the patient has been stable for some hours, we will start a course of

treatment. We tell the patient how to take the drugs and we monitor them for adverse reactions.

And when do you start planning for discharge?

As soon as possible. Hospital beds are in great demand. In some cases we arrange a meeting for the patient and anyone concerned with rehabilitation, such as a social worker, physiotherapist, occupational therapist, and any carers at home.

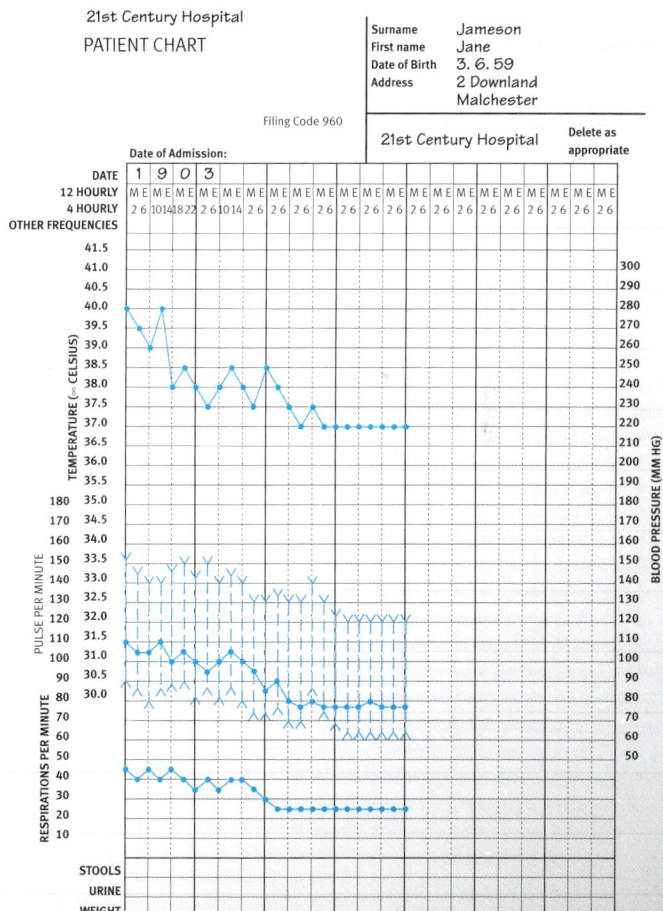

A patient's chart records temperature, blood pressure, pulse, and breathing rate. These must stabilise before the patient leaves the ward.

Diagnose and treat

Last summer I discovered an unusual lump on my back. It was between my shoulder blades, just within reach of my fingers when I stretched over one shoulder. It felt big, scaly, and wrong. This is the story of how that lump became a harmless scar.

Health centre

I phone the health centre straight away. The receptionist asks a few questions before giving me an appointment with my GP in three days' time. I could see another GP tomorrow. However, I've had the same GP for 12 years, so I prefer to see her.

Diagnose

In the waiting room I read a leaflet about the dangers of sunbathing. It reminds me that I should always use sunblock when out in the sun. When I get into the consulting room, the doctor is reading my medical record on the computer screen. She starts by asking me questions to check my lifestyle and **medical** history. To find out about my general health she asks about my:

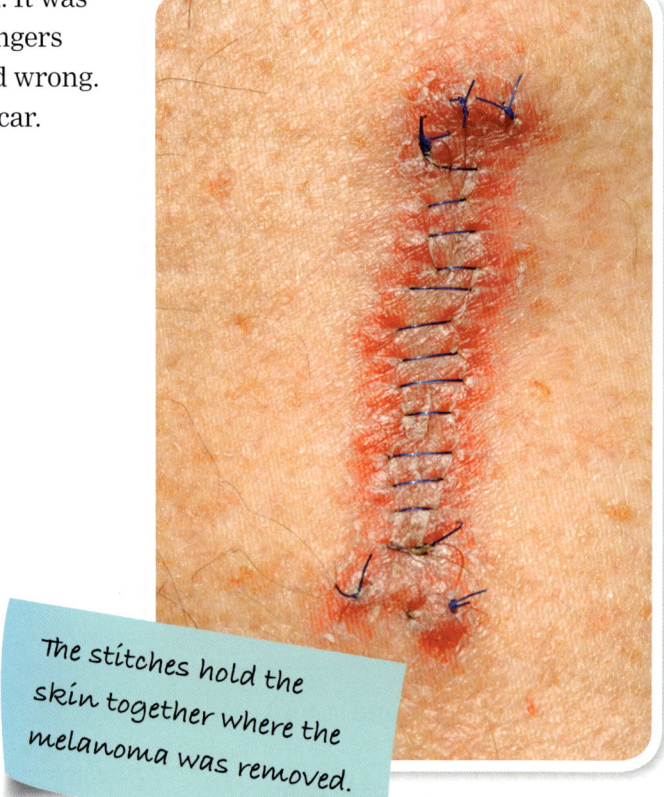

The stitches hold the skin together where the melanoma was removed.

- alcohol intake
- tobacco consumption
- general level of physical activity
- family medical history.

She also asks me:

- about my **symptoms** such as pain or bleeding
- whether I am taking any medication
- if I have had any previous treatments
- if I have any allergies.

This information will help her when she makes a decision about treatment.

My doctor answers all my questions to reassure me.

The specialist explains the procedure to me and any risks. She asks me to sign a consent form if I want to go ahead with the treatment.

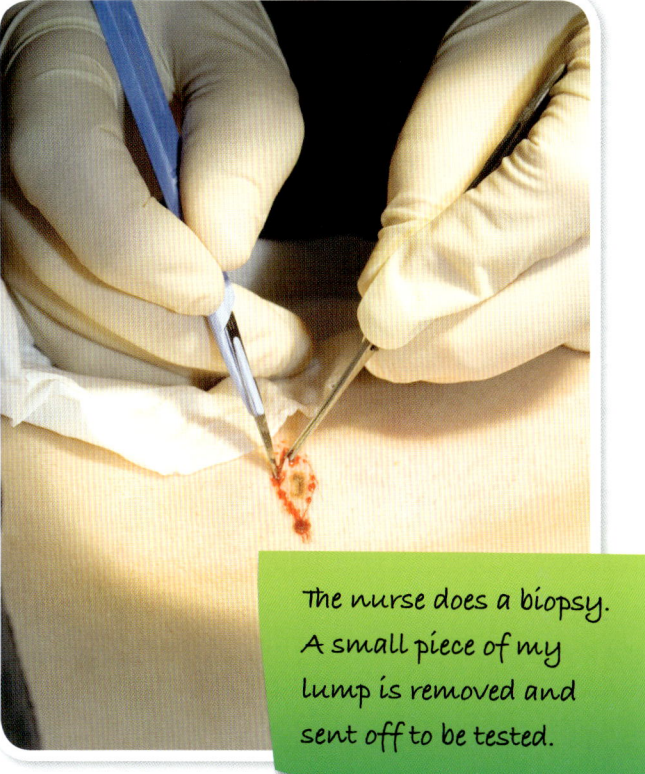

The nurse does a biopsy. A small piece of my lump is removed and sent off to be tested.

Then she takes a good look at my lump and examines it with a special magnifying glass. She asks me when I first noticed the lump, whether or not it itches, and about my sunbathing. She thinks the lump might be a melanoma, a form of skin cancer brought on by too much exposure to sunlight.

Test

The doctor can't be sure of her diagnosis without a proper test. So a nurse at the health centre does a biopsy. She cuts out a small piece of my lump, puts it in a tube, and sends it off to the hospital laboratory. A specialist technician looks at it under a microscope and writes a report to my doctor to confirm that it is cancerous.

Treat

Two days later I discuss my treatment options with my GP. She is going to refer me to a skin-cancer specialist at my local hospital.

Consent and operate

A week later, I'm a day patient at the hospital. I have a long talk with the specialist about what she is going to do, what the scar will look like, the chance of the lump growing back, and what they might do then. She also explains any **risks** associated with the procedure so that I know what could go wrong, the chance of it happening, and how serious it might be. Then I sign the consent form. I have a local anaesthetic so that the procedure doesn't hurt. The specialist removes the lump and some of the surrounding tissue. She then puts in a few stitches to bring the sides of the hole together. A nurse explains how the cancer clinic can support me. Then I make an appointment to come back in three months to make sure all of the lump has been removed.

Antenatal care

Sandy Domoney is a midwife in Cornwall. She is based at a health centre, and also visits her patients at home.

9 weeks

Finding out you're pregnant is a very exciting time. Most people do a home pregnancy test before contacting the health centre. The receptionist books the woman in to see me – often I'll visit at home for this first appointment. At this session I will give information and support so that the pregnant woman can make the right choices for her, and for the baby.

I give information on antenatal screening to test the fetus for several conditions, including **Down's syndrome** and **spina bifida**. The woman and her partner can then decide which tests to have.

With the woman's consent I take a blood sample to be tested for blood group, **anaemia**, infections such as HIV, and immunity to rubella.

She will have about 10 antenatal appointments. The frequency depends on the individual's needs. She can also come to antenatal classes.

12 weeks

I arrange the first ultrasound scan and the results come back to me. I check to make sure the dates are correct. These pictures show how the baby develops – it's amazing how quickly they grow. You can see the baby's heart beating while the scan is being done. At the next appointment you can hear the heartbeat – a very positive experience.

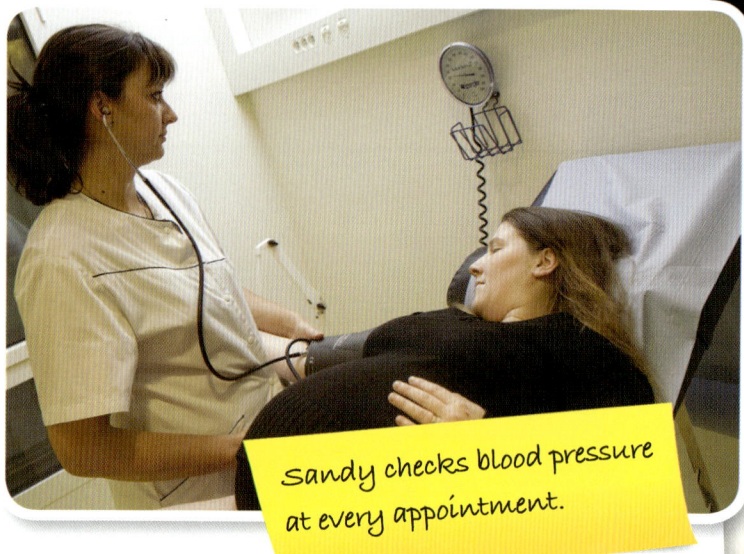

Sandy checks blood pressure at every appointment.

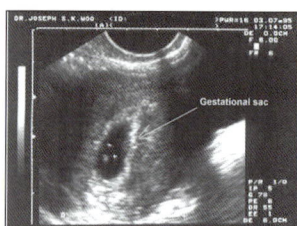

6 weeks

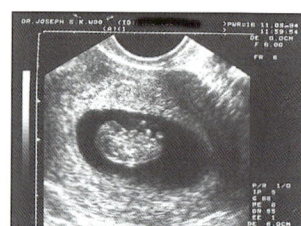

8 weeks

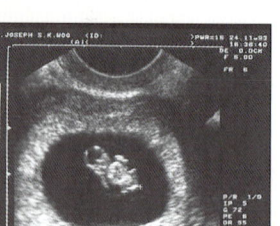

9 weeks

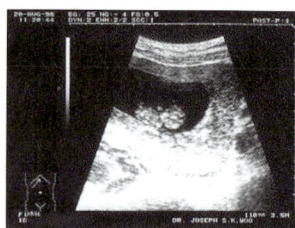

10 weeks

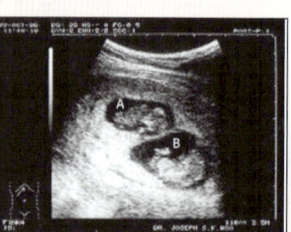

10 weeks – twins

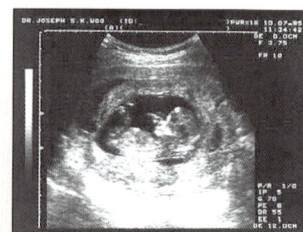

12 weeks

16 weeks

If the woman has decided to have screening for Down's syndrome and spina bifida, I take a blood sample. This is tested for chemicals in the mother's blood to see if there is a chance of the baby having these conditions. If the test is positive more tests are carried out to confirm the diagnosis.

20 weeks

Another scan checks the baby for abnormalities. If there is concern about the position of the placenta, I'll offer another scan later in the pregnancy. We can often tell now if it's a boy or a girl.

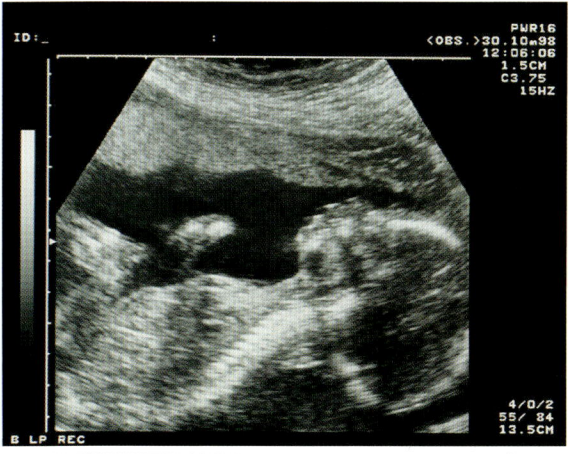

20 weeks

20–42 weeks

I see the woman regularly to check that the baby is growing well and that her weight gain, blood pressure, and urine are normal. I test the urine for protein and sugar. Protein suggests possible infection, and sugar may indicate gestational diabetes. I complete a birth plan after 32 weeks, saying whether she wants to give birth at home or in hospital, and what sort of pain relief she would prefer.

41–42 weeks

If the baby has not been born we offer to start labour using artificial hormones. After 42 weeks the placenta starts to break down slightly. If this continues the baby could be at risk.

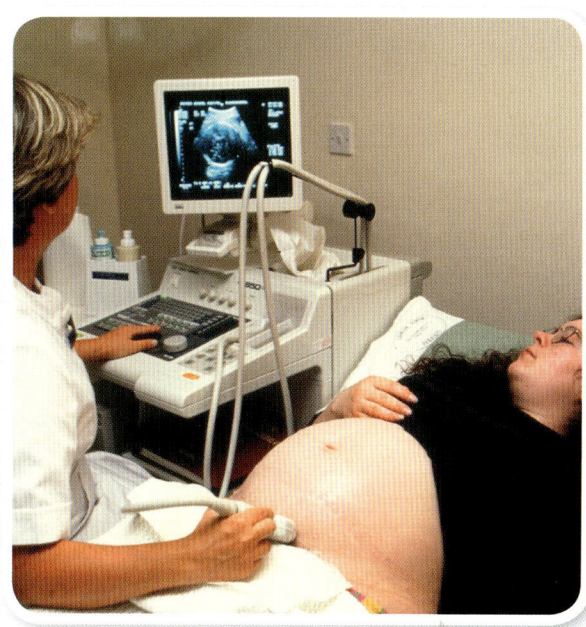

Labour and birth

The woman may have chosen to give birth at home, looked after by my team. If she has chosen a hospital birth she is cared for by hospital midwives unless a problem arises. Then they call in the consultant.

Following birth

I visit the mother and baby at home for up to 10 days to check that the baby is developing properly and the mother is well. It's lovely to see the parents with their new member of the family.

The National Health Service

> The NHS employs more than 1.7 m people. Of these, only about 585 000 are hospital doctors, general practioners (GPs), nurses, and ambulance staff. The others have a wide variety of different jobs.

This section introduces some of the people and organisations involved in healthcare. Teams of skilled people work together to respond to emergencies, provide treatment and aftercare for sick people, and work to prevent disease.

Independent agencies work to protect public health.

About the NHS

Which organisation is the largest employer in Europe? The answer may surprise you. It's the **National Health Service** in the UK.

The NHS was set up in 1948 to provide free healthcare for everybody. You may meet NHS practitioners at:

- the dentist's
- the optician's
- the local pharmacy
- the local health centre.

You can also get help quickly from the NHS by phone or on the Internet. Nurses give telephone advice about treatment or further help. The NHS website has information about common illnesses and how to treat them.

Features of a national health service:

- provides health care for everybody
- provides specialist care that is not always available locally
- monitors national trends
- plans suitable healthcare
- decides where and when resources should be used
- balances providing direct healthcare with organising and managing the service.

How did you become . . . a nurse?

You need a caring attitude to become a nurse.
Below is the route to qualifying:

Nurses are helped by healthcare assistants. To become a healthcare assistant you need a good general education or work experience, and can gain NVQ qualifications on the job.

A **general practitioner (GP)** is your gateway to the NHS. Your local GP can refer you to a hospital, arrange specialist treatment for you, and prescribe medicines.

The career of a doctor can take many paths. Instead of being a GP, doctors may specialise in a particular branch of medicine. Here are some of the specialist fields a doctor might choose:

- paediatrics
- obstetrics and gynaecology
- pathology
- radiology
- ophthalmology
- surgery
- psychiatry
- cardiology
- anaesthetics

Key words

- ✓ **National Health Service (NHS)**
- ✓ **general practitioner (GP)**
- ✓ **paramedic**

How did you become . . . a doctor?

It takes many years to become a doctor but once qualified there is a choice of about 60 specialities. Each needs different skills, but all involve working as part of a team. Below is the route to qualifying:

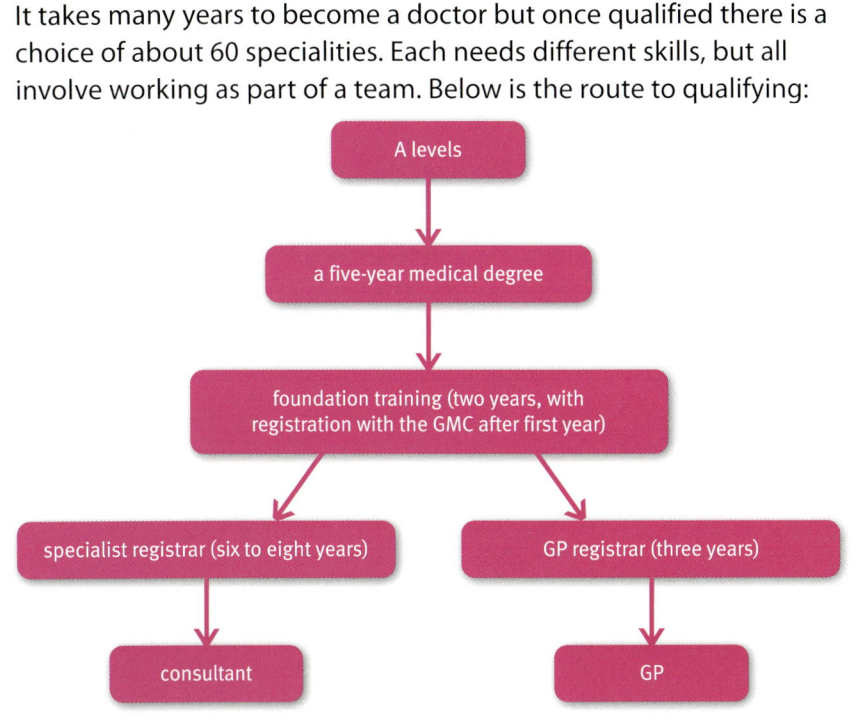

A levels

↓

a five-year medical degree

↓

foundation training (two years, with registration with the GMC after first year)

↓ ↓

specialist registrar (six to eight years)   GP registrar (three years)

↓ ↓

consultant   GP

How did you become . . . a paramedic?

You have to be physically fit and highly skilled to be a **paramedic**. You have to make decisions quickly, use a variety of equipment, and be able to calm and reassure your patients. Paramedics are helped by an emergency care assistant (ECA). ECAs need to have a good general education or relevant work experience. Below is the route to qualifying:

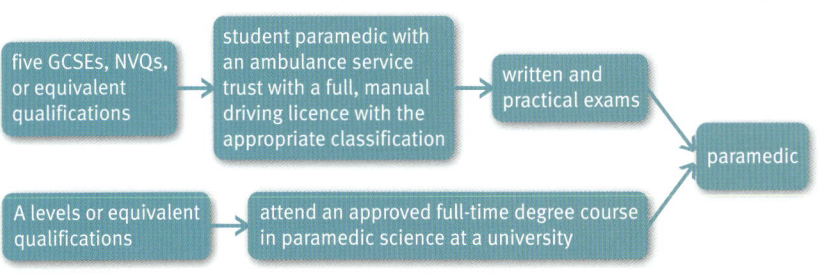

five GCSEs, NVQs, or equivalent qualifications → student paramedic with an ambulance service trust with a full, manual driving licence with the appropriate classification → written and practical exams → paramedic

A levels or equivalent qualifications → attend an approved full-time degree course in paramedic science at a university → paramedic

How the NHS is organised

The NHS is funded by taxpayers and costs over 100 billion pounds a year. Health-service managers make decisions about where to spend this money, balancing treatment with prevention of disease.

Paramedics, GPs, midwives, health visitors, practice nurses, dentists, opticians, and pharmacies all provide **primary healthcare**. This means care from the practitioner who first sees the patient.

If the primary healthcare team cannot meet the patient's needs, the GP may make a referral, usually to a hospital. Local hospitals cannot treat every condition, and some hospitals provide specialised care. Two famous hospitals in Britain are Stoke Mandeville Hospital, specialising in spinal injuries, and Great Ormond Street Hospital for Children.

Patient records

Our healthcare is provided by a team of healthcare professionals. No one person has all the essential information about you so it is important that your personal medical information is recorded and kept up to date. Your patient notes with your unique NHS patient number are stored so that they can be accessed when needed. All members of the healthcare team can then provide you with appropriate care.

How the NHS is regulated

The UK government Department of Health works with independent agencies to produce guidelines for the best ways to deliver healthcare services. Clear rules and regulations cover some areas of work in healthcare. These regulations mean that people offering healthcare services must be properly qualified and must work in conditions that are safe for them and their patients.

Healthcare professionals keep records of how they treat patients. These records can be used as evidence if anything goes wrong. Reviewing the success of different methods allows the agencies to develop the guidelines for good healthcare.

Making choices

The NHS publishes information about healthcare options and the success rates of different treatments. This includes information about the risks and benefits of **vaccinations** such as flu, and the advantages of particular operations or non-invasive treatments. This enables patients and parents to make informed decisions about their own healthcare. Independent agenices are responsible for protecting public health and reducing the impact of:

- infectious diseases
- chemical and radiation hazards
- unhealthy lifestyle choices.

Public health

These agencies constantly monitor threats to public health and study health trends and lifestyle changes in the population. Agencies inform the government about particular dangers, such as the spread of an infectious disease, and advise on the best action.

When a threat to health is identified campaigns target the problem. A campaign is expensive, but it can save lives and save money in the long run, as less money will have to be spent on treatment.

One campaign was about *Chlamydia*, the most common sexually transmitted infection in the UK. People cannot always tell they have the disease so it's easy to pass it on to someone else without knowing. A pregnant woman can pass the infection on to her baby during birth. The baby may have an eye infection or pneumonia.

If untreated, *Chlamydia* can cause serious health problems. The government launched this health-education campaign to raise awareness of this sexually transmitted disease.

Questions

1 Give two characteristics of the National Health Service.

2 Give one advantage and one disadvantage of contacting the NHS by phone or through the website instead of visiting a GP when you feel unwell.

3 Name three different providers of primary healthcare.

4 Name two fields you could specialise in if you became:
   a a doctor
   b a nurse.

Postnatal care

Ellie is like any other child born in the UK. As soon as she was born, Ellie was in the hands of healthcare professionals. They have continued taking an interest in her ever since.

APGAR score

Within a minute of being born, the **midwife** checks Ellie's **APGAR score**. This gives a score of 0, 1, or 2 for five different signs.

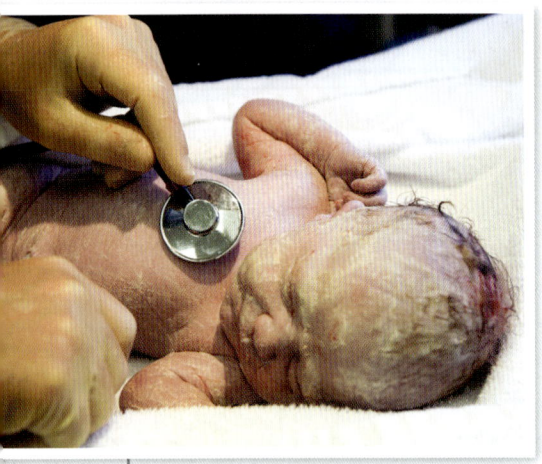

Ellie gets her APGAR check as soon as she is born.

Sign	Score of 2	Score of 1	Score of 0
activity	moving actively	flexing arms or legs	no motion
pulse	above 100 beats/min	below 100 beats/min	no pulse
grimace	sneezes, coughs, turns away	grimaces	no response
appearance	pink or yellow all over	pale hands or feet	blue–grey all over
respiration	good, crying	slow, irregular	not breathing

Ellie's APGAR score is 9. She gets 1 for only grimacing when the midwife touches the inside of her nostril with a thin tube.

APGAR score	Assessment
7 or more	satisfactory progress
from 4 to 6	might need medical help
3 or less	requires immediate attention

Development checks

Ellie's health visitor is responsible for checking her development. To start with, she weighs Ellie and measures her height every week. She also looks for important signs of her development at six months, 18 months, three years, and before she goes to school. For example: How does Ellie play with her toys? How active is she? How does she interact with other people?

Ellie passes this development test easily.

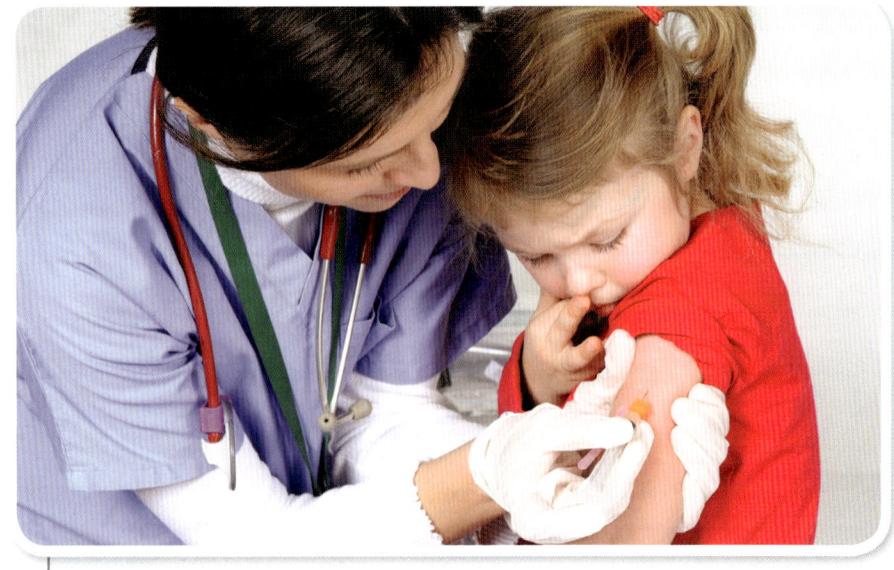

Ellie is vaccinated at the clinic.

Growth charts

The health visitor plots Ellie's weight and length data on graphs called **growth charts**. Here is Ellie's weight chart for her first year of life. Each plotted point shows her weight when the health visitor weighed her.

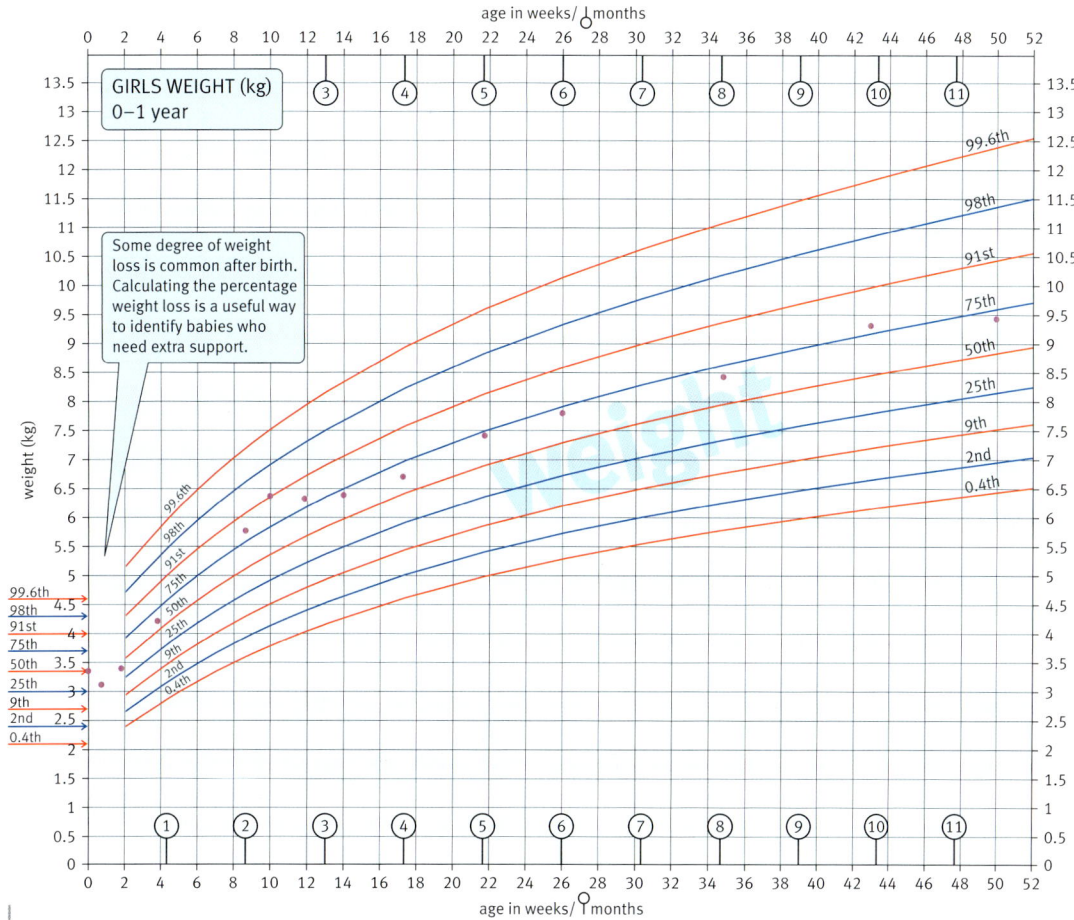

Ellie's weight chart shows how her weight varied in her first year.

The chart has curves that are numbered. These are called centile lines and show the range of weights of most children. At any particular age 75% of babies will weigh less than 75th centile, and 25% of babies will weigh more. It is quite healthy for a baby's weight to be anywhere within the centile lines on the chart.

The chart shows that for the first 10 weeks Ellie put on weight rapidly. The health visitor worried that this weight gain was too rapid, so she decided to monitor Ellie's weight more closely with more frequent checks. Ellie caught a bad cold at week 10 and lost a little weight. Her appetite didn't recover until week 12. For the rest of her first year Ellie's weight stayed close to the curve for the 75th centile.

Clinic visits

Ellie meets a wider range of healthcare professionals on her regular visits to the local health centre. As well as the health visitor, she meets nurses who give her vaccinations and she also has tests for hearing and vision.

Key words
- ✔ midwife
- ✔ APGAR score
- ✔ growth chart

Questions

1 What are the five signs of the APGAR test?

2 Name three different healthcare professionals that Ellie meets.

3 Write out a job description for a health visitor.

The female reproductive system

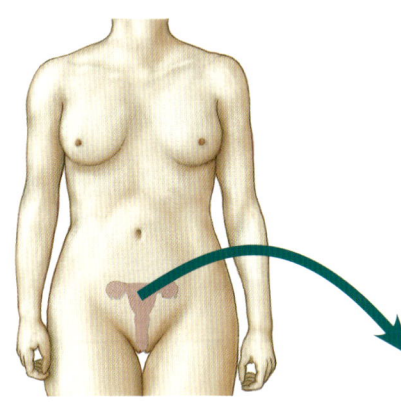

Uterus or womb: here a fertilised ovum develops into a baby. Part of the lining of the uterus is shed each month if the woman is not pregnant. This is called the menstrual cycle.

Ovaries: contain developing eggs or **ova**. When a girl is born each of her ovaries has about a million immature ova. Of these, only about 500 develop during her life. One ovum a month is released between puberty and **menopause**.

Fallopian tubes: tubes with funnel-shaped ends. An ovum passes into the Fallopian tube after it has been released from the ovary.

Cervix: a ring of muscle surrounding the lower end of the womb. Its central opening allows sperm or menstrual blood to pass through. During childbirth it dilates (widens) so that the baby can pass through.

Vagina: a passage leading up to the womb. It secretes mucus. It widens during sexual intercourse and childbirth.

Human **reproductive systems** have one main function: to produce babies.

THE SCIENCE

Key words
✓ cervix
✓ uterus (womb)
✓ Fallopian tubes

Questions

1 Describe the journey of an ovum after being released from the ovary, if a woman is not pregnant.

2 **a** What is ovulation?
 b Approximately how many days after the start of menstruation does ovulation usually happen?

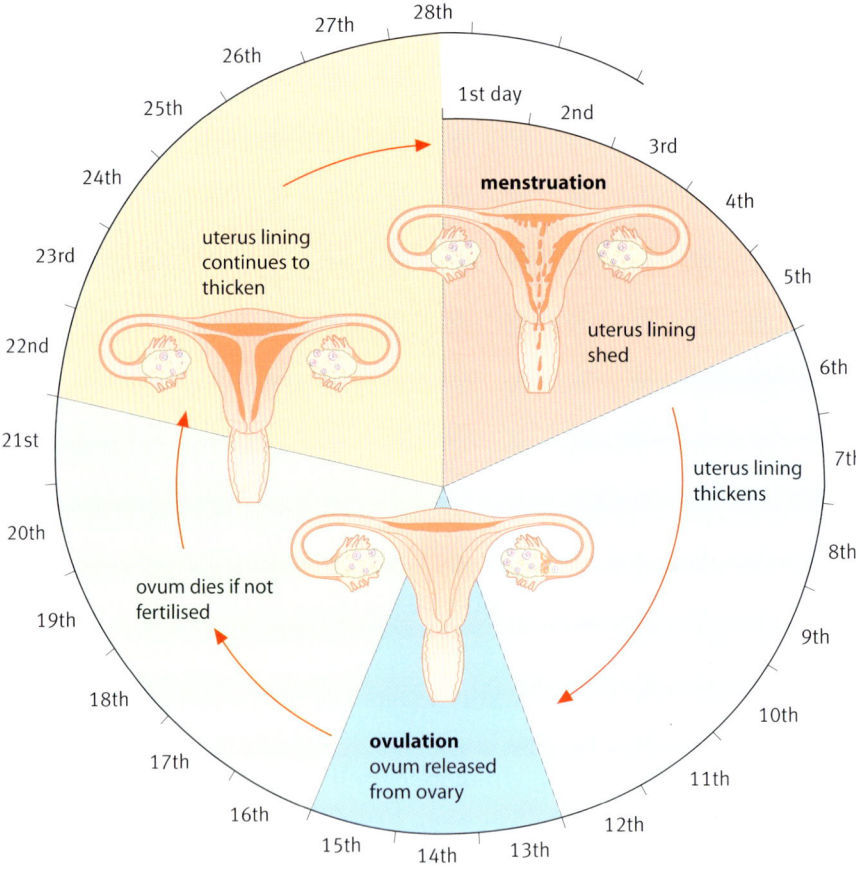

The **menstrual cycle** is controlled by **hormones** carried in the blood. Menstruation starts at puberty and stops between 45 and 55 years old, when a woman can no longer have children. This time in her life is the menopause.

Pregnancy

Fertilisation: the start of pregnancy

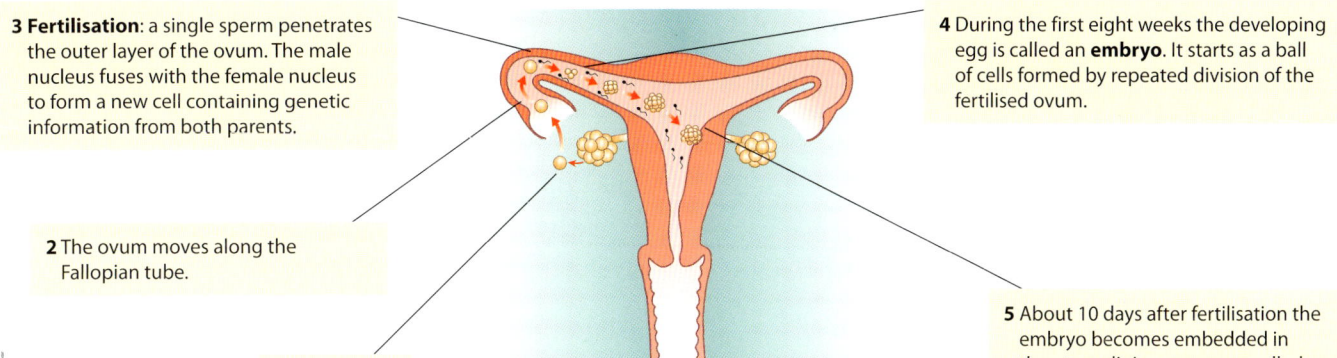

3 **Fertilisation**: a single sperm penetrates the outer layer of the ovum. The male nucleus fuses with the female nucleus to form a new cell containing genetic information from both parents.

2 The ovum moves along the Fallopian tube.

1 Ovulation

4 During the first eight weeks the developing egg is called an **embryo**. It starts as a ball of cells formed by repeated division of the fertilised ovum.

5 About 10 days after fertilisation the embryo becomes embedded in the uterus lining, a process called **implantation**.

Pregnancy is the period between fertilisation and the birth of the new baby.

First trimester (0–3 months)
The fetus's toes and fingers are distinct with tiny nails, but may still be joined by webs of skin. Ears, eyelids, and teeth buds have formed. The fetus is recognisable as a human being.

Second trimester (3–6 months)
The fetus grows rapidly and moves more vigorously. Genital organs are visible.

Third trimester (6–9 months)
The fetus becomes increasingly mature. After about week 28, its organs have grown enough that it could just survive with expert care if born prematurely.

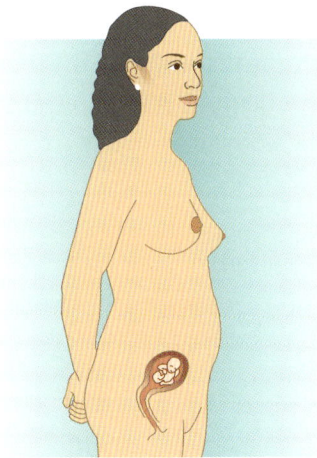

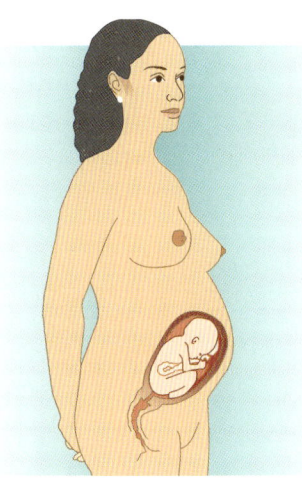

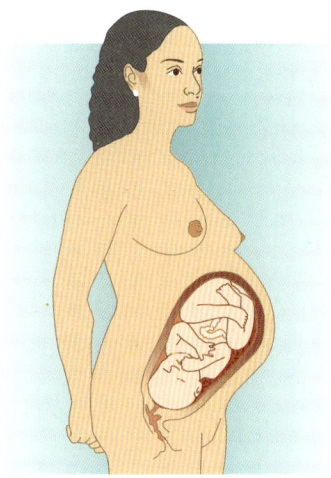

The mother's breasts become tender and enlarge. The area around the nipples darkens. Morning sickness is common. She begins to gain weight.

The mother's heart rate increases and her uterus enlarges. From about 20 weeks on, she can usually feel the baby move.

The mother's skin stretches over the abdomen. She feels slight contractions of the uterus, which become more intense as the birth approaches. The enlarged uterus may press on the bladder, increasing the need to urinate. She may feel tired and breathless. Back pain and **heartburn** are common.

Key word
✓ **pregnancy**

THE SCIENCE

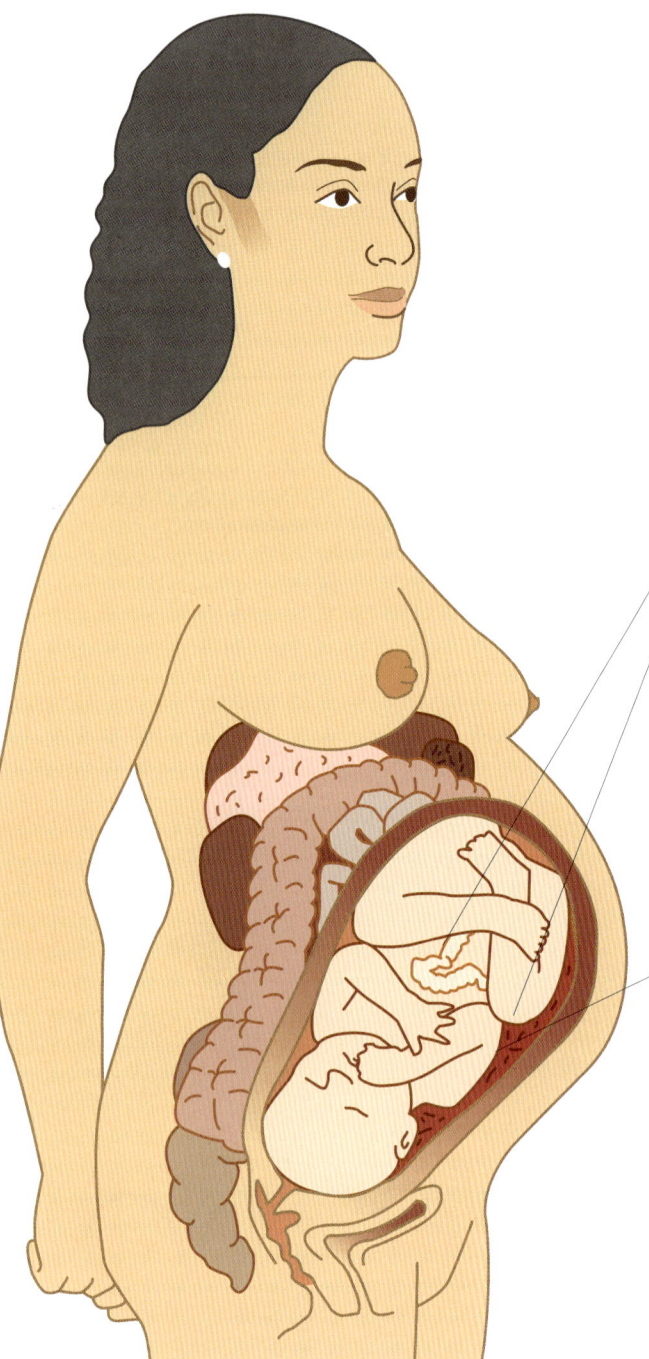

At first the embryo obtains its nutrients and oxygen directly from fluids in the uterus. Later an organ called the placenta forms. Here oxygen and nutrients are exchanged between the mother's blood and the developing baby's blood.

From eight weeks until birth, the developing baby is called a **fetus**.

Umbilical cord: connects the fetus to the placenta.

Placenta: this exchanges materials between the mother's and the baby's blood. Their blood does not mix but materials are exchanged across a thin membrane.

The placenta also produces female hormones that prevent further pregnancies. They also ensure that the uterus grows at the same rate as the baby and that the breasts are ready to produce milk soon after the baby is born.

Amnion: a bag within the uterus filled with **amniotic fluid**. This clear liquid cushions the fetus, protecting it from knocks. The fluid is swallowed by the fetus, absorbed into its bloodstream, and excreted as urine.

Key words
- placenta
- amnion
- amniotic fluid

Questions

1 What is the difference between an embryo and a fetus?

2 What is the amnion and what are its functions?

3 Draw a flowchart showing the process of pregnancy from fertilisation to just before birth.

A healthy pregnancy

For the majority of women, pregnancy doesn't stop life going on as normal. However, there are a couple of health problems that can sometimes arise.

Pre-eclampsia

Pre-eclampsia affects about 10% of pregnancies in the UK, and is more common in first-time pregnancies.

In the early stages of pregnancy pre-eclampsia has no symptoms. It can only be detected by regular antenatal checks of the mother's **blood pressure** and urine. Health practitioners also watch for a slower-than-normal growth of the baby.

Later in the pregnancy the condition can become acute, causing the mother's blood pressure to rise to dangerously high levels. This can cause fluid retention, swelling, and protein in the urine. Without treatment, the woman may develop increasingly severe symptoms including fits.

She will be admitted to hospital for bed rest, where doctors and midwives can give her drugs to lower her blood pressure and keep a close watch on the health of her and her baby. At present, the only way to treat women experiencing severe symptoms is to deliver the baby early, by **caesarean section**.

Gestational diabetes

Gestational diabetes affects about 5% of pregnancies in the UK.

Hormones from the baby interfere with the mother's ability to control glucose levels in her blood. This is detected by measuring the amount of glucose in her blood. Often there are no other symptoms. The condition is usually treated by controlling the mother's diet. Gestational diabetes can affect the baby's liver, giving the baby **jaundice**.

The majority of women have a healthy pregnancy.

Questions

4   What are the symptoms of pre-eclampsia? What is the treatment?

5   What are the symptoms of gestational diabetes? What is the treatment?

Key words

- ✓ blood pressure
- ✓ pre-eclampsia
- ✓ gestational diabetes

IVF stands for *in vitro fertilisation*. It means that the egg is fertilised by sperm in a glass container. Children conceived this way are sometimes called test-tube babies. There are five stages in the IVF process.

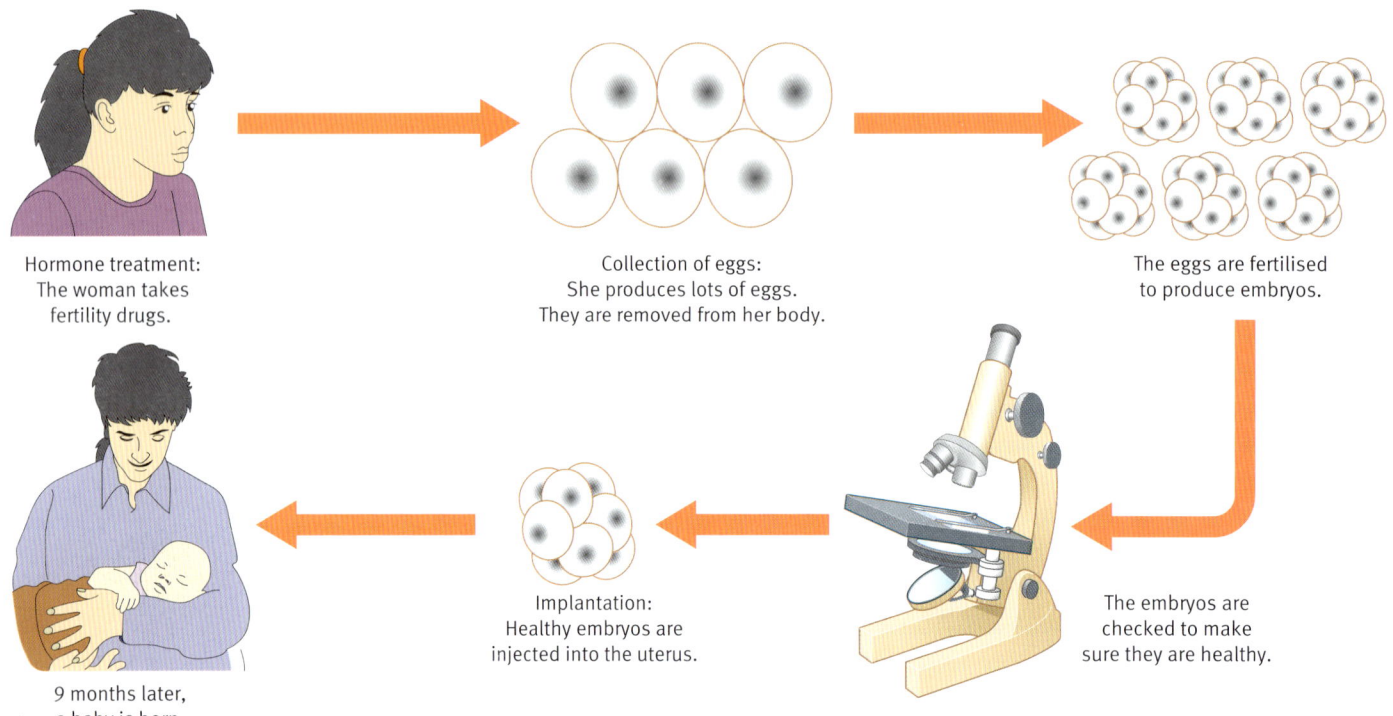

Hormone treatment: The woman takes fertility drugs.

Collection of eggs: She produces lots of eggs. They are removed from her body.

The eggs are fertilised to produce embryos.

Implantation: Healthy embryos are injected into the uterus.

The embryos are checked to make sure they are healthy.

9 months later, a baby is born.

Stages of IVF treatment.

Before people take part in IVF treatment they are given **counselling**. This helps them to understand the procedures that they will undergo. They are also encouraged to think about the possibilities of both success and failure.

Stimulation

The woman has **hormone treatment** to stimulate her ovaries to ripen and release several eggs instead of one. This requires a series of injections in her abdomen at precise time intervals over a period of 10 days. This can be uncomfortable and sometimes painful. In about 3% of women it can result in over-stimulation, which can be dangerous to her health.

Egg collection

Ultrasound is used to check the number of eggs developing and their exact position in the body. When the eggs are mature, they are collected from the womb or Fallopian tubes using a tube inserted through the vagina. The procedure is generally done under a local anaesthetic, with the woman sedated. Several eggs are collected – often up to eight.

Fertilisation

Sperm that has been collected from the male donor is then added to the eggs. If the eggs are **fertilised** they produce embryos and grow by cell division. After a few days the embryos can be viewed under a powerful microscope. One cell of each embryo is collected and tested for genetic abnormalities. The embryos are not damaged by this procedure. They just produce more cells and continue to grow.

Implantation

The process of transferring the embryos from the laboratory into the woman's uterus is risky. The risks are reduced by giving the woman drugs and hormones that prepare the uterus lining. The woman should not wear perfume or use scented soaps immediately before the procedure because this can affect the embryos. Usually two (or sometimes three) embryos are placed in the uterus, through the vagina, using a long tube on a syringe. The woman is then advised to rest for one or two days.

Pregnancy

If the transfer of the embryos is successful, the woman becomes pregnant. This will need to be confirmed with a blood test. Drugs and hormones will be given daily to ensure the pregnancy continues. A frequent complication is that more than one baby may be produced, resulting in a **multiple birth**. The condition of the fetus(es) is monitored regularly by ultrasound scanning.

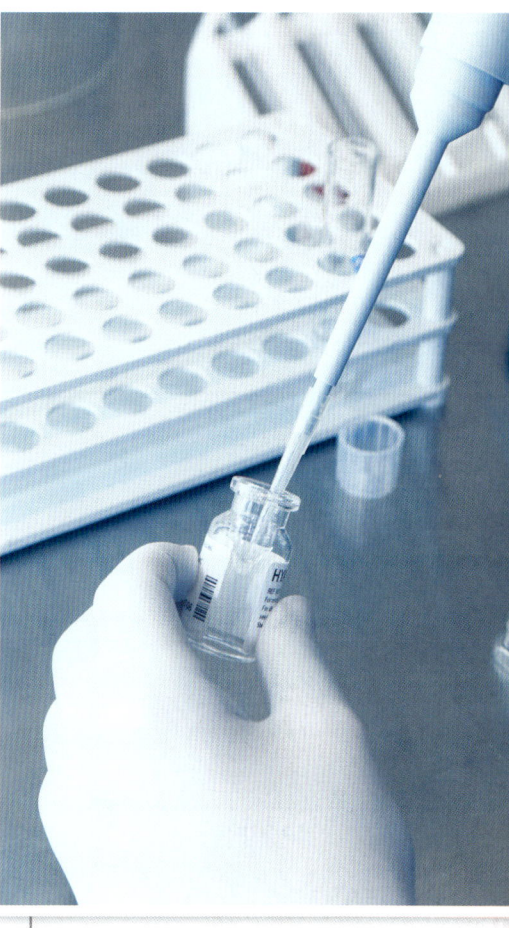

The sperm are added to the eggs in the glass jar.

IVF treatment can often result in more than one baby.

Key words

- ✔ **counselling**
- ✔ **hormone treatment**
- ✔ **fertilisation**
- ✔ **implantation**
- ✔ **multiple birth**

Questions

1 Describe the stages of IVF.

2 Explain why one cell of the embryo is removed.

Clinical tests

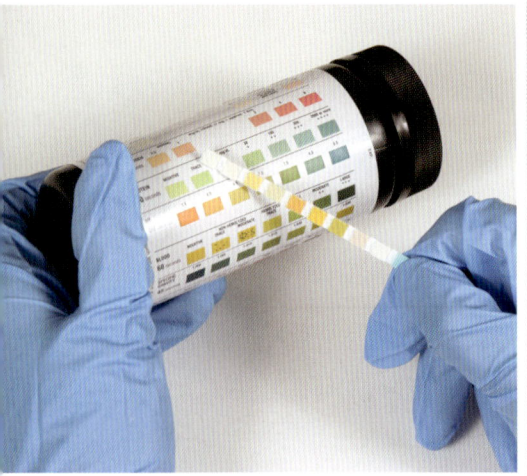

A urine test. The specially prepared strip is dipped into the urine sample and then compared against the colour chart. Each coloured square on the strip indicates the level of a different compound present in the urine.

THE SCIENCE

CHECK SAFETY

Always wear surgical gloves when handling body fluids.

A pregnant woman can expect to have some tests when she visits the antenatal clinic. These can provide information about the health of both the mother and her fetus.

Testing urine

Regular monitoring of levels of protein and glucose in the mother's urine can give valuable warning of complications such as pre-eclampsia and gestational diabetes. The procedure is very straightforward.

Sampling blood

Positive results from a urine test often need to be confirmed by further tests made on a blood sample. These have to be done in specialist laboratories, usually in hospitals. A small sample of blood will be taken at the antenatal clinic and sent off to the hospital for analysis. Blood samples are usually taken from the arm, just below the elbow, as follows:

- Apply a pressure collar around the upper arm.
- Find a vein just under the surface of the skin.
- Sterilise the skin surface using an alcohol swab.
- Insert the needle of the syringe into the vein.
- Draw off the blood into the syringe.
- Remove the syringe and apply pressure to the arm to stop the bleeding.
- Transfer the blood to the sample tube.
- Label the sample tube with the patient's details.

Sample tubes have colour-coded lids to indicate the type of test that needs to be performed at the laboratory.

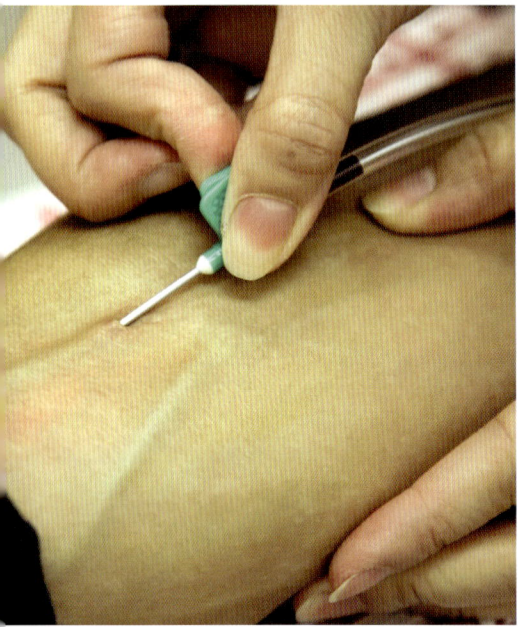

The needle must be carefully inserted into the vein.

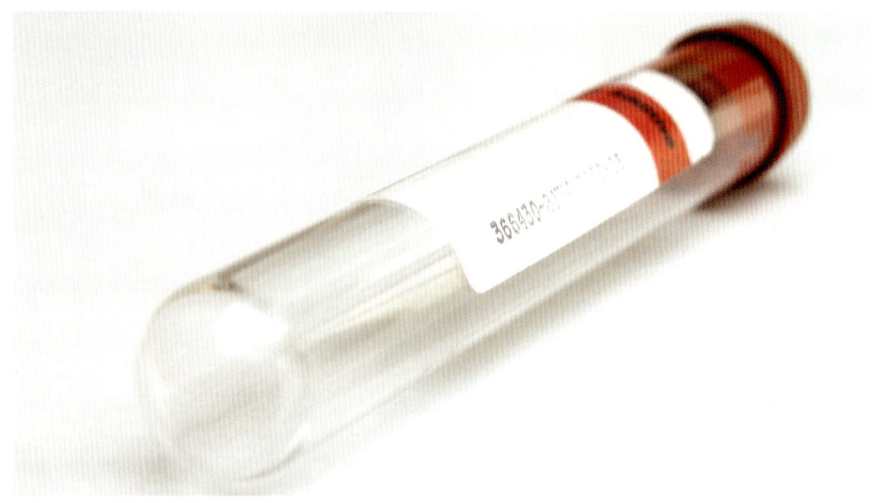

An empty blood-sample tube. Note the label for the patient's details. Some sample tubes are supplied with all of the air removed. This type of tube is connected to the needle once it is in the patient's arm. The blood from the vein flows straight into the tube and a syringe is not needed.

Measuring blood pressure

Sudden rises in blood pressure can indicate the onset of pre-eclampsia, so blood pressure is measured regularly during pregnancy. Electronic monitors are portable and easy to use. With these monitors pregnant women can do their own daily checks at home if necessary.

Amniocentesis

Sometimes tests are carried out during pregnancy to see whether the fetus is developing normally.

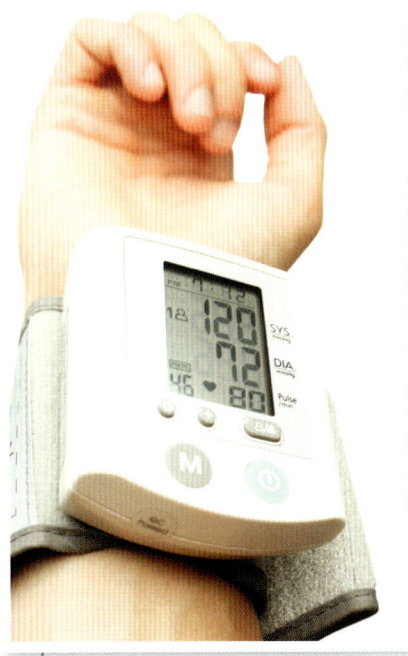

An electronic blood-pressure meter on a patient's wrist.

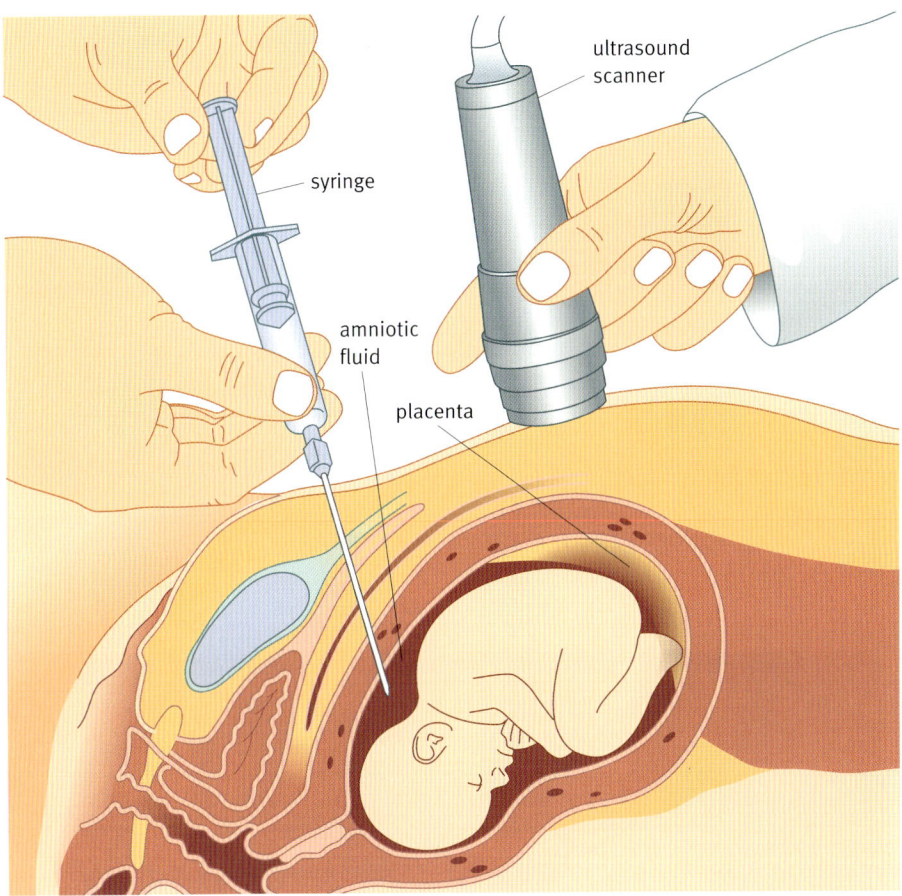

ultrasound scanner

syringe

amniotic fluid

placenta

Some tests need to be performed at a hospital instead of a clinic. **Amniocentesis** is one of them. Using an ultrasound scan as a guide, the doctor inserts a needle into the mother's uterus. A sample of amniotic fluid is then removed from around the fetus. The amniotic fluid undergoes a series of tests. This procedure carries a small risk to the fetus, so it is only carried out when there is a good reason to investigate how the fetus is developing.

Questions

1 Describe how to do a urine test.

2 Describe how to take a blood sample.

3 Why don't all pregnant women have an amniocentesis?

Ultrasound scanning

THE SCIENCE

Ultrasound scanning uses sound waves to make an image of the inside of the body. The technique is also known as **sonography**. The GP may refer a patient for an ultrasound scan if they have an unexplained lump, for example.

The sonographer spreads a gel on the patient's skin and moves a scanner over the skin surface. The scanner sends very high frequency sound waves through the body. Where the waves meet boundaries between different tissues, such as between muscle and fatty tissue, they are reflected in a particular way.

The scanner picks up these reflected waves and a computer turns the signals into a picture. The sonographer can see the heart beating and the arteries expanding rhythmically. They can recognise organs such as the liver, kidneys, and pancreas on the greyscale image. The scan will reveal whether the organs appear normal, or whether further investigation is needed.

As well as taking images externally, small ultrasound probes can be placed into body openings such as the mouth, ear, or vagina, to get a clearer picture.

Ultrasound scans are carried out on pregnant women to check the fetus is developing normally. There are other imaging techniques that can be used to see inside the body, such as X-rays. X-rays are not used on pregnant women because they can harm a developing fetus. Ultrasound is much safer.

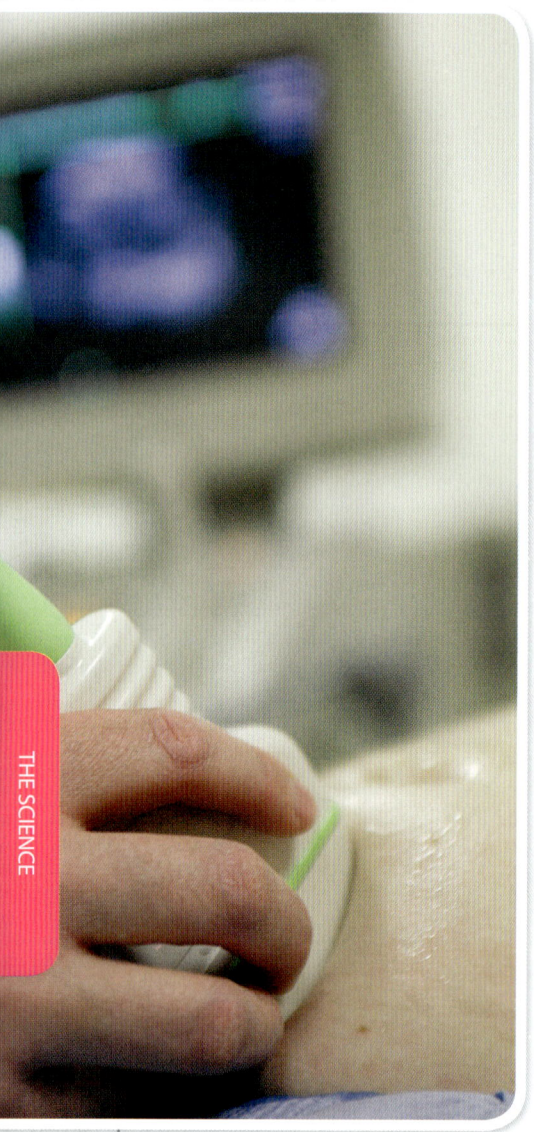

The scanner is moved over the skin. Without the gel, air gaps would spoil the image.

The sonographer can see the internal organs. Moving the scanner around gives a different view.

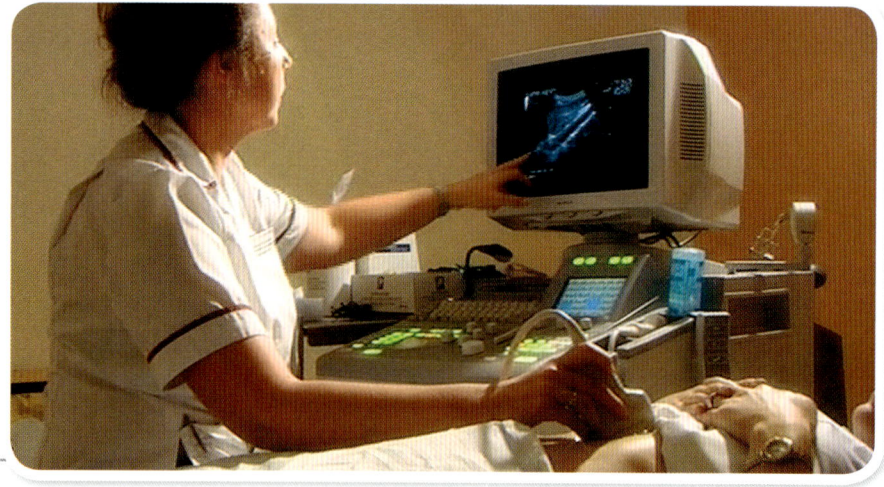

Key word

- ✓ **ultrasound scanning**

Questions

1 What type of waves does ultrasound use?

2 Why is ultrasound the imaging method used for pregnant women?

Birth

Birth takes place at about 40 weeks. There are three stages of labour, shown below.

First stage
Contractions of the uterus force the baby's head onto the cervix. The cervix dilates to 10 cm and its mucus plug is discharged. The amnion may break so that amniotic fluid escapes from the vagina, known as 'breaking of the waters'.

Second stage
Birth contractions become stronger and closer together. The mother feels a strong urge to push with each contraction until the baby is born.

Third stage
Further contractions of the uterus push the placenta and umbilical cord (the 'afterbirth') out of the mother's body.

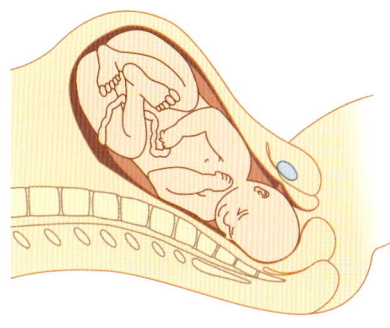

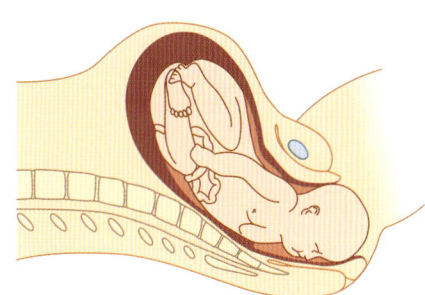

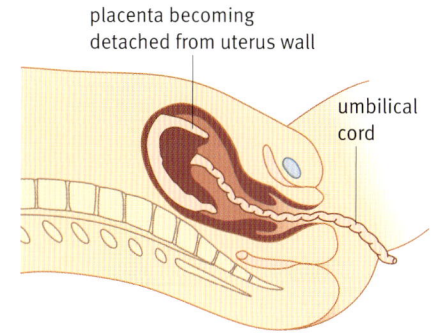

placenta becoming detached from uterus wall

umbilical cord

THE SCIENCE

Pain relief during labour

Some mothers experience little pain during childbirth and manage the pain using natural techniques such as relaxation and controlled breathing. For others, drugs such as nitrous oxide (in "gas and air") or pethidine are needed for pain relief. An **epidural** is an injection of anaesthetic into the epidural space around the spinal cord.

Question

1 Make a flow chart showing all three stages of labour.

Gas and air

The woman inhales gas and air through the mask.

tube to cylinder containing mixture of oxygen and nitrous oxide

Gas and air is a mixture of nitrous oxide (dinitrogen oxide) and oxygen. It softens labour pain but does not remove it completely. The woman can control the amount she inhales. Too much might make her feel light-headed and unable to concentrate.

Epidural

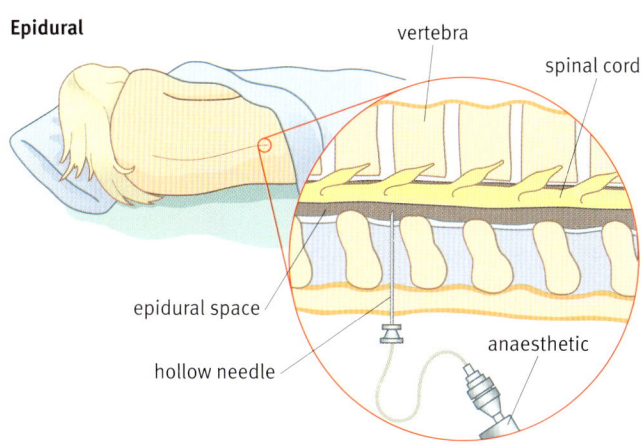

vertebra

spinal cord

epidural space

hollow needle

anaesthetic

An epidural is an injection into the epidural space around the spinal cord. It can provide complete pain relief but the woman cannot move around. She needs a catheter and a drip, and is hooked up to a fetal heart monitor.

Measuring blood pressure

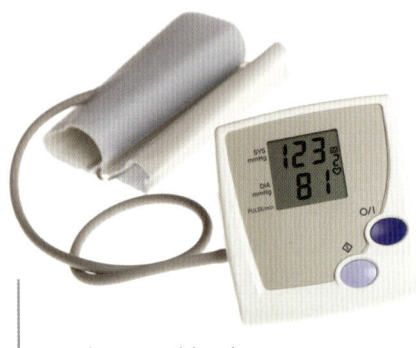

An electronic blood-pressure meter.

Equipment

Blood pressure is often measured with an electronic **sphygmomanometer**. This consists of:

- an inflatable sleeve, called a cuff
- a box containing the electronics
- a tube to carry pressurised air from the box to the cuff.

Procedure

1. Place the cuff around your arm, just above the elbow.
2. Connect the tube to the box and switch it on.
3. Sit at a table and place your forearm on it. Relax.
4. Press the start button. The box pumps air into the cuff, inflating it.
5. A valve in the box slowly releases air from the cuff, registering when a pulse is first detected.
6. The remaining air in the cuff is released and the box measures the point when the pressure is too low for a pulse to be detected.

The reading

- Blood pressure is measured in **millimetres of mercury (mm Hg)**.
- There are two numbers, for example, 120/80 mm Hg ('120 over 80').
- The first figure is the highest pressure at which the pulse can be detected – the **systolic blood pressure**.
- The second figure is the lowest pressure at which the pulse can be detected – the **diastolic blood pressure**.
- Pulse rate is measured in **beats per minute (bpm)**.

Interpreting the reading

This chart shows possible diagnoses based on the blood pressure of a pregnant woman.

Systolic pressure (mm Hg)	Diastolic pressure (mm Hg)	Possible diagnosis and treatment
< 100	< 60	low: dehydration; drink more fluids blood pooling in the feet; wear pressure stockings
100 to 120	60 to 80	normal: no action required
120 to 140	80 to 90	high: reduce salt and fat in diet; stop smoking or drinking alcohol
> 140	> 90	too high: possible pre-eclampsia; bed rest and drugs required

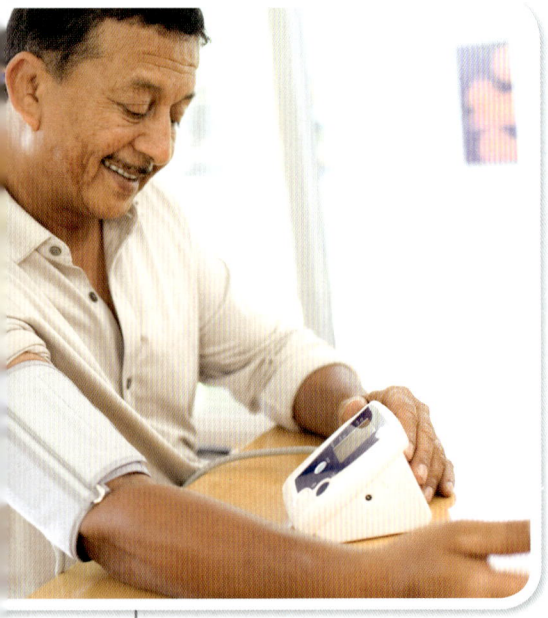

This man is measuring his own blood pressure.

Testing urine

Equipment
- sterile container to hold the urine sample
- test sticks and colour chart (usually on the side of the stick bottle)
- surgical gloves.

Procedure

Collecting the specimen
- The best time for the patient to collect a sample of urine is first thing in the morning.
- They should insert the container into the urine stream after it has started. This prevents contamination of the sample by substances on the surface of the urethra.

Testing
Always wear surgical gloves when handling body fluids.

1 Take a fresh test strip out of the bottle, holding it by the white end.

2 Dip the other end of the stick into the sample of urine for the time specified by the manufacturer. This is typically a minute.

3 Hold the end of the stick against the colour chart on the bottle.

4 Find the best colour match, then read off the level for each substance tested.

Tests available include:
- the amount of glucose in the urine
- protein or blood in the urine
- whether a woman is pregnant.

Interpreting the reading
- Glucose present may be a sign of diabetes.
- Protein present may be a sign of kidney damage or disease, or a urinary tract infection.
- Blood present may be a sign of disease in the kidney, urinary system, or bladder.

The results should be reported to the patient's GP.

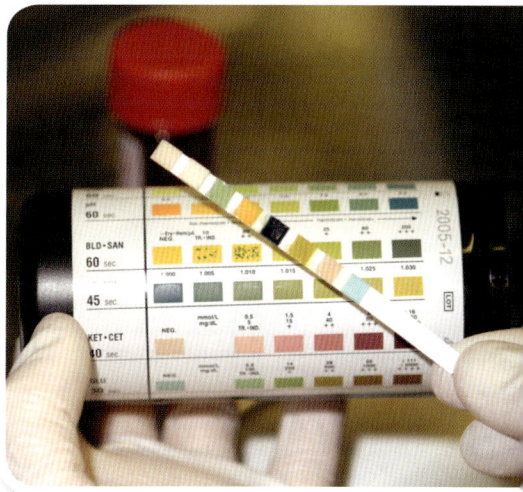

Each stick can test for several different substances in the urine sample. You need to hold each band against its own colour chart on the bottle.

PROCEDURES & TECHNIQUES

Module Summary

People and organisations

- about local organisations that provide healthcare for your community, what they do, and how they affect your community (for example, health centres, hospitals, opticians, and dentists)
- that the National Health Service makes healthcare available for all citizens, monitors national health trends, and plans and allocates resources for suitable healthcare including specialist care
- that people who work in healthcare have scientific and technical skills and they include doctors, nurses, nutritionists, pharmacists, opticians, and dentists
- that hospital A&E departments practise for major incidents to make sure they have staff and equipment available and know what resources they need
- that it is a key role of GPs to refer people to other specialist services
- that regular contact between people working in healthcare and their patients allows them to build trust and be aware of an individual's medical history
- that some organisations provide education and public information about health
- that people working in healthcare have to work within health-and-safety guidelines and other regulations that control the work they do
- that patients have a right to informed consent for any diagnostic procedure or treatment because all procedures and treatments carry some risk that is weighed against the benefits gained
- that personal medical information must be recorded, stored, and made available to make it easier to treat the patient, to provide information about different treatments, and as evidence if things go wrong.

The science

- the main parts of the female reproductive system, what they do, and how they change during pregnancy and birth
- about the main stages in IVF
- about gestational diabetes and pre-eclampsia – conditions that can occur during pregnancy and that are hazardous for mother and child
- how an APGAR score and growth charts are used in the postnatal care of a child, with development tests and visits to a clinic
- how to interpret weight and height data shown on an infant growth chart.

Standard procedures

- how paramedics and A&E departments use a triage system to prioritise patients
- why people working in healthcare need to know about our medical or lifestyle history before treatment begins
- the steps in taking a blood sample
- how a urine test, blood test, or blood-pressure monitoring could lead to diagnosis of gestational diabetes or pre-eclampsia
- how blood tests can lead to the diagnosis of anaemia and birth defects such as Down's syndrome and spina bifida.

Review Questions

1 Many doctors work for the National Health Service (NHS).

a Describe the job of two other health practitioners who work for the NHS.

b Explain the role of the NHS in the UK.

2 The NHS has a vaccination programme for babies and young children. Explain why money is spent on publicity for the programme.

3 Drew breaks her leg. She is taken to the hospital accident and emergency (A&E) department.

a Why do A&E staff use triage to decide how quickly to treat Drew?

b An X-ray is taken to show the damage to the bone. X-rays are ionising radiation, which can damage body cells. Explain why the X-ray is taken despite this risk.

c Staff decide that Drew needs an operation to set the broken bone. They ask her to sign a consent form. What do they need to do **before** she signs the form?

d Describe **two** pieces of information about Drew's medical history that the staff will need to record before the operation.

e Explain why details of Drew's treatment at A&E need to be recorded and stored.

4 Anna is pregnant.

a Her GP is worried that she might have gestational diabetes. He decides that she needs a blood test.

   i Explain, in detail, how to take a blood sample from a patient.

   ii What substance should Anna's blood be tested for?

   iii Explain why Anna's GP is concenred about gestational diabetes.

b Explain why Anna visits the midwife regularly during her pregnancy.

c Anna goes to the hospital for an ultrasound scan. Describe the information that can be obtained from the scan.

5 A midwife helps Freda to give birth to Ellie.

a Explain why the midwife gives the baby an APGAR score as soon as she is born.

b Explain why the baby's weight is measured at the clinic regularly during her first year.

This is the growth chart for Ellie (you can see a larger version on page 67).

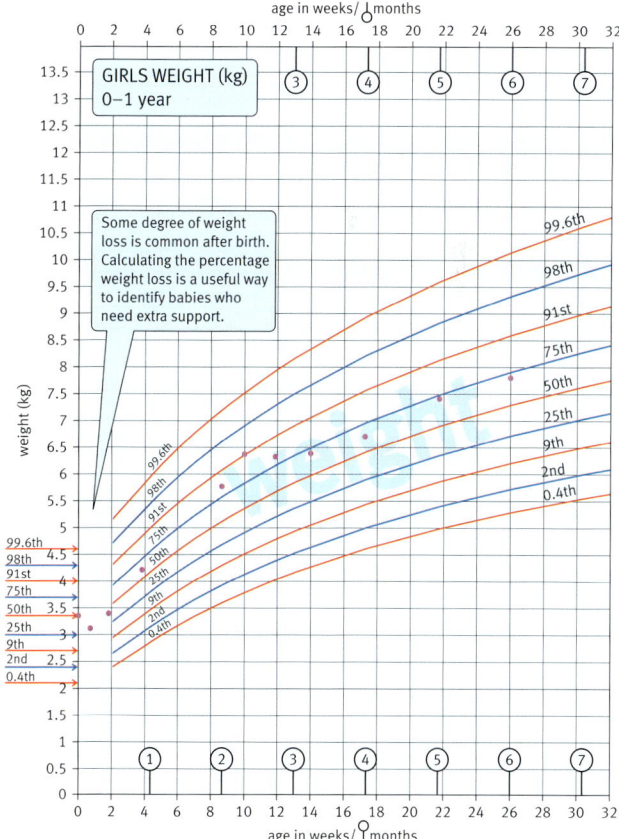

c Describe how Ellie's weight changed over the first six months after she was born.

d When Ellie was six months old, which centile line was she closest to?

collar

skin sterilisation

select vein

blood sampling

insert needle

withdraw sample

label sample

triage

Down's syndrome

spina bifida

birth defects

blood tests

anaemia

diabetes

lifestyle / medical history

specialists (by referral)

glucose

protein

urine tests

pharmacists

doctors

people

dentists

nurses

nutritionists

midwives

opticians

organisations

dentists

NHS

health centres

national plans

hospitals

resources

opticians

all citizens

education

public health information

A&E departments

WORKING IN HEALTHCARE

biology of human reproduction

- hazardous conditions
  - gestational diabetes
  - pre-eclampsia
- female reproductive system
  - parts
    - amniotic fluid
    - uterus
    - Fallopian tubes
    - placenta
    - amnion
    - cervix
  - changes
    - pregnancy
    - birth
- IVF
  - counselling
  - monitoring
  - fertilisation
  - collection of eggs
  - hormone treatment
  - implantation
  - possibility of multiple births

standard procedures

interpreting data
- APGAR
- growth charts (height + weight)
- risks and benefits of treatments

regulations
- health and safety
  - informed consent
- record-keeping
  - storing personal information

A3 Monitoring and protecting the environment

Why study the environment?

At the beginning of the 21st century, people across the world are experiencing the effects of global climate change and the impact of 200 years of industrial activity. This topic shows you how scientists work to measure and assess the health of an environment. Careful monitoring shows us when an area is changing. This helps us decide when to take action.

What you already know

- organisms are affected by changes in their environment

- changes in the environment can be measured using living and non-living indicators

- living indicators include lichen and mayfly larvae

- non-living indicators include nitrate levels, temperature, and carbon dioxide levels

- repeating measurements leads to a better estimate of the true value

- how to calculate the mean from a set of data

- how to identify outliers in a set of data.

The Science

In this module you have a chance to practise some of the techniques used by environmental scientists. Some methods are used in the field and others in the laboratory. If you are interested in a career in environmental protection, you will see some of the work that is done and how it is managed, standardised, and regulated.

Find out about

- how observations can be recorded

- collecting and keeping samples

- the quantitative and qualitative tests scientists do

- how to evaluate data in scientific reports

- how environmental information can help us to spot pollution events and understand climate change.

Cape Farewell

The Cape Farewell project is an unusual collaboration of scientists, artists, and educators. The people involved are concerned about climate change and are enthusiastic about explaining some of the important results of a changing climate.

Several expeditions in the Cape Farewell project have been to the Arctic Ocean. On these particular expeditions, scientists collect information about the Arctic environment. Artists experience the environment, see how it is changing, and work alongside the scientists.

Collecting information regularly is part of monitoring any environment. Environmental scientists can track how an environment is changing by comparing data collected on different expeditions and data collected by other scientists. When they see that the changes are happening very quickly, or causing irreversible damage, they can try to take action to protect an environment.

Cape Farewell boat in the Arctic Ocean.

Arctic animals have evolved to blend into their surroundings.

What would you measure in the Arctic?

Temperature

The Arctic and the Antarctic are the coldest places on Earth. In these places the temperature is usually so low that any freshwater is frozen as ice or snow. One of the big changes in our climate is increasing global temperature. In the Arctic and Antarctic, changing temperatures will have a big environmental impact and should be easy to detect.

Carbon dioxide

One of the causes of global climate change seems to be the increasing amount of carbon dioxide and other greenhouse gases in the atmosphere. Carbon dioxide dissolves in water and makes water more acidic. It dissolves more in cold water than in warm water. Measuring the pH of Arctic waters might tell us something important about carbon dioxide levels.

Much of the landscape of the Arctic is ocean. The seawater contains huge numbers of tiny plants (phytoplankton) and tiny animals (zooplankton) that provide food for many of the other animals in the environment. The phytoplankton use carbon dioxide when they photosynthesise. If carbon dioxide levels are rising this could make them grow more quickly. These plankton are at the base of the food chain and they live entirely in the ocean. Observations and measurements of plankton populations help us to understand the Arctic environment.

Scientists sample water for phytoplankton.

Salinity

As the freshwater locked in ice and snow melts, it dilutes the seawater. So the saltiness, or salinity, of the seawater in the Arctic is also an important indicator.

Scientists discuss their observations and data on the boat.

Why are artists involved?

The artists in the Cape Farewell project have created many works of art including sculptures, videos, songs, poems, and installations. Exhibitions and presentations of this artwork explain aspects of climate change and may inspire people to take action to reduce human impact on environments.

Artwork from the Cape Farewell project: Ice lens by Heather Ackroyd and Dan Harvey.

Working in environmental protection

Over 5 million people in England and Wales live and work in properties that are at risk of flooding from rivers or the sea.

PEOPLE & ORGANISATIONS

Hydrologists study water flow and suggest how to prevent flooding and protect sensitive environments.

Many different kinds of people work in environmental protection, including:

* technicians
* scientists
* engineers
* lawyers
* educators
* journalists
* managers.

If you are interested in environmental protection, you could choose to study in many areas of science, for example, biology, biochemistry, and microbiology. You could focus on a particular environment, such as marine, aquatic, or forest, and focus on how change affects people or places. Some courses are very practical, others more theoretical.

You could look for a career in a private company, or working for a government agency or charity. There is work in environmental protection in the UK and all over the world.

The UK government and international organisations such as the United Nations Environment Programme (UNEP) employ people to decide global and local priorities for action to protect the environment. They also monitor the work that others do. For example, Defra, the Department for the Environment, Food, and Rural Affairs in the UK, sets targets for water quality, biodiversity, air quality, marine health, and land management. Defra works with other UK government departments to make sure these national goals are met.

The Environment Agency in the UK works to protect and improve the environment in line with the priorities set by government. Private companies also provide environmental services, such as environmental impact assessment or environmental monitoring.

More environmental work is done by non-governmental organisations, or charities such as the World Wide Fund for Nature (WWF). Charities also campaign to raise awareness and influence policy on the issues they think are most important.

Working for the WWF

Jon works for the WWF, on the Climate Change Team. Jon and his colleagues work to persuade businesses and governments across the world to take action that will:

- reduce emissions of greenhouse gases
- help vulnerable people adapt to **climate change**
- where possible, protect natural habitats from the effects of climate change.

How did you become ... an environmental scientist?

Jon became programme manager in the Climate Change Team at the WWF in 2008. He studied Ecology at University, and later studied computing and environmental policy. He has worked as a zookeeper, a science teacher, and for a private company carrying out environmental impact assessments. Members of the team have degrees in politics, economics, law, and journalism. They share a commitment to protecting the natural world and vulnerable people who depend upon it for their livelihoods.

Scientists working for the WWF monitor the effects of climate change.

Questions

1 Name two organisations that work to protect the environment.

2 Give two ways in which charities can raise awareness of the issues that are important to them.

Key word
- ✓ **climate change**

In England and Wales the Environment Agency is responsible for the protection and improvement of the environment. Scientists working for the Environment Agency regularly monitor air, water, and land quality. They work with industry and local communities to keep pollution to a minimum.

Monitoring water

Scientists from the Environment Agency are also responsible for maintaining and improving water quality, both freshwater and seawater. They have responsibility for the freshwater in lakes, rivers, and streams, and the groundwater held in rocks. They also make sure that inland and coastal waters are safe for wildlife and recreation. They may use indicator organisms as a sign of water quality. For example, mayflies are an indicator species for a healthy river environment.

The Drinking Water Inspectorate (DWI) oversees the quality of the tap-water supply in England and Wales. It checks up on the standards of the water supply companies. The checks include tests for bacteria and chemical contaminants and the look and taste of the water. The DWI can take action if a water supply company's standards are not good enough.

Monitoring the land

Testing soils for poison is another job for analytical scientists. In the past, new housing and factories were built on 'greenfield' sites. However, there is now pressure to reuse old industrial sites, or 'brownfield' sites. It is important for the planners to know whether or not a brownfield site is contaminated with poisons.

Monitoring the air

Smoke, dust, pollen, and poisonous gases are some of the pollutants that may be in the air you breathe. The main contributions to air pollution in the UK are from power stations, transport, and industrial activities. Levels of pollutants in the air have reduced since 1990. This is due to laws regulating emissions. Tighter controls have led to the development of cleaner fuels and industrial technologies. There are targets for even greater reduction of some pollutants.

The Environment Agency and local authority scientists record the presence of different pollutants in the air from monitoring stations all over the country.

This Environment Protection Officer describes her job as extremely varied. 'In the morning I could be meeting a Managing Director to discuss the redevelopment of a contaminated land site. In the afternoon I might be investigating fly-tipping or water pollution.'

This sensor can be installed to monitor the quality of drinking water from the mains supply. It measures properties such as pH, chlorine content, colour, and cloudiness.

This table shows part of the results of an air-quality survey carried out by the Environment Agency at Merthyr Tydfil in South Wales, from December 2002 to April 2003. Local residents had complained of smells from a nearby landfill site. Environmental scientists from the Agency measured several airborne pollutants at hourly intervals (or more often) for 118 days. In the table, the results are compared against maximum permitted values. These are Air Quality Strategy objectives (set by the UK government) or World Health Organization guidelines.

Pollutant	Averaging time	Recorded maximum ($\mu g/m^3$)	Maximum permitted value ($\mu g/m^3$)
particulate matter (< 10 micrometers)	24 hour	54.2	50
SO_2 (sulfur dioxide)	15 min	59.9	266
NO_2 (nitrogen dioxide)	1 hour	200	200
H_2S (hydrogen sulfide)	30 min	4.3	7 (annoyance)
H_2S (hydrogen sulfide)	24 hour	3.1	150 (human health)

A further study of the air quality in this area between November 2005 and May 2006 found that the levels of particulate matter and nitrogen dioxide were now lower and that hydrogen sulfide was at the 'annoyance' level less than 1% of the time.

Sometimes environmental scientists take their labs with them. The Environment Agency Wales uses this mobile monitoring station to check air quality.

Questions

These questions are about the air-quality survey at Merthyr Tydfil in 2002–2003.

1 What triggered the survey at this location?

2 What pollutants have been studied?

3 How does the Environment Agency decide if the levels of pollutants are harmful?

4 Have the scientists found any hazards?

5 What advice would they give to the people managing the landfill site?

6 Why is it useful to repeat the investigation after a few years?

Good laboratory practice

Working in a laboratory

There are many European and national regulations designed to protect people at work, the general public, and the environment. Laboratories have to comply with these regulations. In order to work safely and accurately it is necessary to have a well-organised laboratory. Good laboratory practice depends on:

- following **health-and-safety** regulations
- regular **maintenance** and checking of equipment
- up-to-date training for staff.

It is the responsibility of every worker in a laboratory to observe the health-and-safety regulations.

Health-and-safety regulations

A special health-and-safety officer is usually responsible for creating a set of health-and-safety regulations for a laboratory. It is the responsibility of every worker to observe these regulations. Health-and-safety regulations vary for different laboratories according to the hazards, but they might include:

- instructions on doing risk assessments
- rules about using protective clothing
- standard safety procedures to be followed
- rules about tidiness, including how to label and store materials and dispose of waste
- special rules about handling hazardous chemicals, or working alone
- how to report accidents or equipment faults
- what to do in an emergency.

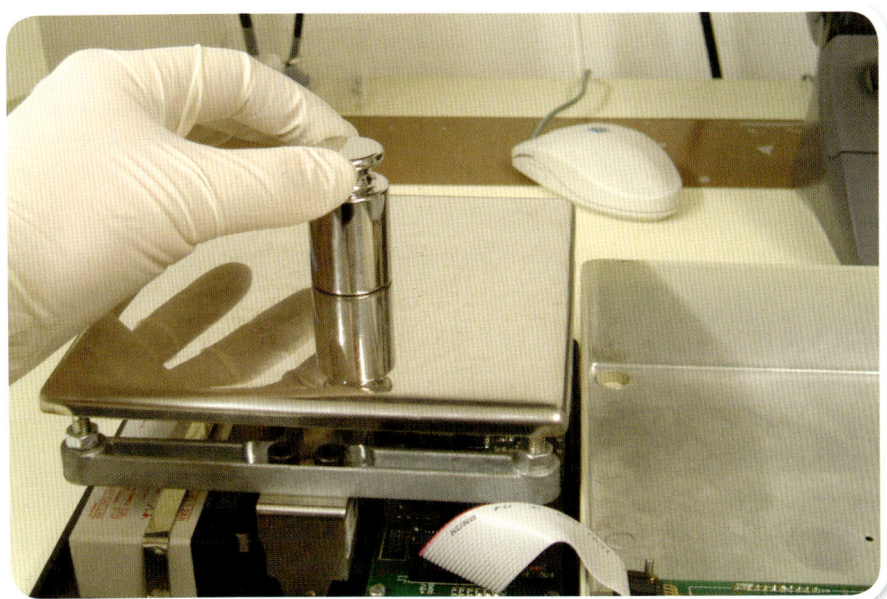

This electronic balance, like all equipment, must be maintained correctly.

Looking after equipment

It is important for good laboratory practice to keep equipment properly maintained. For glassware, this involves efficient cleaning and drying after use, and suitable storage. A dirty pipette does not deliver the correct volume of liquid. Laboratory instruments are only accurate if they are properly maintained. Many need frequent cleaning and adjustment, together with regular visits from a service engineer.

The manufacturer's instructions should always be followed when using laboratory equipment. Special training may be needed before using some complicated pieces of equipment, such as a **spectrophotometer**.

Staff training

Staff must be properly trained to use the laboratory equipment safely and accurately. New equipment may require new training. It is also important for scientific staff to keep up to date in the latest developments in their field. Good laboratory practice depends on staff being given opportunities to attend suitable training and **professional-development** courses.

Question

1 Explain why each of the following is an important part of good laboratory practice:
   a rules about wearing protective clothing
   b that equipment is left clean after use
   c that staff have been trained to use equipment correctly.

Key words
- ✓ **health and safety**
- ✓ **maintenance**
- ✓ **professional development**

Defining "quality"

The International Organization for Standardization (ISO) produces documents describing international standards for a huge range of items and procedures. The British Standards Institution (BSI) also produces information to help standardise procedures between organisations. The ISO or BS standards for laboratory tests provide a framework that guides the work of analytical laboratories.

Proficiency tests

Analysts use **proficiency tests** to check the quality of their work. A group of laboratories each receives identical samples to analyse. They send in their results and receive back a confidential report containing all the results. The analysts can see how well their results compare with the group. The schemes are designed to be helpful. The scheme organiser, who has plenty of experience of the tests used, often gives advice to a laboratory having difficulties.

Accreditation

Analytical laboratories must show that they can do the job properly. Laboratories can apply to the United Kingdom Accreditation Service (UKAS) for **accreditation**. Laboratories that meet the required standards can display the UKAS symbol with a list of the tests they are accredited to carry out. The list is the 'scope of accreditation' for the laboratory.

Water being analysed in a laboratory.

000

The UKAS accreditation mark.

What would you expect UKAS to check?

A UKAS accreditation shows that a laboratory can repeatedly produce accurate and precise results for analytical tests. This gives clients confidence in the quality of their work. But what does UKAS look for? It looks for more than just good results in a proficiency test.

UKAS also looks at documentation. It checks:

- records and documents kept to describe the procedures used for sampling, field testing, and laboratory methods
- records of all the checks and assessment of the methods
- records of the staff carrying out the work
- notes on the training programmes for the staff.

It also looks at laboratory processes by:

- testing the competence of at least one sampler and one field analyst
- looking at the management of the laboratory including staffing and supervisory arrangements.

As part of this, they observe one complete survey. They ensure that the methods used match the laboratory documentation. A survey includes:

- sampling and field testing
- laboratory analysis and reporting
- the way samples are transported back to the laboratory.

They also undertake a close study of the analytical methods used. This includes assessing:

- the written method
- how the analysis is carried out
- the suitability of equipment and procedures
- the proficiency test results and other assessments
- how observations and results are recorded
- how calculations are made from raw data
- how reports are written
- what records are kept.

Questions

1 Why are there international standards for some laboratory tests?

2 Explain why laboratories take part in proficiency tests.

3 What would happen if a laboratory had a poor result in a proficiency test?

Key words
- proficiency tests
- accreditation

River sample being collected.

River sample being analysed.

THE SCIENCE

Key words
✓ pH
✓ indicator organisms

Monitoring river water in the UK

The Environment Agency is responsible for monitoring the quality of our rivers and lakes. It checks 7000 sites each year. Monitoring involves recording observations and measurements regularly over a long period of time. Scientists use **a system of common practice and procedures** for collecting and analysing samples. They collect samples at fixed times in the year. Some of the analysis is done 'in the field' as the samples are collected. Other tests are done back in the laboratory.

What would you measure in a river?

Simple observations can tell us a lot about the quality of a river. One set of observations covers the **aesthetic** qualities of the river – the colour and smell of the water, and the amount of litter, foam, or sewage present.

Other measurements could be:
*   oxygen concentration – this affects animals that live in the water
*   nutrient status – the level of nitrate and phosphate in the water
*   chemical quality – the presence of any polluting chemicals, for example, oil or toxic metals
*   **pH** – the acidity of the water could tell us if there are pollutants present.

Snails can tell stories

Insects, snails, shrimps, and worms can tell scientists a lot about water quality. Some creatures, known as **indicator organisms**, are used to check levels of pollution. In the best-quality water, there are lots of different indicator organisms. When water contains only a few organisms and they are types that can survive in polluted water, this shows that the water quality is poor.

Tools and techniques

Some equipment used to collect samples for monitoring is very simple:
*   rulers to measure the depth of water
*   nets to catch invertebrates
*   jugs to collect water
*   tubes to measure how clear the water is.

Some tests of water quality are done with electronic meters. Scientists put probes in the water and the meter gives a reading, for example, oxygen concentration and pH.

Evidence in colour

Test kits

Scientific detectives use tests of many types. Some tests used for water and soil make use of chemicals that change colour. Special strips can be dipped into water to measure the levels of nitrates and other pollutants. Kits are available to check water from ponds, aquaria, swimming pools, or tap water. These **test kits** are all easy and quick to use.

Coloured results

Test kits give an easily seen coloured result. There are at least three steps in every test:

- add the kit chemicals to the sample being tested
- match the colour formed against the kit colour chart
- read the test result from the chart.

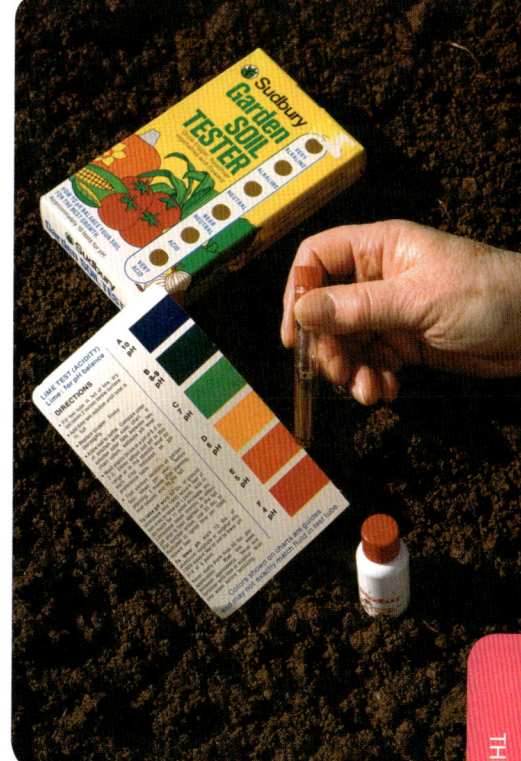

Testing soil using a simple kit: the colour in the test tube can be compared with a colour chart to find the pH of the soil.

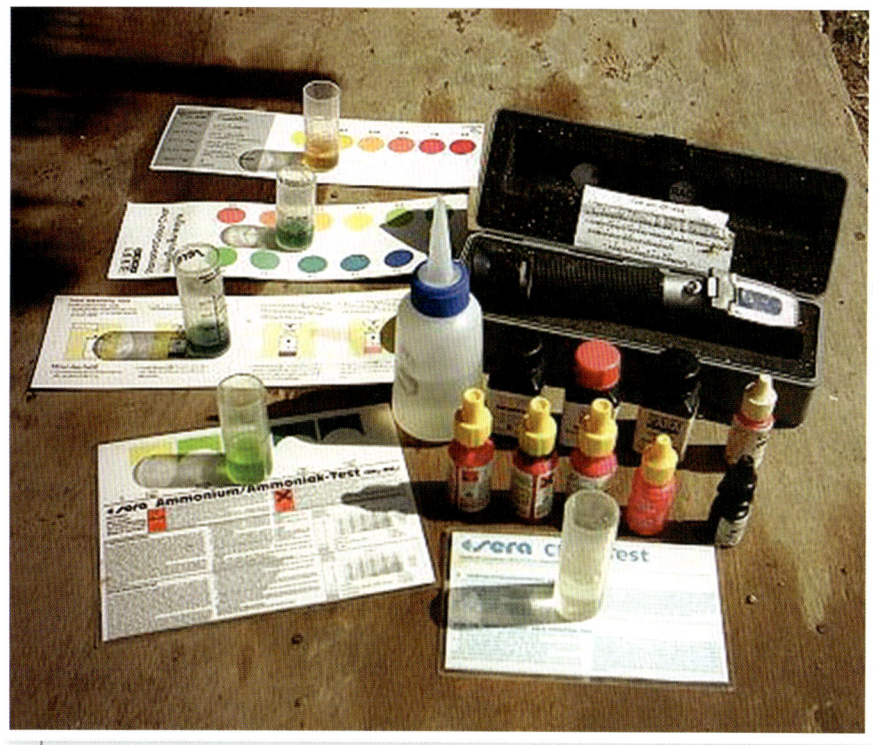

Colour matching being used to test water.

Carrying out tests

Testing soil

Soil is an important part of many environments. It is a mixture of:

- non-living ingredients – air spaces, particles of rock, water with dissolved minerals
- parts of living things – plant roots, rotting leaves and stems, microorganisms, and small invertebrates.

Some assessments of soil are quite simple. Soil samples are weighed, dried in an oven, and weighed again. The percentage of water in the soil is then easy to calculate.

$$\text{percentage of water} = \frac{\text{mass before drying} - \text{mass after drying}}{\text{mass before drying}} \times 100\ (\%)$$

To find the percentage of air in a sample, add 100 cm³ of water to 100 cm³ of soil in a measuring cylinder. The added water fills the air spaces, so the percentage of air in the sample can be calculated using this formula.

$$\text{percentage of air in sample} = 200 - \text{volume of soil and water mixture (cm}^3\text{)}$$

Other assessments are more complicated and involve mixing the soil with distilled water then testing the solution for dissolved minerals.

For a valid comparison of two soil samples it is important that the samples:

- are collected in a similar way – for example, with an auger of known diameter and at a particular depth
- are kept in a sealed container for transport to the laboratory
- are tested in exactly the same way in each laboratory – with trained staff using the same reagents and the same equipment to make the same measurements
- give the same results if tested again.

Standard operating procedures

Analytical scientists use a system of common practice and procedures. These are carefully written instructions explaining exactly what to do. These instructions mean that if someone carried out the same test twice they would expect to get the same result – the measurement is **repeatable**. They also mean that if anyone else carried out the test they would expect to get the same results – the measurement is **reproducible**. The written instructions also describe how the results should be presented. This may be as tables, graphs, or drawings.

Standard operating procedures often require measurements of mass or volume. Balances or graduated glassware used to measure mass or volume must be properly **calibrated**.

A researcher collecting a soil sample from a field. The sample will be tested for pollutants, to make sure the area is safe for visitors.

THE SCIENCE

Standard reference materials

Equipment is calibrated using a known mass or volume. Special weights are used to check electronic balances regularly. These 'known' materials are called **standard reference materials**.

There are many different sorts of standard reference material. For example, the minerals present in drinking water or soil can be measured using a variety of tests. Analytical laboratories can buy pre-tested samples of water or soil from accredited suppliers and check their own procedures by testing the samples and comparing their results.

These weights are standard reference materials. They each have a known mass and must be kept clean, dry, and unscratched.

Key words
- ✓ repeatable
- ✓ reproducible

Questions

1 The mass of a soil sample was 26.2 g before drying and 21.2 g after drying. Calculate the percentage of water in the sample.

2 What four things can be done to ensure that the results of a soil test are reliable?

Analysing and evaluating results

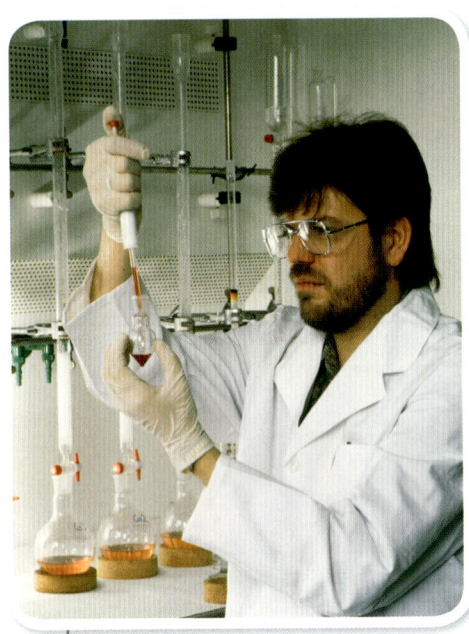

There are random errors even when measuring equipment is used correctly. Using a pipette incorrectly could lead to systematic errors.

Questions

1. Use the formula below to calculate the area of a strip of grass measuring 5 metres 50 cm by 80 cm.
   A (m^2) = l_1 (m) $\times$ l_2 (m)
   where A = area,
   l_1 = length of longer side,
   l_2 = length of shorter side.

2. Pick the outlier from this series of temperature readings taken every 3 hours at an environmental site:
   10, 15, 20, 24, 25, 13, 20, 12 (all °C).

3. What range of values would the following measurements lie between?
   a 79.5 ± 0.5 cm
   b 3.560 ± 0.005 g
   c 5000 ± 100 items.

Formulae

Mathematical formulae are given for any calculations needed in standard operating procedures. Results are calculated by substituting the measured values (in the correct units) in the given formula. For a simple example, the following formula might be given for calculating the **area** of a rectangle:

$$\text{area in m}^2 = \text{length of longer side in metres} \times \text{length of shorter side in metres}$$

Alternatively, it might be written in symbol form:

$$A \text{ (m}^2) = l_1 \text{ (m)} \times l_2 \text{ (m)}$$

where A = area, l_1 = length of longer side, and l_2 = length of shorter side.

Repeating measurements

If you take several measurements of the same quantity you will probably get different results. This is because:

- you used the equipment differently
- there were differences in the equipment itself.

It is better to take several measurements and use these to estimate the **true value**. The true value is what the measurement should really be.

Measurement uncertainty

All measurements have some **uncertainty**. Analytical scientists often give a result showing the range of possible values. For example, the purity of a drug may be given as 99.1 ± 0.2%. This indicates that the actual value lies between 98.9 and 99.3%. It shows the level of **uncertainty** for that value. Two types of possible error contribute to the level of uncertainty.

Random error

Random errors can happen during any analysis. They might happen during transfer of materials using graduated glassware, when making subjective judgements (for example, colour matching), or in estimating between two whole graduations on a scale. Random errors cause the same measurement repeated several times to give slightly different values. Skilled workers can reduce but not eliminate random errors.

Systematic error

Systematic errors can result from incorrectly calibrated or incorrectly used measuring instruments (such as balances, thermometers, or graduated glassware). For instance, measurements made at a steady, but wrong, temperature show a systematic error. Systematic errors give values that are consistently higher (or lower) than the true value when the same measurement is repeated several times. Proper training and high-quality, well-maintained equipment reduce systematic errors.

Outliers

Sometimes a set of results just does not look right. One of the values seems out of step with all the other values. This value is called called an **outlier**. An outlier may be due to an error in measurement, or it may be due to a genuine extreme result. Outliers need to be carefully considered before the data is analysed. For example, measurements of nitrate concentration on five samples of the same river water gave the results (all in mg/l):

- 23.4
- 23.8
- 23.2
- 25.1
- 23.5

The value of 25.1 mg/l is out of line with the other results. A decision needs to be taken whether to:

- calculate the best estimate of the nitrate concentration in the river water as the mean (average) of all five values
- ignore the value 25.1 mg/l, and find the mean of the remaining four values
- make some more measurements.

The decision depends on whether the outlier is thought to be incorrect, or an extreme result within the expected range.

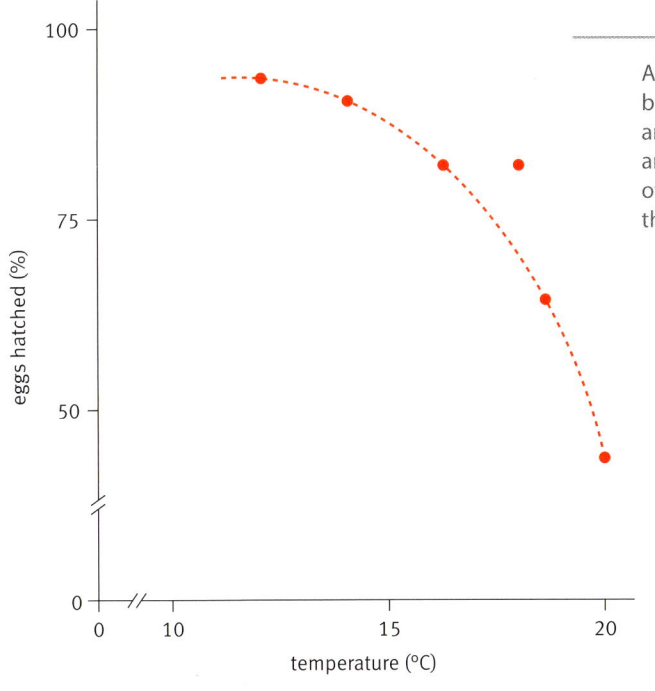

A graph showing the relationship between the hatching of mayfly eggs and temperature. One measurement, an outlier, has been ignored. A 'curve of best fit' has been drawn through the remaining points.

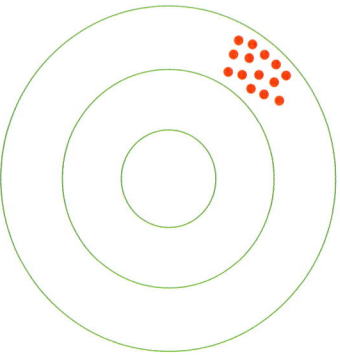

precise, not accurate

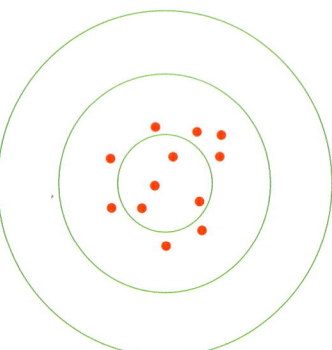

accurate, not precise

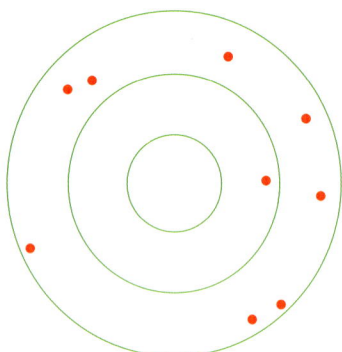

inaccurate and imprecise

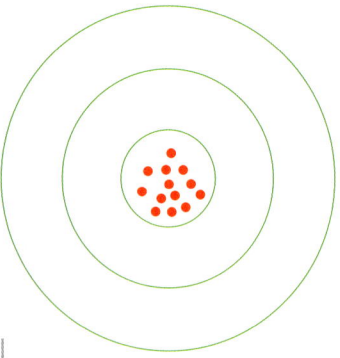

precise and accurate

Accuracy and precision are not the same thing.

Significance of findings

Scientists need to be able to interpret the significance of the results and state what they show. They must draw conclusions from their investigations that are:

- **valid,** meaning that the procedures and tests were suitable for the question being investigated
- **justifiable**, meaning that the conclusions are supported by sound, reproducible evidence.

Accuracy and precision

A test is often repeated to give a number of measured values. The mean is then calculated to find the best estimate of the true value.

Accuracy is a measure of how close results are to the true value. An accurate meter gives results that you can depend on. You can test the accuracy of a meter by using it to measure standard reference materials.

Precision is a measure of the spread of measured values. A big spread of values means there is a greater uncertainty than there is for a small spread. If you perform the same test on a sample time after time and get the same result, your work is precise. It is still possible that there is a systematic error in your method, but your work is repeatable.

Questions

Three groups of four students are asked to measure the length of a metal rod. The rod is passed around and each student independently measures its length. The table shows their results.

Group	Student 1	Student 2	Student 3	Student 4
A	10.2 cm	10.1 cm	9.9 cm	10.1 cm
B	12.2 cm	12.3 cm	11.8 cm	12.0 cm
C	10.3 cm	8.0 cm	12.2 cm	11.4 cm

The manufacturer of the rod is a very reputable firm. A report from the manufacturer certifies the length to be 10 cm long ±0.1%. Answer the following questions. You may find it helpful to draw a number line and to plot the results.

4 Which group of students has results that are:
   a accurate but not precise?
   b precise but not accurate?
   c both precise and accurate?

5 Suggest how the students have got different results when measuring the metal rod.

Collecting samples

Representative samples

Analysts work with **samples**. The size of the sample needed depends on what tests need to be carried out. Samples of any item must be **representative**. In other words, there should be enough of the sample to give a picture of the whole item. The sampling method depends on the nature of the sample.

The composition of a homogeneous item is the same throughout, like a solid milk-chocolate bar.

The composition of a heterogeneous item is not all the same, like a chocolate bar made in layers.

Labelling samples

All samples should be clearly **labelled**. Labels are likely to include information about the time, date, and place the sample was taken. It will also need to include any other information that distinguishes the sample from other samples taken at the same time and place.

Preservation of samples

It is important to prevent any changes or **deterioration** of the sample. Some samples have to be analysed immediately. Others can be stored under suitable conditions.

Avoiding contamination

Accurate analytical results depend on protecting samples from **contamination**. Environmental scientists investigating a possible pollution incident put all samples into sealed bags or bottles to avoid contaminating samples. The scientists may have to appear in court to give evidence about possible routes of contamination.

Key words
- accuracy
- precision
- sample
- representative
- label
- deterioration
- contamination

THE SCIENCE

Scientists sampling water for microscopic plants called phytoplankton. Each sample is labelled to record when and where it was collected.

Avoiding tampering

Some samples become evidence in legal actions or prosecutions. So for some samples it is important to have 'chain of custody' evidence and tamper-proof seals. This means it can be shown exactly who had access to the samples.

Planning

Taking samples needs careful planning. There are many factors to be considered, such as:

- where, when, and how to collect samples
- how many samples, and how much of each to take
- how to store and transport the samples to the laboratory
- how to avoid contamination
- how to safeguard the samples from **tampering**.

The stream water is heterogeneous – samples may vary from one part of the stream to another. The time of year when samples are collected affects the results. Some samples may need to be examined for living organisms immediately. Samples for chemical analysis must be bottled, stored, and transported back to the laboratory. This is a complex sampling problem that needs careful planning.

THE SCIENCE

Questions

1 What would be a representative sample of the solid milk-chocolate bar?

2 What would be a representative sample of the chocolate bar made in layers?

3 How could you find out if soil in an area was heterogeneous or homogeneous?

4 The Environment Agency wants to check the stream in the photo for chemical pollutants over the course of a year. Plan a suitable sampling scheme.

5 Suggest reasons why it can be difficult to obtain a representative sample when analysing:
   a air in a busy street
   b soil from land contaminated by a waste tip
   c water from a reservoir.

Key word
✓ tampering

Recording visual information

Written descriptions

Records can be made of evidence from visual examination in various ways. A written description is a quick method of recording key details. Written descriptions of evidence should be as short as possible to give an accurate and clear record.

Written descriptions have the advantage that they can concentrate on the most important visual information. They can also record details of movement and change over time, which is not possible in a sketch or still photo. Notes might include the flight pattern of a bird, or the changing colour of a test strip.

Drawings

Drawings can be used to record visual information. The type of drawing can vary from a quick sketch used as a memory aid to a detailed, accurate, measured drawing as a permanent record. Both can be useful.

Field about 30 m x 50 m
7 large trees
thick vegetation
small pond ~ 2 m x 3 m,
full of weed
surrounded by reeds
and boggy area
no buildings
slopes down to pond
slope is south facing
4 samples taken

Written notes recording observations about a plot of land.

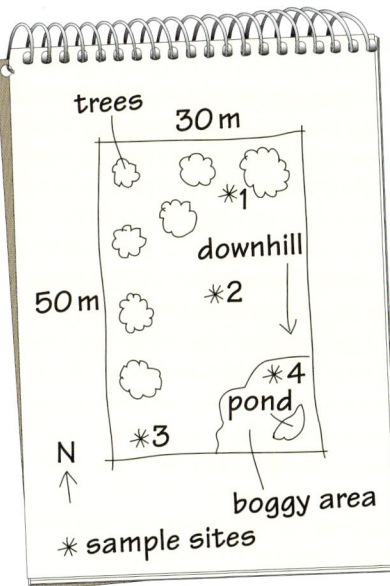

A sketch is a quick way to record places where samples were taken. It can also record key features of the site being investigated. The writing, or annotation, has been used to increase the amount of information recorded.

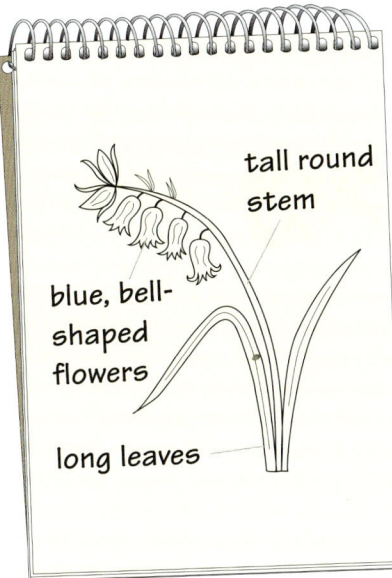

Botanists find drawings can sometimes be more useful aids to plant identification than photographs. A detailed drawing such as this one of a bluebell can record and label all the key features clearly.

Photographs and videos

Photography, whether still or video, can be a quick and accurate way to record information. It can be used for all sorts of visual information, from close-up or telephoto shots to remote 24-hour surveillance. The level of detail recorded varies with the quality of the camera. Image-enhancement techniques can sometimes be used to gather extra details from poor-quality photos.

Satellite images

For some environmental studies, images of large areas as seen from a satellite in orbit round the Earth are useful. These images can show temperature differences between different areas of ground or show up populations of plankton in the sea. Images captured using other wavelengths of radiation, such as ultraviolet or infrared, may show useful details not seen with visible light. Computer enhancement and analysis makes images like these easier to interpret.

Identifying features

Sometimes evidence of certain key features is all that is needed. For example, an environmental officer needs to know whether a certain sort of mayfly nymph is present in a sample. The key feature to look for is the three hairy tails. There is no need to record other details of every creature found.

In other instances it may be important to compare the similarities and the differences between two samples or two images.

Whatever the source of the image, there are key features that affect how useful the image might be to a scientist. These include:

- the **sharpness of focus** of the image
- the **contrast** between different parts of the image
- the **magnification** of the image
- the **depth of field,** which is the distance between the nearest and farthest objects that appear in sharp focus in the image.

A mayfly nymph showing the distinct three hairy tails.

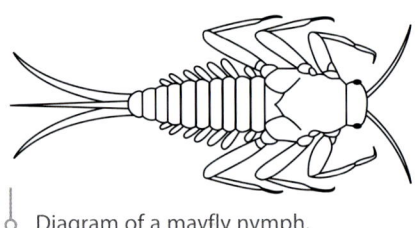

Diagram of a mayfly nymph.

THE SCIENCE

Key words

- ✓ **sharpness of focus**
- ✓ **contrast**
- ✓ **magnification**
- ✓ **depth of field**

Questions

1 You notice an unusual tree when you are out and want to know what type of tree it is. You have a camera, a notebook, and a pencil. What details would you record, and what method would you choose to record them?

2 Write a short description of part of the school grounds that would help another person to locate it.

3 Look at the photo and the diagram of the mayfly nymph. Say which is most useful and in what situations.

Measuring colour

Colour chemistry

Many chemicals are coloured. Looking at the colour is a simple way of checking which compounds are present. Colour is also a clue to how much of a coloured compound is present. Sand, for example, is mainly silicon dioxide, which is colourless, but sand is usually yellow because it contains iron(III) oxide. Dark-yellow sand contains more iron(III) oxide than pale-yellow sand.

Colour can help identify rocks and minerals.

Colour change

Litmus is red in acid but blue in alkali. Litmus is a good acid–base indicator because it changes colour just as the pH changes from acidic to alkaline. It is used as a **qualitative** test to tell if a solution is acidic or alkaline. A qualitative test will tell you only that a chemical is present or that it is absent.

Universal indicator solution is a mixture of acid–base indicators, chosen to show a range of colours depending on the pH of the test solution. Matching the colour to a colour card gives the result as a pH value.

This is a **semi-quantitative** test: it gives an approximate measure of the degree of acidity or alkalinity.

Key words
- litmus
- universal indicator
- qualitative
- semi-quantitative

Key word

- ✓ intensity

Questions

1 Is the test for iron(III) compounds in the photo on this page qualitative or quantitative?

2 Name one thing a soil test kit might test for.

Colour concentration

Chlorophyll in algae is green. The more chlorophyll in a water sample, the deeper the green. The **intensity** of colour can be used to estimate how much algae is present. Many test kits use colour intensity to give a semi-quantitative result. Colour intensities are compared with printed colour cards.

More precise, **quantitative** tests can be done using a colorimeter. This is an instrument that measures depth of colour exactly. Colorimeter tests might need to be done in a laboratory rather than in the place where samples are collected.

Producing colours

If the main product of a test is not coloured, it is reacted with other chemicals to give a coloured result.

Soil analysis using a semi-quantitative test. The green colour can be compared with a colour chart to find the approximate pH.

Very dilute solutions of iron(III) sulfate are almost colourless. Adding potassium thiocyanate gives a deep-red solution. This shows that there was an iron(III) compound in the original solution.

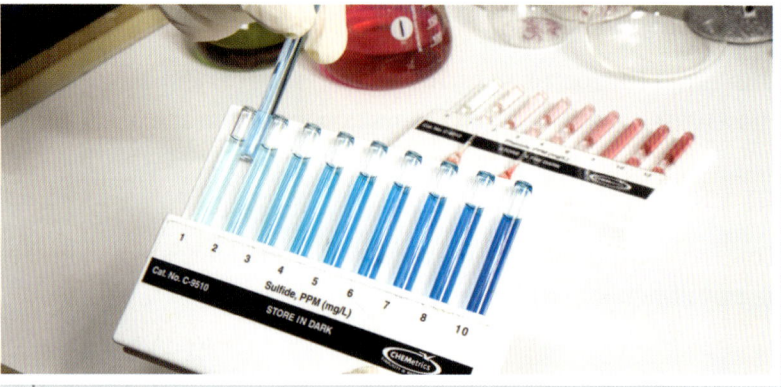

Testing waste water for the pollutant hydrogen sulfide by colour matching.

Turbidity

What is turbidity?

Turbidity is an important measure of water quality. It is caused by very small particles of solids **suspended** in the water but not **dissolved** in it. If you filter turbid water you can remove the particles to weigh or analyse them. Chemicals that are dissolved in the water cannot be separated out by filtering.

When the turbidity is high, light doesn't pass easily through it. This means that aquatic plants may not get enough light to photosynthesise and grow. Turbid water also warms up more quickly in sunlight than clear water. This may help some organisms to survive but can upset others.

High levels of turbidity can be dangerous.

What does it tell you?

Turbidity is a sign of water contamination. High turbidity can be caused by stirring up mud or sand from the bottom. Alternatively it can be caused by blooms in populations of tiny plants. In drinking water, turbidity can be a sign of hazardous contaminants such as sewage.

How is it measured?

Turbidity is measured in deep water by lowering in a Secchi disc on a marked line. When you can no longer see the black and white sectors clearly, you read off the length of the line. The longer the line is, the less turbid the water.

In the laboratory you can use a turbidity tube to measure the turbidity of a water sample, as described on page 111. You can also assess turbidity in the laboratory by measuring the amount of light detected at 90° to a light beam shone through the water. The more turbid the water is, the more the light will be scattered by the particles suspended in it.

Key words

- ✓ **turbidity**
- ✓ **suspended**
- ✓ **dissolved**

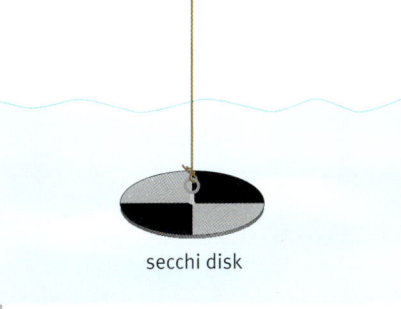

secchi disk

A Secchi disc can be used to measure turbidity.

Questions

1. What causes turbidity?

2. Describe two environmental problems that can be caused by high turbidity.

Principles

Data collected as evidence must be reliable and fit for purpose. Choose a balance with a level of accuracy fit for the purpose of the task, and follow the manufacturer's operating instructions. Use a notebook to record all readings.

Procedure to weigh by difference

1 Check that the balance is clean and reading zero.

2 Place the sample in a suitable clean, dry weighing vessel on the balance platform. Record the mass.

3 Transfer the sample to another container. Weigh the weighing vessel and record its mass.

4 Calculate the mass of the sample from the difference between the two measurements.

Procedure to weigh direct

1 Check that the balance is clean and reading zero.

2 Place a suitable empty weighing vessel on the balance platform. Set the display to zero. (This is known as taring the balance.)

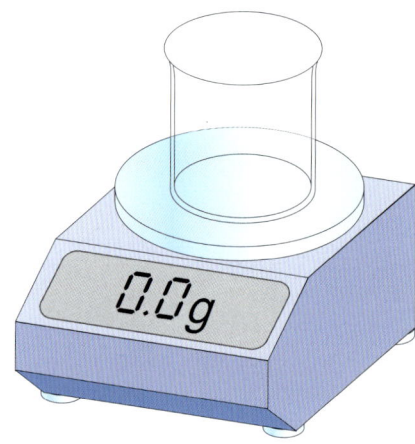

3 Place the sample in the weighing vessel on the balance platform.

4 The reading on the balance is the mass of the sample.

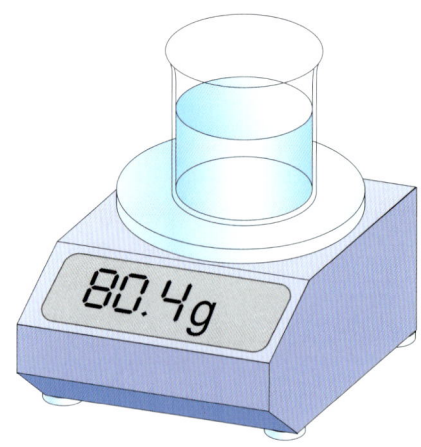

Good laboratory practice: using a pipette

Principles

Glassware for measuring volume is designed to be used in a specific way. Variations in procedure lead to errors, making data less accurate. The errors can be minimised with good practice.

Procedure

1 Check that the pipette is clean, not chipped, and has a clear mark. Attach a suitable filler.

2 Place the tip of the pipette below the liquid surface. Draw up liquid to just above the mark.

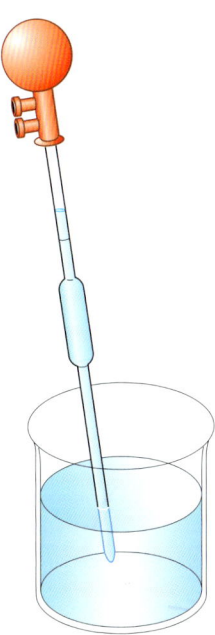

3 Hold the pipette above the liquid. With your eye level with the mark, let liquid out until the bottom of the **meniscus** is on the mark.

4 Touch the pipette tip against the side of the container to remove any drops.

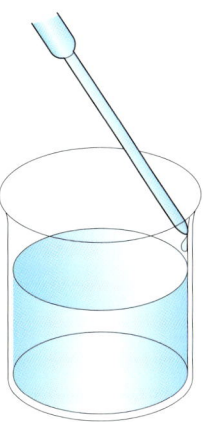

5 Place the pipette tip just touching the side of the receiving vessel and let the liquid drain out. Allow an extra three seconds after it stops flowing. Do not blow out the last drop.

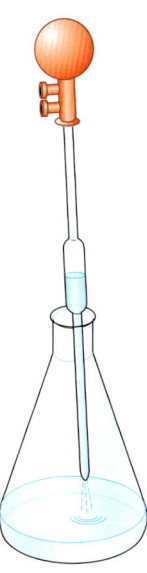

Reading a linear scale

Principles

Many measuring instruments have a **linear scale**. This is a series of equally spaced lines, or **graduations**. You use the scale to read off a value. When a measurement falls between two graduations, it has to be estimated.

Procedure

1 Look at the scale you are reading and decide what the distance between two graduations represents.

2 Record the nearest graduation that is less than the length you are measuring.

3 Estimate the extra distance beyond this graduation and record this part of the measurement.

4 To show the uncertainty of the measurement, decide on the range that you are sure the length falls in. Record this as a ± extra value.

An example is shown below, for measuring the length of a needle.

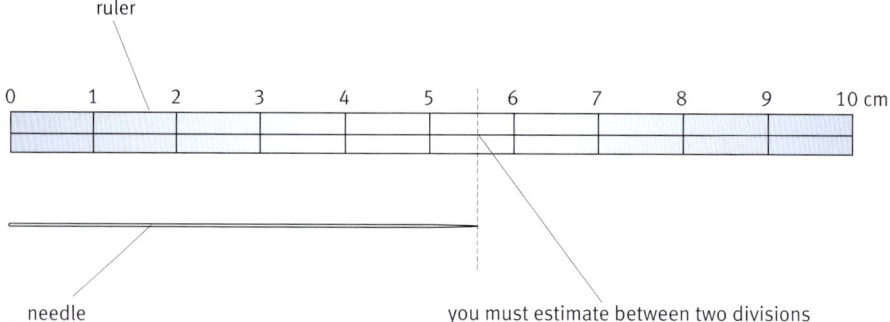

ruler

0   1   2   3   4   5   6   7   8   9   10 cm

needle   you must estimate between two divisions

The length is between the 5-cm graduation and the 6-cm graduation. A good estimate would be 5.6 cm. The length of the needle is 5.6 cm ± 0.1 cm.

Kick sampling

Principles

Biological samples are collected from rivers in spring and in autumn. Following a set procedure, scientists collect samples by holding a net in the flowing river. Then they kick the stream bed upstream of the net for a fixed length of time (30 seconds to 3 minutes). This technique can be used to collect indicator organisms to find out about pollution levels in the water. The organisms found are compared against a chart showing the organisms that would be expected in water with various levels of pollution.

Equipment

- net with a least one straight side to the opening
- white tray

Procedure

1 Choose a suitable point in the river where it is safe to stand.

2 Hold the net with the handle vertical and the bottom edge of the net close to the stream bed downstream of your feet.

3 Kick the stream bed gently but firmly to dislodge stones and riverbed-living invertebrates.

4 Collect the dislodged material in the net over a fixed length of time.

5 Tip the contents of the net into a white tray.

6 Identify, count, and record the organisms you have found.

Interpreting the results

The number and type of organisms give an indication of the condition of the environment you have sampled. Remember to look for indicator species that are typical of the environment.

Principles

Universal indicator solution contains a mixture of compounds that change colour at different levels of acidity or alkalinity. When universal indicator solution is added to a sample, the colour it turns shows the pH of the sample. This is a semi-quantitative test.

Procedure

1 Put about 2-cm depth of the sample solution in a test tube.

2 Add three drops of universal indicator solution. Shake gently.

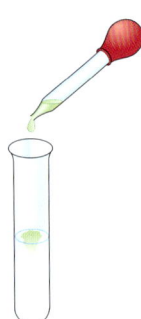

3 Compare the colour of the solution with the printed colour chart and decide the nearest match.

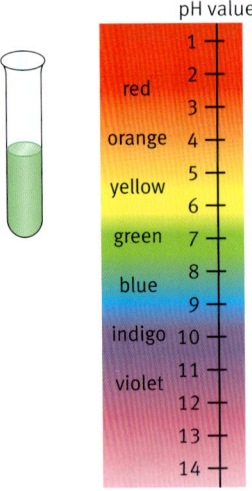

4 Use the chart to estimate the pH value of the sample solution.

Measuring turbidity with a turbidity tube

Principle

Very small particles of solids suspended in water make it look cloudy or turbid. If there are more contaminants, the turbidity is higher. These particles are not dissolved in the water, and may settle to the bottom if the water is left undisturbed. You cannot see through turbid water as easily as through clear water because it does not transmit light as easily.

Equipment

* turbidity tube: this a graduated tube with a disc at the bottom, with a pattern of black on white shapes similar to a Secchi disc

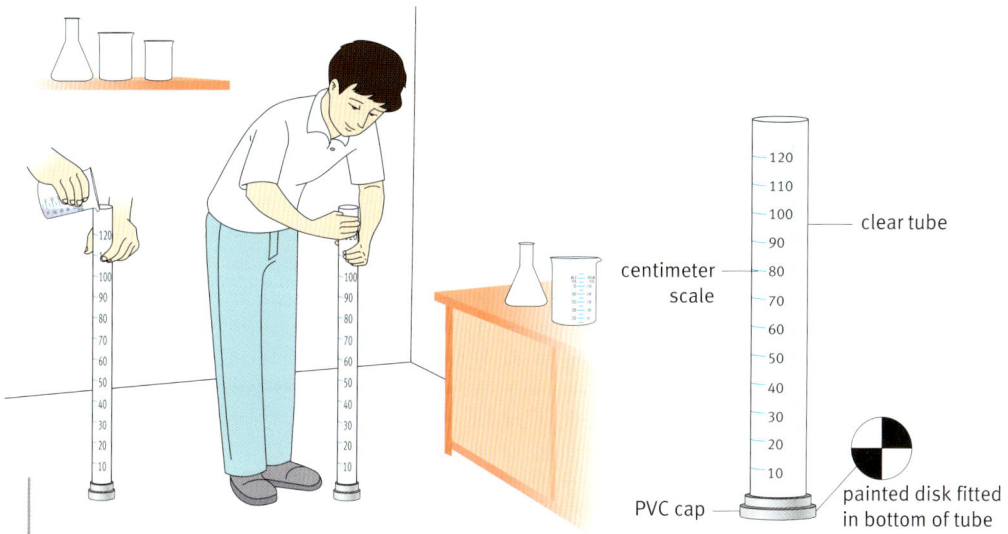

Water sample being measured using a turbidity tube.

Procedure

1 Collect a sample of water in a large jug.

2 Stand the turbidity tube vertically and look into the top.

3 Pour the water sample into the tube very slowly, looking down at the black-on-white pattern.

4 Stop pouring when you can no longer see the pattern.

5 Read the depth of water from the scale on the graduated tube.

Interpreting the results

The turbidity of the water sample is inversely proportional to the depth of water in the turbidity tube:

* The deeper the water is, the lower the turbidity.
* The shallower the water is, the higher the turbidity.

PROCEDURES & TECHNIQUES

Module Summary

People and organisations

- that people with scientific and technical skills are employed in environmental protection
- that organisations gather scientific data that is used to make decisions about environmental protection, including monitoring industrial sites, checking water pollution and flood risks, monitoring air quality, and protecting wildlife
- that there are regulations affecting the work of people in environmental protection, including workplace health-and-safety regulations
- that public laboratories have a system of accreditation for their work that checks their accuracy and precision and gives clients confidence in the data.

The science

- how environmental scientists can use samples of living organisms found in the oceans to gather information about climate change
- how indicator organisms give information about levels of pollution in freshwater (such as ponds, streams, rivers, and canals)
- that information about environments can be recorded in different ways, including written descriptions, drawings, photographs, and videos
- that the key features of an image used to monitor an environment are sharpness of focus, contrast, magnification, and depth of field
- why a given measurement may not be the true value of the quantity being measured and why several measurements of the same quantity may give different values
- how random errors and systematic errors make measurements uncertain
- the difference between accuracy and precision of data
- the difference between quantitative, semi-quantitative, and qualitative tests
- the difference between dissolved and suspended solids in a water sample
- how to interpret data about indicator organisms in freshwater
- how to calculate an area and why a calculated area has a greater uncertainty than the measured lengths
- how to make sense of data on measurements of dissolved or suspended solids in a water sample
- how to make sense of data from test kits used in environmental monitoring and indicator solutions used to determine pH.

Standard procedures

- that using common procedures and standard techniques increases the repeatability and reproducibility of results, which makes data more valuable
- that good laboratory practice depends on all equipment and instruments being maintained, calibrated, and checked as well as training and professional development of staff
- that proficiency tests make sure the people working in laboratories carry out standard procedures effectively
- how representative samples are collected, labelled, stored (to avoid deterioration, tampering, and contamination) and prepared for analysis
- how to take readings from a linear scale, and how to estimate readings between graduations
- the colour of litmus in acidic and alkaline solutions, and how Universal indicator is used
- how to assess the quality of a water sample by measuring its turbidity.

Review Questions

1 Ali works as an Environmental Protection Officer for a local council. Describe four different responsibilities he might have.

2 British Waterways sends samples of canal water to a laboratory to be analysed for pollution.
   a Explain how the samples should be collected and stored.
   b Explain why the laboratory uses standard procedures to analyse the samples.
   c Explain why the laboratory looks at which living organisms are present in the water.

3 Gary is a scientist who is interested in climate change.
   a Explain what is meant by climate change.
   b Explain why his research includes sampling organisms from the oceans.

4 Fred uses a microscope to study a pollen sample. Here is the image he sees. The scale has 100 μm between divisions.

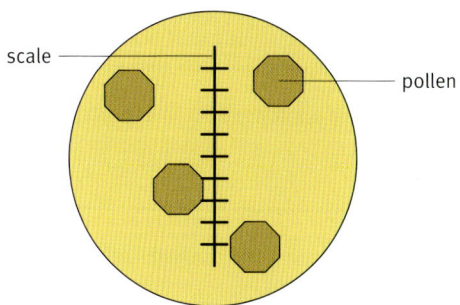

scale — pollen

Fred consults this table to identify the pollen.

Pollen source	Diameter (μm)
ash	50
beech	200
chestnut	800

Which tree provided the pollen? Justify your answer.

5 Gillian studies orchids growing in a field. She counts 8 of them inside a 5.0 m² area. The field is 20 m long and 30 m wide.
   a Estimate how many orchids are in the field.
   b Harry repeats Gillian's measurement. Give reasons why his estimate is different.
   c Explain how Harry and Gillian can obtain a better estimate of the number of orchids in the field.

6 Irene uses both litmus and Universal indicator to test the pH of a water sample.
   a What are the colours of litmus in:
      i acid?   **ii** alkali?
   b Explain why Universal indicator is a better test of pH than litmus.
   c Universal indicator is a semi-quantitative test. Explain what this means.

7 An Environment Agency report on water quality includes the information in the table below.

Water-quality indicator	Current standard	Total number of tests in 2010	Number of tests not meeting the standard
Water leaving water treatment works			
E. coli bacteria	0/100 ml	20255	1
Turbidity	1 NTU	20243	9

   a Explain what 'current standard' means.
   b Approximately how many tests are carried out each day to test for *E. coli*?
   c How many tests for *E. coli* did not meet the standard? Suggest what the water company should do when it finds the water is not meeting the standard.
   d Explain what is meant by 'turbidity' and how it can be measured.

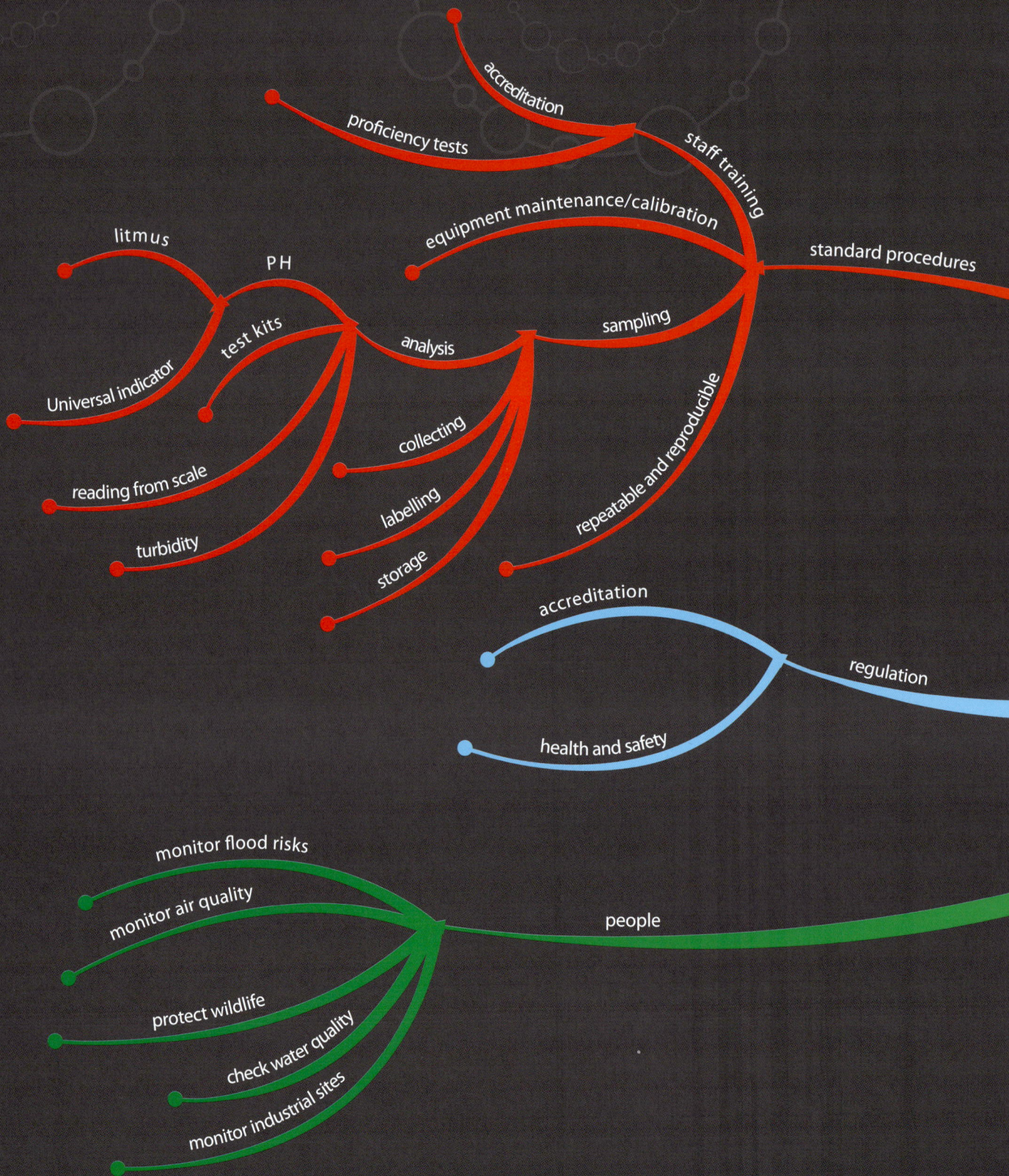

accreditation

proficiency tests

staff training

equipment maintenance/calibration

standard procedures

litmus

PH

test kits

analysis

sampling

Universal indicator

reading from scale

collecting

labelling

storage

repeatable and reproducible

turbidity

accreditation

regulation

health and safety

monitor flood risks

monitor air quality

people

protect wildlife

check water quality

monitor industrial sites

WORKING TO MONITOR AND PROTECT THE ENVIRONMENT

science
- recording
  - videos
  - sound
  - drawing
  - writing
- measuring
  - errors
    - precision
    - accuracy
    - random
    - systematic
- testing
  - quantitative
  - semi-quantitative
  - qualitative
- sampling
  - soil
  - air
  - water
    - dissolved solids
    - suspended solids
  - living organisms
    - climate-change evidence
    - freshwater pollution evidence

interpreting data
- test kits
- indicator organisms
- PH indicators
- calculated areas
- dissolved/suspended solids in water

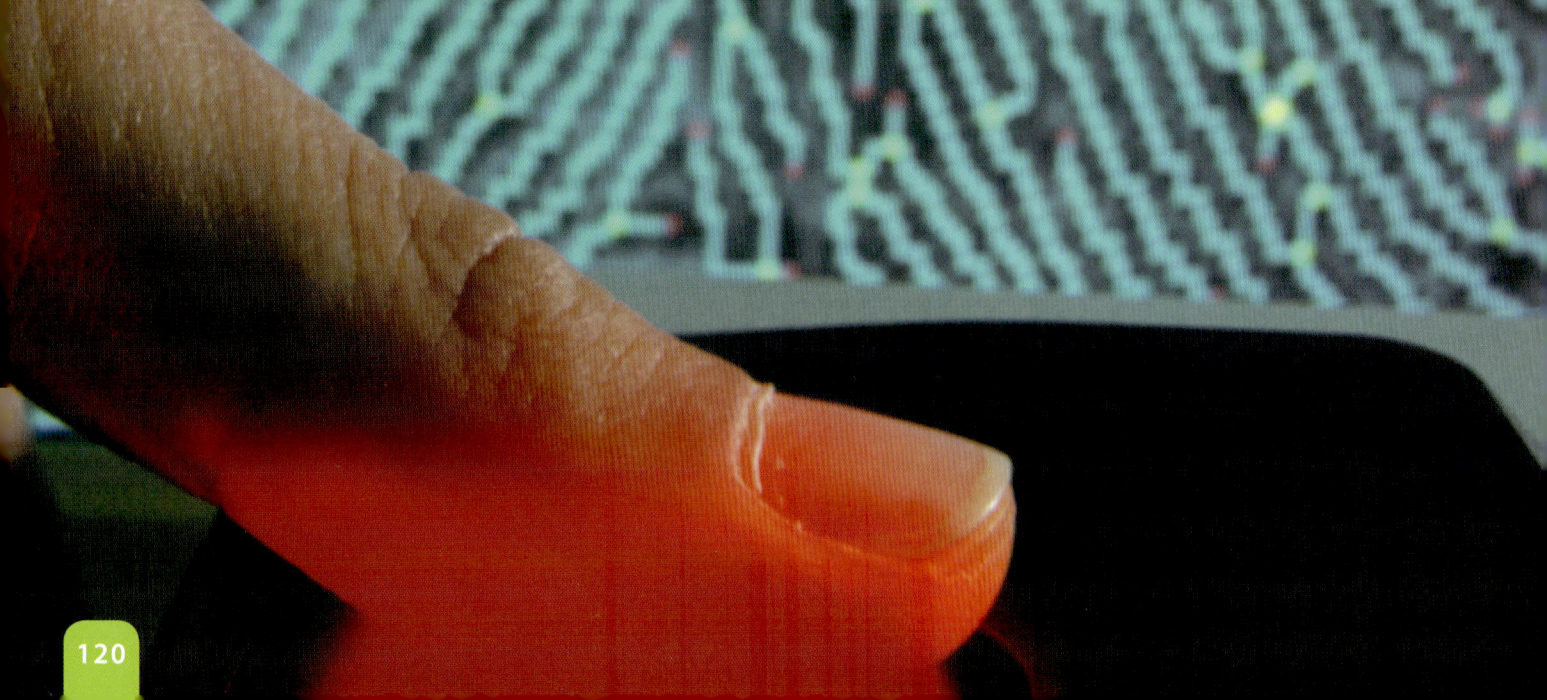

A4 Scientists protecting the public

Why study how scientists protect the public?

Popular TV programmes show how science can help to detect criminals. Forensic scientists search for the clues that link suspects to the scene of a crime. They find the evidence that can lead to convictions in court. People who work in environmental health may not be celebrities but their work is very important. They help to make sure that the food we buy is safe and that we are not cheated by being sold goods that are not up to standard.

What you already know

- the importance of careful observation, counting, and measurement in the laboratory or in the field
- the need for standard procedures when collecting and analysing samples to be used as evidence
- the use of colour and test kits in analysis.

Find out about

- how colourimetry is used to measure the concentration of food additives
- how chromatography can help to detect unsafe chemicals in food and drink
- how scientists gather evidence from crime scenes
- how to use microscopy to look for tiny clues
- how to examine and analyse evidence.

The Science

Scientific detectives often have to sift and sort tiny samples in their search for clues. This means that their methods of analysis have to be very sensitive. The scientists have to know how to use microscopes to examine very small specimens. They also apply techniques such as colourimetry and chromatography to identify and measure tiny traces of chemicals.

Solving crime

Evidence from a team of forensic scientists helped convict a murderer in a recent police investigation.

Evidence of identity

When the victim's body was found, it could not be identified immediately. Forensic scientists used DNA analysis to match the body with a tooth kept by the victim's mother. This confirmed the identity of the victim.

Evidence from fibres

Police found a black shoe at the crime scene. It had 350 fibres trapped in its Velcro strap. Using forceps, a forensic scientist picked them off one by one and 'taped' them on sticky tape. By looking at the fibres through a low-power microscope the scientist identified fibres from the victim's clothes, confirming that it was the victim's shoe.

The police held a suspect. A pair of socks, a red sweatshirt, and a curtain were found in his van. A scientist rechecked all the fibres picked off the shoe. Four dark-red fibres were found. Examination with a high-power microscope showed that these came from the sweatshirt. One multicoloured cotton fibre from the shoe was found to match the curtain. Blue fibres from the victim's trousers were also found on the sweatshirt.

Forensic scientists use light microscopes for the preliminary examination of hairs and fibres.

Additional tests

The scientists needed more evidence to be quite sure of their matching of fibres. They used chromatography to analyse the coloured pigments in the fibres. Over 11 months, 128 thin-layer chromatography tests were carried out. This painstaking, time-consuming work confirmed that the fibres matched and provided more evidence of the link between suspect and victim.

Evidence from hairs

Hairs removed from the curtain and sweatshirt were sent for DNA profiling. There were 40 hairs and all but one matched the suspect. A single hair gave a DNA profile that matched the victim. This vital evidence linked the suspect to the victim.

After 18 months of careful investigation, a man was sentenced to life imprisonment for kidnap and murder. Forensic scientists had analysed over 500 items. The evidence that led to the conviction came from several fibres and a single hair.

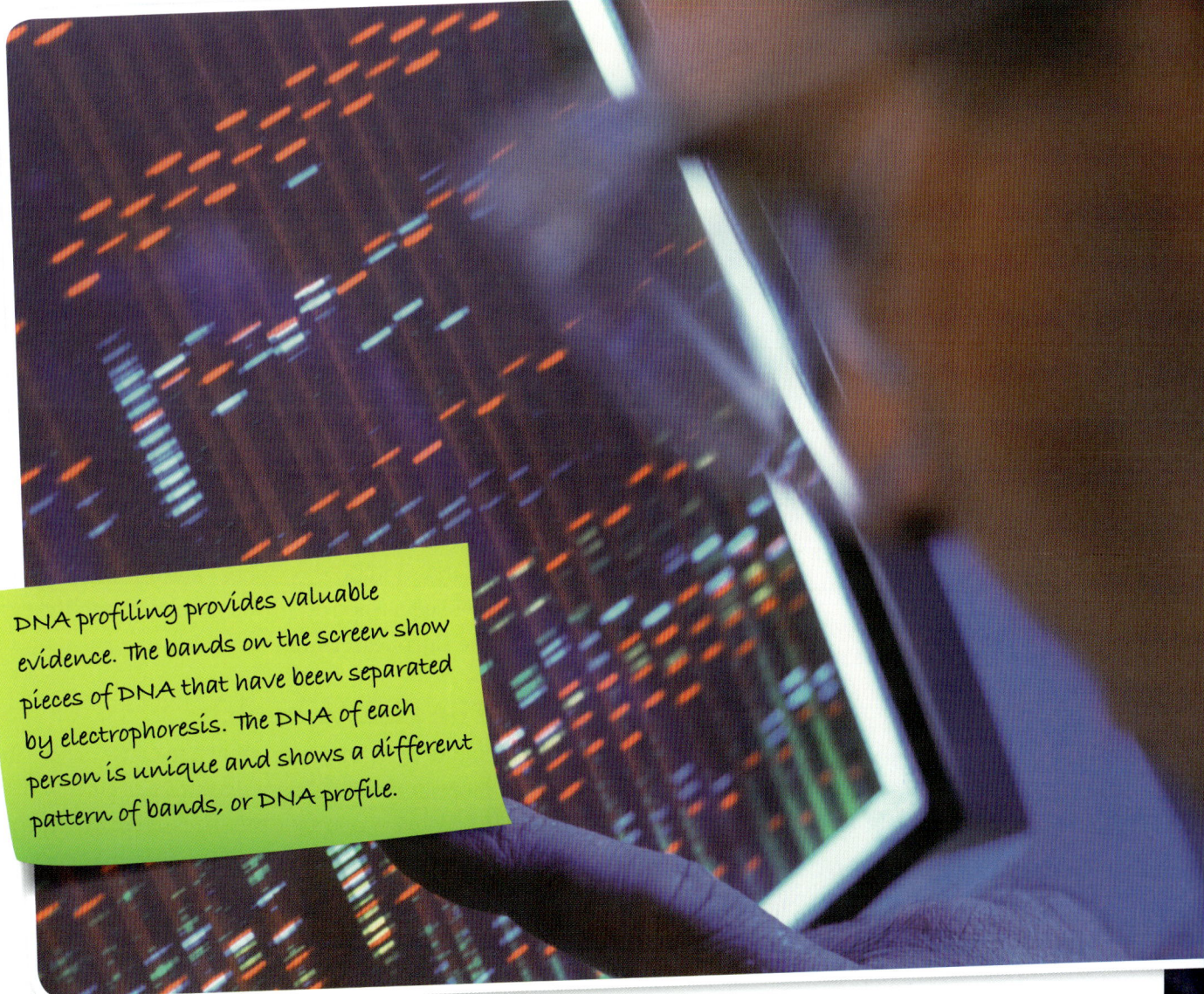

DNA profiling provides valuable evidence. The bands on the screen show pieces of DNA that have been separated by electrophoresis. The DNA of each person is unique and shows a different pattern of bands, or DNA profile.

Dangerous dyes

Chilli and other spices in a market.

The danger

In 2005 a UK government agency called the Food Standards Agency issued a warning advising people not to eat foods contaminated with a dye that is not safe to eat. The bright-red dye was present in a batch of chilli powder used to make a pepper sauce. Manufacturers included the sauce as one of the ingredients of other foods including soups, sauces, and ready meals. Over 400 products were found to be affected.

Colourful but not for food

Analysts used chromatography to show that there was a type of colouring called a Sudan dye mixed with the chilli powder. The Sudan dyes are synthetic red dyes that are normally used for colouring solvents, oils, waxes, and petrol, as well as shoe and floor polishes. They are not allowed to be used as food additives in the UK and the rest of the European Union.

The risk to health

Sudan dyes have been shown to cause cancer in laboratory rats. This research suggests that they are likely to harm people too. The Sudan dyes are not considered safe to eat because they may cause cancer.

In the 2005 incident the amount of dye in any one meal was tiny. The risk to health was small but even so the Food Standards Agency had to act fast to get the contaminated products removed from sale.

Protecting consumers

The Food Standards Agency worked with local authority environmental-health departments to stop any products containing Sudan dyes from being sold in the UK. The agency required companies that were producing and selling the contaminated products to make sure these products were quickly withdrawn and recalled.

It is now the rule that all cargos of dried and crushed or ground chilli powders coming into any country in the European Union must have a certificate. The certificate has to show that they have been tested and found to be free of Sudan dyes. Local authority health departments and port officials collect random samples for analysis to check up on imported foods.

What's in a name?

No-one knows for sure how the red dyes came to be called 'Sudan dyes'. The dyes were discovered in the 19th century. Maybe their fiery red colour made people in Europe think of hot countries such as Sudan. In 2005 the government of Sudan was angry that the name of its country was linked to dyes that cause cancer. Sudan's ambassador wrote to the UK government agency, asking it to change the name to stop further harm to Sudan's reputation.

Chilli powder – you want to be sure that it is free of dangerous dyes.

The skills of scientists protecting the public are used in many different fields of work. They usually work in teams to analyse evidence and draw conclusions that are supported by careful observation and measurements. Scientists working in consumer protection make sure products meet certain safety standards. Forensic scientists help provide evidence in criminal investigations.

Setting the standards

Nearly all organisations set standards for the quality of their work and their products. Some standards are set by national and international organisations. Many products sold in the European Union (EU) must meet certain general product safety requirements. These European directives set standards for:

- food, covering additives, composition, contaminants, labelling, and safety
- consumer goods, covering many different types of household goods
- pharmaceuticals, covering all over-the-counter medicines
- agricultural products, covering animal feeds, fertilisers, and pet food.

Checking the standards

Local authority Trading Standards officers are responsible for protecting the local community from illegal or unfair trading practices. They make sure that products on sale in the shops comply with safety standards.

The Trading Standards Service enforces over 80 Acts of Parliament for the protection of consumers. Samples are sent to public analysts when testing is needed to check if health-and-safety standards are being met.

A Trading Standards officer checks the calibration of petrol delivery pumps.

LGC is the UK's largest independent analytical laboratory. Its work covers food and agriculture, oil and chemicals, pharmaceuticals, environment, healthcare, life sciences, and law enforcement.

Public analysts

Public analysts are scientists who specialise in chemical analysis and other tests necessary to find out if food, drugs, and consumer products meet the standards required by law. Public analysts must know about food technology, have skills in chemical analysis, and have a good understanding of the law. Their work might include:

- analysing foods for additives, contaminants, and composition
- giving advice on labelling food
- checking that consumer products comply with safety standards
- examining food and water for microorganisms that may cause disease, and providing advice and research in this area
- interpreting regulations, both UK and EU
- being an expert witness in legal court proceedings.

Much of the testing and analysis is carried out by laboratory technicians. Public analysts are responsible for the accuracy of the work and the conclusions reached.

Public analysts' laboratories have to be UKAS accredited (see pages 96 and 97) for the tests that they carry out. They also regularly take part in proficiency testing schemes.

Questions

1 Give three examples to show why the public expects to be protected by Trading Standards.

2 Give an example of a person with scientific expertise whose job is to protect consumers. Explain why this person needs to be a scientist.

3 Explain why it is essential for public analysts' laboratories to have UKAS accreditation.

Crime-scene investigators

Crime-scene investigators take photographs and gather **forensic** evidence such as fingerprints, footprints, hairs, and fibres from the scene of a crime. They also take fingerprints from people who might have been at the scene. They place evidence into bags and send it to forensic laboratories. It is important that the evidence is carefully labelled, does not become contaminated, and cannot be tampered with at any stage.

Crime-scene investigators must have patience and pay close attention to detail. They have to work accurately under pressure. They also need to be able to communicate well with police officers, onlookers, and victims. They sometimes have to give evidence in court.

The crime-scene investigator is wearing a paper suit, overshoes, and gloves, to make sure the evidence is not contaminated.

Forensic scientists

Back in the laboratory, forensic scientists examine the evidence gathered during investigations. This work can help in the investigation by providing important evidence. They may sometimes attend crime scenes, or be called as expert witnesses during a trial. Detailed knowledge and skills are often required, so senior forensic scientists usually specialise in a particular area. Specialist areas include:

- firearms
- image enhancement
- fingerprints
- accident investigation

- documents and handwriting
- medical sciences
- electronic data recovery
- fire investigation

A young burglar took off his shoes and socks before breaking in. He put the socks over his hands – no fingerprints that way. But he did not realise that bare footprints are as unique as fingerprints. Result? A two-year prison sentence.

How did you become.... a forensic scientist?

Forensic science is a popular career choice. However, this means that there is a lot of competition for jobs. Applicants with good qualifications are much more likely to be successful. It is also useful to gain laboratory work experience where possible. This could be in a local hospital or in a school as a laboratory technician.

You need analytical, teamwork, and communication skills to be a forensic scientist. Personal qualities include integrity, objectivity, and an enquiring mind.

The flowchart shows the qualifications needed to become an assistant forensic scientist or a forensic scientist.

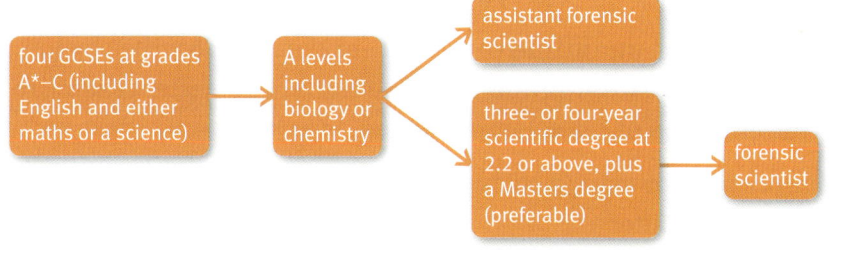

A forensic scientist taking a sample from blood-stained jeans during a criminal investigation.

In the past most forensic scientists in the UK were employed by the government-owned Forensic Science Service. Now there are a variety of different employers including privately owned companies and specific police forces.

Key word
✓ forensic

Questions

1 Suggest reasons why a crime-scene investigator must:
   a have patience
   b pay close attention to detail
   c work accurately under pressure.

2 Give three examples to explain why scientists are needed to help enforce the law.

Colourimetry

Colourimeters measure colour intensity and are used to give quantitative results.

Measuring colour

Colour matching by eye can be used to estimate concentrations of coloured compounds in solution but sometimes more precise measurements of colour **intensity** are needed. Colourimetry is a technique used to find out the **concentration** of a coloured chemical in solution by measuring the colour intensity. The instrument used is a **colourimeter**.

Colourimeters

A colourimeter compares the intensity of light passing through the coloured solution with the intensity of light passing through the colourless solvent. Less light passes through the coloured solution.

The meter gives a **quantitative** measure of the light absorbed by the solution. A **calibration graph** is used to convert the meter reading to the concentration of the coloured solute.

The procedure for using a colourimeter involves several steps:
- fit a suitable filter
- put in a sample of pure, colourless solvent, and set the meter to zero
- put in a series of standard reference solutions, and use the meter readings to plot a calibration graph
- put in the test solution and note the reading
- use the calibration graph to calculate the concentration of the test solution.

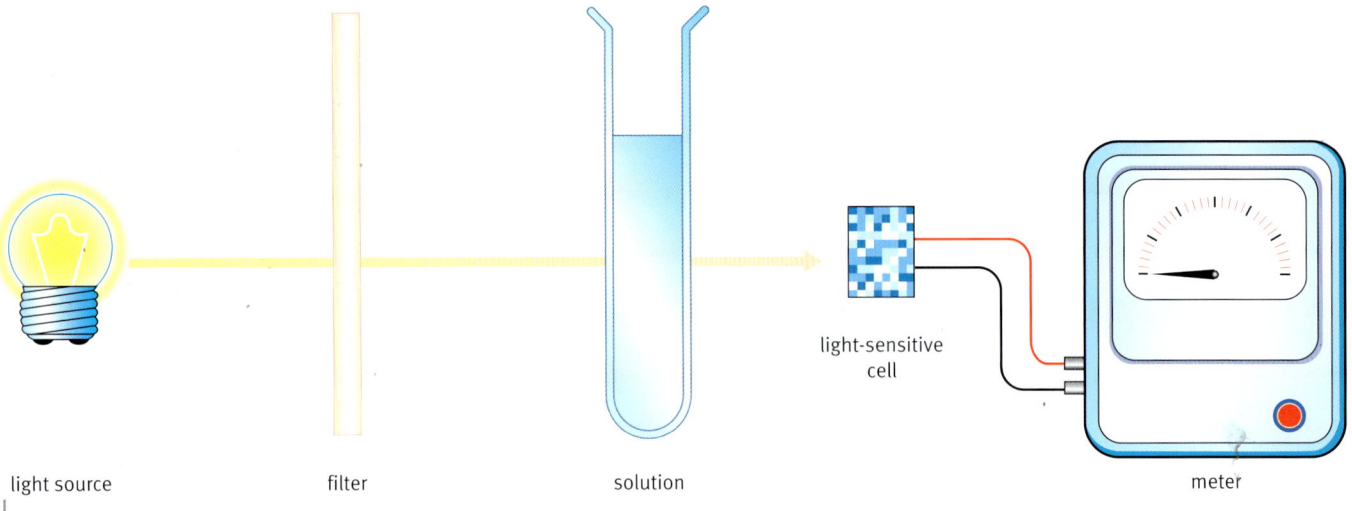

light source   filter   solution   light-sensitive cell   meter

The main parts of a colourimeter.

Strengths and weaknesses of colourimetry

Uncertainty

If people make judgements about a colour and its intensity by comparing colours, two people may not get exactly equivalent results. There would be a high measurement uncertainty. A colourimeter has the advantage of giving quantitative, consistent results.

Range

The range of concentrations that can be determined depends on the compound being analysed. Colourimeters can only measure **absorbance** in a narrow range, so solutions have to be at dilutions suitable for that range. The eye can detect differences in intensity over a much wider range of concentrations.

Sensitivity

The lowest concentration of a chemical that can be measured accurately by a colourimeter, called its sensitivity, depends on how strongly the chemical absorbs light (how highly coloured it is). Weak or colourless solutions are sometimes converted to more intensely coloured compounds for measurement.

It is possible to measure the absorbance of very dilute solutions of copper sulfate with a colourimeter by mixing them with ammonia solution first. The deep-blue solution that forms has a more intense colour than the original copper sulfate.

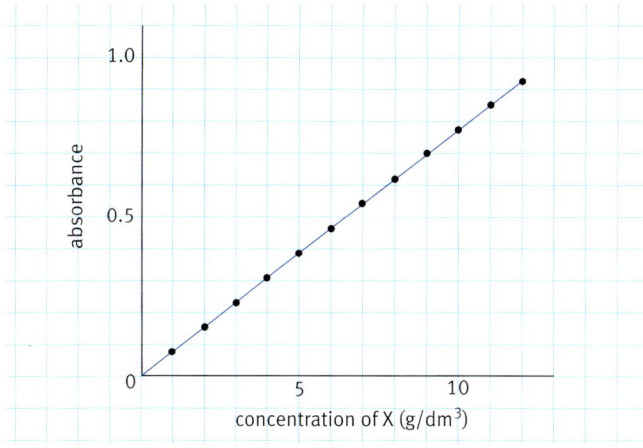

A calibration graph plotted from data obtained from standard reference solutions of compound X.

Key words

- ✓ **intensity**
- ✓ **concentration**
- ✓ **colourimeter**
- ✓ **quantitative**
- ✓ **calibration graph**

Questions

1 Use the calibration graph to find the concentrations of solutions of compound *X* that had absorbency readings of:
   a 0.50
   b 0.20

2 Explain why the colour intensity of a glass of your favourite orange squash helps you tell if it is the strength you like.

3 Explain why the colour intensity might not be the same for the nicest strength of a different brand of orange squash.

Chromatography

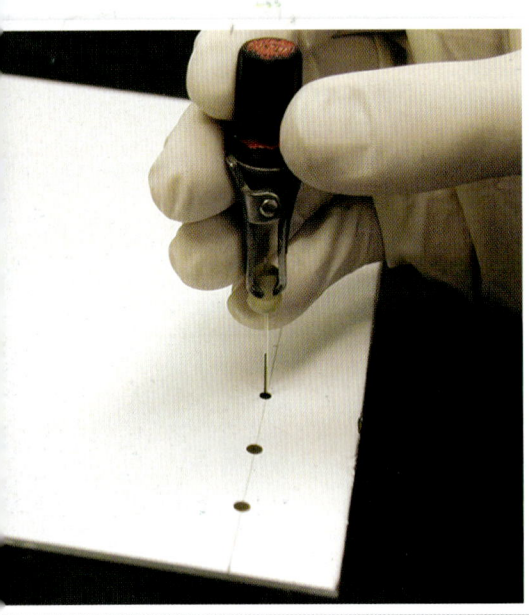

Separation by chromatography

Chromatography separates and identifies chemicals in mixtures. The technique depends on the movement of a liquid through a fixed medium. The liquid is the moving, or **mobile, phase**. The fixed medium, which does not move, is the **stationary phase**.

A small drop of the test solution is put on a chromatography plate and allowed to dry. If the solution is dilute, this is repeated several times. Small concentrated spots are needed. The separation is not so good if the spot spreads too much.

THE SCIENCE

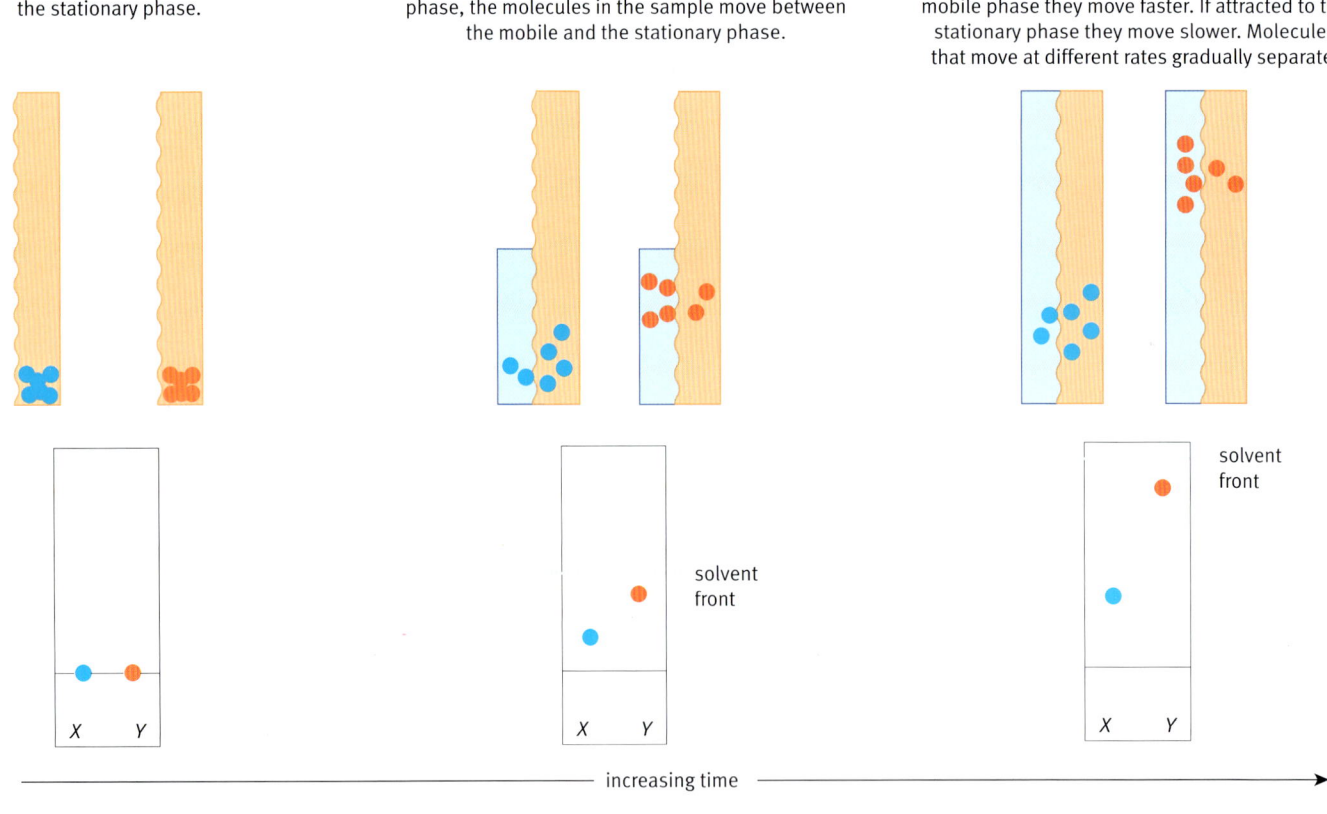

The sample is added to the stationary phase.

As the mobile phase travels through the stationary phase, the molecules in the sample move between the mobile and the stationary phase.

If the molecules in the mixture are attracted to the mobile phase they move faster. If attracted to the stationary phase they move slower. Molecules that move at different rates gradually separate.

solvent front

solvent front

increasing time

Key

● = substance *X*

● = substance *Y*

stationary phase

mobile phase

Paper chromatography

In paper chromatography the stationary phase is the water trapped in the fibres of the paper; the paper just acts as a support. The mobile phase may be a single **solvent** or mixture of solvents. Some chemicals (**solutes**) dissolve well in water (to make an **aqueous solution**) while others are more soluble in **non-aqueous** solvents. The choice of solvent depends on the mixture being separated. Paper chromatography is an easy method of separation though it is now not often used for analysis.

Thin-layer chromatography

In thin-layer chromatography a thin layer of an absorbent solid is coated onto a glass or plastic sheet. The solid acts as the stationary phase. As in paper chromatography, various solvents can be used as the mobile phase. Thin-layer chromatography is quicker than paper chromatography. It is used to analyse dyes in food products and clothing fibres, and to test for controlled drugs, especially cannabis.

Developing the chromatogram

When the solvent approaches the top of the paper or thin-layer plate, the **chromatogram** is removed from the tank and allowed to dry. The position of the **solvent front** is always marked. Coloured spots are marked directly, in case their colour fades. 'Invisible' spots, on special thin-layer plates, are sometimes viewed and marked under ultraviolet (UV) light. Otherwise it is necessary to **develop** the chromatogram by spraying it with a chemical that reacts with the spots to form coloured compounds that can be seen.

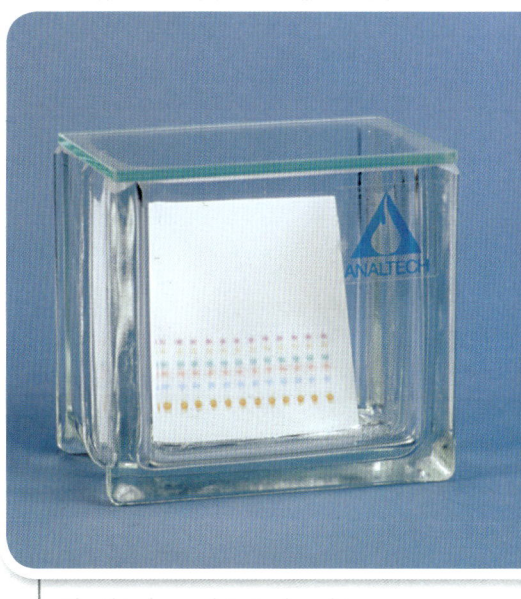

The thin-layer plate is placed in a tank containing solvent. Mixtures are separated by the movement of the solvent (the mobile phase) through the thin layer (the stationary phase).

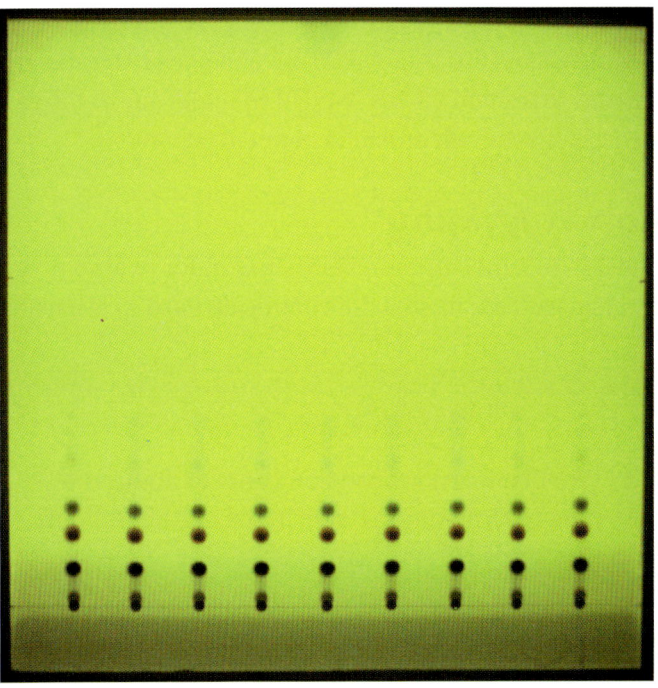

'Invisible' spots seen under a UV lamp. The glow of the special fluorescent material in the thin layer is reduced by the compounds in the spots, which look dark.

Key words
- chromatography
- mobile phase
- stationary phase
- solvent
- solute
- aqueous solution
- non-aqueous
- chromatogram
- develop

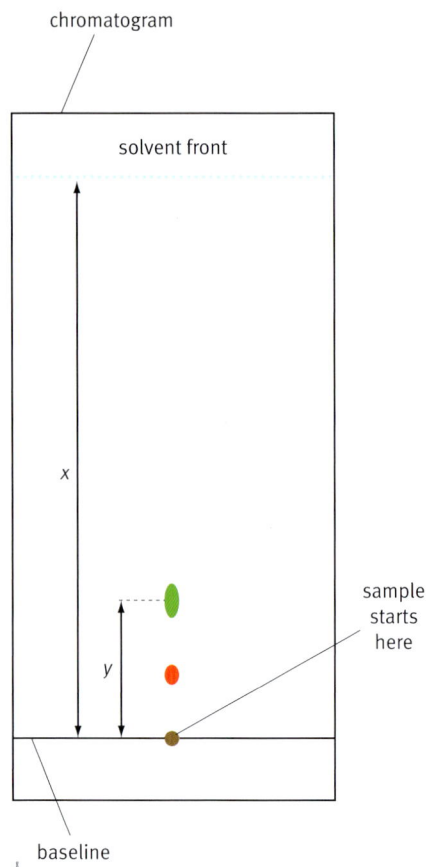

chromatogram

solvent front

x

y

sample starts here

baseline

The retardation factor (R_f) of a spot can be calculated by measuring the distance it travelled and the distance travelled by the solvent.

Key words

- ✔ **retardation factor (R_f)**
- ✔ **qualitative**

Reference materials

One way of using chromatography to identify the chemicals in a sample is to run the **sample** alongside standard reference materials. Spots of compounds suspected of being present in the sample are run on the same chromatogram. The positions of spots from the sample can then be directly compared with the spots from the reference materials.

Retardation factors

A chemical may also be identified by its **retardation factor (R_f)**. It is calculated using the formula

$$R_f = \frac{\text{distance moved by chemical } (y)}{\text{distance moved by solvent } (x)}$$

The advantage of recording retardation factors is that, if the same mobile and stationary phases are used, the retardation factor remains constant. This means that retardation factors can be compared with those of reference materials that were not run on the same chromatogram.

Qualitative analysis

Both paper and thin-layer chromatography are good methods of **qualitative** analysis, to find what chemicals are present in a mixture. Thin-layer chromatography can be used for a wider range of chemicals and usually gives quicker results and better separation than paper chromatography.

Quantitative analysis

Generally thin-layer and paper chromatography give only a rough idea of *how much* of a chemical is present, estimated from the intensity of the coloured spot. Sometimes spots are removed from a paper or thin-layer plate chromatogram and extracted with a solvent to make quantitative measurements of exactly how much of each chemical is present.

Two-way chromatography

Sometimes the chemicals in a mixture are very similar, for example, the coloured pigments in leaves. This makes the chemicals hard to separate

Questions

1. Paper chromatography is used to separate a mixture of a red and a blue compound. The blue compound is more soluble in water, while the red compound is more soluble in the non-aqueous chromatography solvent. Sketch a diagram to show the chromatogram that you would expect.

2. What conclusions can you come to about the nine test mixtures shown in the chromatogram on page 133?

3. Calculate the R_f values of the red and green spots in the diagram on this page.

by simple chromatography. The R_f values in any one solvent are all very similar. So, a single solvent is not enough to separate the chemicals. One method of improving separation is to use two-way chromatography. This method uses two different solvents.

First, a baseline is marked in pencil on a paper or plate. A small drop of the mixture is put on the baseline, towards one end, and allowed to dry. The paper or plate is placed in a solvent and left until the solvent front is about 1 cm from the top. The position of the solvent front is marked in pencil before it dries. This is labelled 'solvent front 1'. At this stage the chemicals are not well separated but are smeared across the chromatogram.

When the paper, or plate, is dry it is rotated by 90 degrees, and placed in a second solvent. As before the spots must be just above the solvent surface. Once the second solvent is about 1 cm from the top, the paper or plate is removed, and the position of the solvent front is marked, and labelled 'solvent front 2'.

It is unlikely that chemicals with similar retardation factors in the first solvent also have similar retardation factors in the second solvent. This means that the spots move by different amounts and all the components in the mixture are separated.

You cannot use reference materials with two-way chromatography because the result would be a confusing mixture of spots. Two-way chromatograms are analysed using the R_f values for retardation factors of all the chemicals in the two solvents.

Questions

4 With the help of figures 1 and 3 estimate the R_f values (a) for the green chemical and (b) for the yellow chemical
   i   with solvent 1
   ii with solvent 2.

5 Use your answers to question 4 to explain why two-way chromatography is needed to bring about a complete separation of the four chemicals.

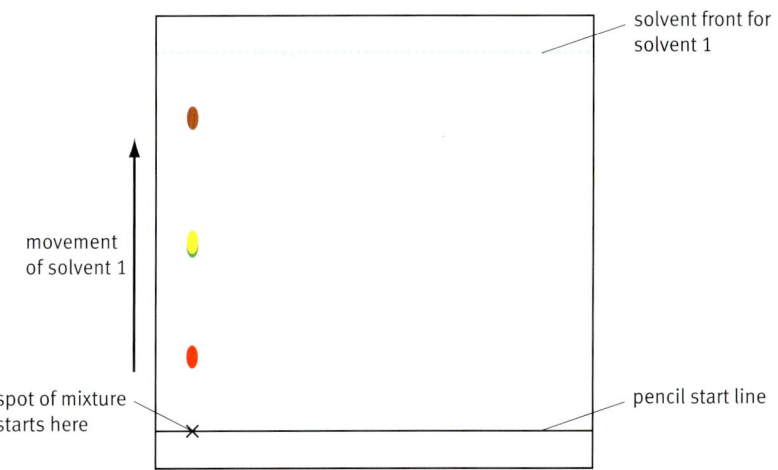

Figure 1: The chromatogram after the first stage of two-way chromatography to separate a mixture of four chemicals.

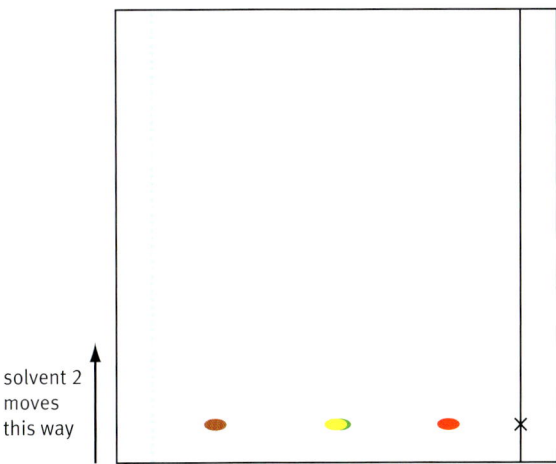

Figure 2: The chromatogram rotated through 90 degrees ready to place in the second solvent.

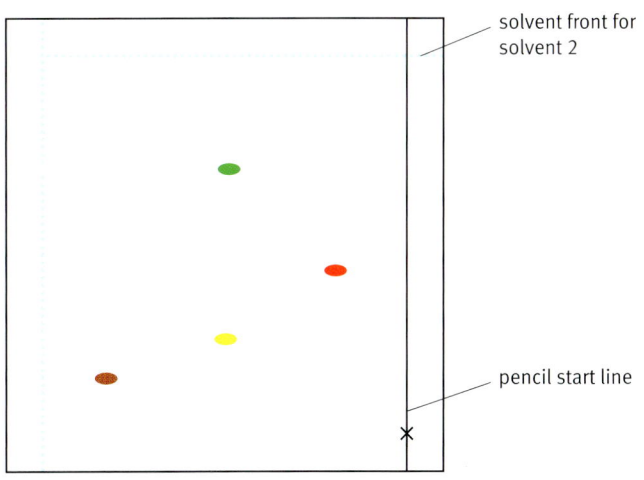

Figure 3: The chromatogram after the second solvent has moved up the paper.

Microscopic examination

To match fingerprints, the expert compares the important features (whorls and loops) and makes measurements of their relative positions.

THE SCIENCE

Key words
- ✔ **objective lens**
- ✔ **eyepiece lens**
- ✔ **magnifying power**
- ✔ **resolving power**
- ✔ **depth of field**
- ✔ **micrograph**

Question

1 A microscope is fitted with an eyepiece lens of magnification ×10. What is the magnifying power of the microscope using:
   a an objective lens of magnification ×10?
   b an objective lens of magnification ×40?

Visual information – comparing images

Modern fingerprint analysis uses automatic recognition software to compare prints with those on a database. A fingerprint expert then carefully compares the likely matches to make a final identification. In photographs of faces, key facial features can be measured to confirm a match. A light microscope is used to compare small samples like hairs or fibres as it gives more detail.

The compound light microscope

Microscopes are used to see things that are too small to be seen by eye. A light microscope has two magnifying lenses. One is near the object being looked at, called the **objective lens**. The second is near the eye, called the **eyepiece lens**.

Magnifying power

The eyepiece lens magnifies the image formed by the objective lens. The **magnifying power** of the microscope is therefore found by multiplying the magnifying powers of the two lenses. This can be expressed as the formula

$$\text{magnification of microscope} = \text{magnification of eyepiece lens} \times \text{magnification of objective lens}$$

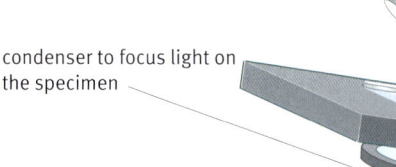

eyepiece with a magnifying lens

objective lenses of different magnification

coarse focus

fine focus

condenser to focus light on the specimen

stage with stage clips

mirror to collect light

arm

base

The main parts of a light microscope.

Resolving power

Magnification is only useful if the image makes it possible to see the fine detail. A microscope that lets you see details that are very closely spaced has a high **resolving power**. The resolving power of light microscopes is limited by the nature of light. In the very best light microscopes, the minimum distance between points that can be seen as distinct is 0.0002 mm (0.2 μm). This limit means that the useful magnification of the best light microscopes is about ×1500. Many light microscopes will not be useful for samples smaller than a micrometre (1 μm). Transparent samples that are smaller than this will also be difficult to see with a light microscope.

Depth of field

The light microscope has to be focused exactly onto the structure being examined. Objects above and below appear blurred. This is because a light microscope has a very narrow **depth of field**.

Preparing specimens

Samples need to be mounted onto a glass slide. Temporary slides, to be looked at straight away, can be prepared quickly. A longer and more complicated treatment is needed to make permanent slides. Some material needs squashing, smearing, or slicing first to give a layer thin enough for examination. Coloured stains can be used to reveal certain features.

Recording information

An image seen through a light microscope might be recorded as a description, drawing, or by still or video photography. A microscope image recorded in the form of a photograph is known as a **micrograph**. It is important to know the magnifying power used when interpreting microscopic images.

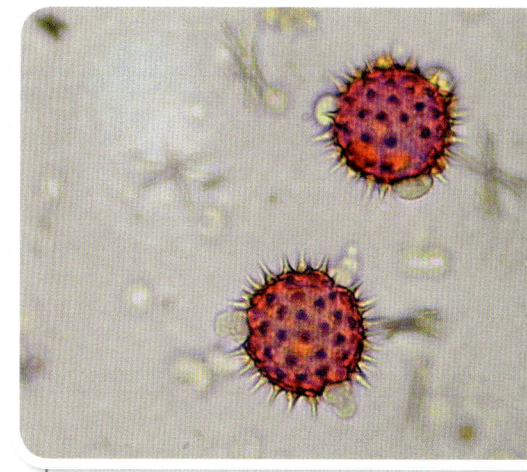
A light micrograph of (stained) sunflower pollen. (Magnification ×900.)

 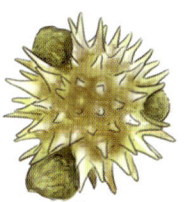
Sunflower   Calendula

Petunia   Hibiscus

Reference illustrations can be used to identify and interpret an unknown sample.

A light micrograph of silk fabric (the colour is due to special lighting). (Magnification ×100.)

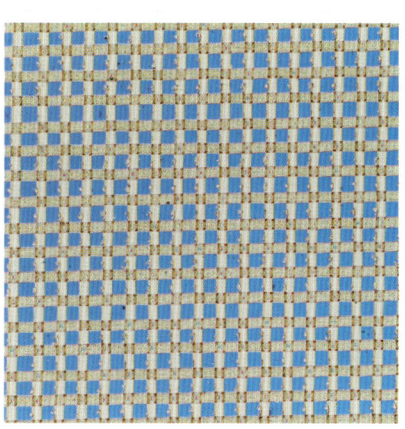
A light micrograph of a synthetic fabric (Trevira). (Magnification ×100.)

Questions

2 Use a written description or an annotated sketch to record the main features of sunflower pollen. Include an estimation of the size of a pollen grain.

3 Make a table to compare the features that you can see in the micrographs of the silk and synthetic fabric. Remember to count and measure the features you can see.

The electron microscope

An **electron microscope** uses a beam of electrons instead of light to produce an image. A vacuum pump has to remove all the air from the inside of an electron microscope to stop air molecules getting in the way of the beam of electrons.

Glass lenses focus the rays in a light microscope. In an electron microscope the electron beams are focused by powerful magnets.

Magnifying and resolving power

Electrons are so tiny that an electron beam can pick out parts of an object that are very close together. This gives electron microscopes a very high resolving power. The high resolution makes higher magnifications possible with electron microscopes. Some electron microscopes have a magnifying power of $\times 500\,000$.

Depth of field

Electron microscopes have a much greater depth of field than light microscopes. This is particularly useful in scanning electron microscopes (SEMs) where the surfaces of solid specimens are examined.

This SEM is used to study engine parts in racing cars. It can be used to look at metal surfaces and has better resolution, magnification, and depth of field than a light microscope.

THE SCIENCE

Types of electron microscope

- **Scanning electron microscopes** (SEMs) use a fine beam of electrons to scan the surface of a specimen and generate electrons from its surface. These secondary electrons are collected by a detector and processed to form an image on the screen.
- **Transmission electron microscopes** (TEMs) pass a beam of electrons right through the sample. Electrons passing through the sample are captured as an image on the screen, showing dark and light areas.

Benefits and costs of electron microscopy

Electron microscopes have a much higher magnifying power than light microscopes, so can show much greater detail. However, electron microscopes are expensive and the equipment is not portable. Specimens for electron microscopy have to be dried and specially mounted to survive in the vacuum. This is a complex process and it means that electron microscopes cannot be used to look at living specimens or biological material in its natural state.

- For an SEM, biological materials must be coated with a very thin layer of gold or carbon. Coating is not necessary for metals or minerals.
- A TEM needs very thin specimens. They are sometimes treated with chemicals to show up different microstructures.

Recording information

Images from electron microscopes are usually recorded as a micrograph. They are sometimes digitally coloured afterwards. The colours are not the true colours of the specimen. The false colours make it easier to see different structures. These structures can then be compared with reference images to help interpret the micrograph. It is also very important to know the magnification when gathering information from an electron micrograph.

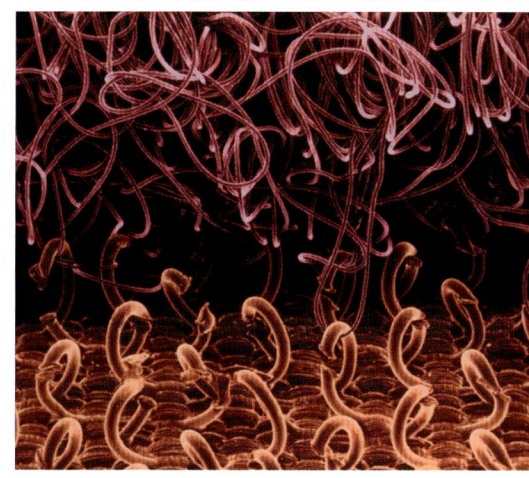

A coloured SEM micrograph of Velcro. The good depth of field means that the whole image is in focus. (Magnification ×7.)

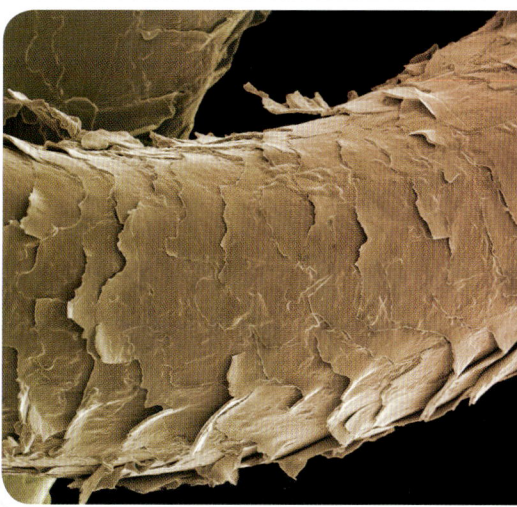

A coloured SEM micrograph of a human hair. (Magnification ×380.)

Questions

1 Look at the top micrograph above. Estimate the area of the hooked Velcro in the picture in mm^2.

2 Give the approximate number of hooks per mm^2 on the Velcro sample.

3 Write a short report of what you can see in the lower micrograph. Include details of all the main features.

Key word
- ✓ **scanning electron microscope (SEM)**

Electrophoresis

Separation by electrophoresis

Electrophoresis, like chromatography, is another technique used to separate and identify chemicals. It uses an electrical voltage, instead of a moving solvent, to separate out mixtures. The technique is particularly useful for small quantities of large biological molecules, such as fragments of **DNA**. In solution, at the correct pH, these molecules become electrically charged.

A small amount of the sample solution is placed on a solid support formed from gel. The support is in a tank full of a solution that keeps the pH constant. An electrical voltage is applied to **electrodes** placed at each end of the tank. This causes charged particles to move through the gel. The negative electrode attracts molecules with a positive **charge**. The positive electrode attracts molecules with a negative charge. The speed of each particle depends on how big its charge is, and its size.

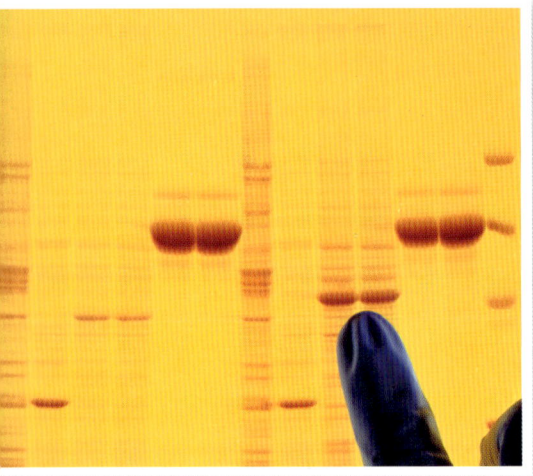

DNA fragments after separation by electrophoresis. Several different samples (across the picture) have been run on the same gel. Each sample shows many bands – each is a different sized fragment of DNA.

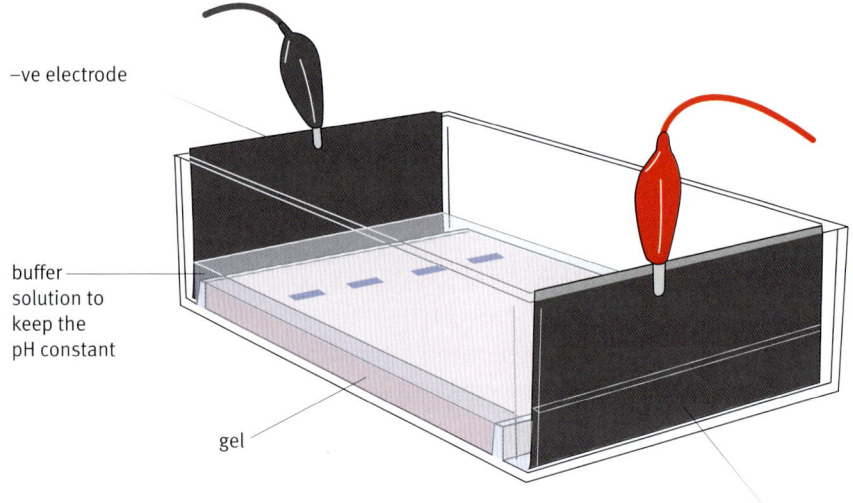

−ve electrode

buffer solution to keep the pH constant

gel

+ve electrode

When an electric voltage is applied between the electrodes, positive particles move towards the negative electrode and negative particles move towards the positive electrode.

Electrophoresis of DNA fragments

DNA fragments for electrophoresis are dissolved in a special solution at a pH that means that they have a negative charge. A gel is used as the solid support. The negatively charged DNA fragments all move towards the positive electrode. The speed that they move through the gel depends on their size. After a suitable time the electric current is switched off. The gel is removed from the apparatus and soaked in a marker or stain that binds to the DNA. This reveals separate bands of DNA showing how many different sized fragments are present in the sample.

Key words
- ✓ **electrophoresis**
- ✓ **DNA**
- ✓ **electrodes**
- ✓ **charge**
- ✓ **DNA profiling**

DNA profiling

Everyone, except identical twins, has a unique sequence of subunits in their DNA. **DNA profiling** is a way of looking at some of these differences in DNA. Tiny pieces of tissue can provide enough DNA for a test. Several different techniques are used to make a DNA profile, but one step in the procedure is to separate out the fragments of DNA. Electrophoresis is the best way to do this. The pattern of fragments obtained is called a DNA profile. It can be used to match up samples of DNA with a high degree of certainty.

Examples of the use of DNA profiling include:

- forensic investigations – identifying murderers, rapists, and other criminals
- checking parentage – in people and other animals
- wildlife protection – by detecting illegally traded animals
- testing food for genetically modified ingredients.

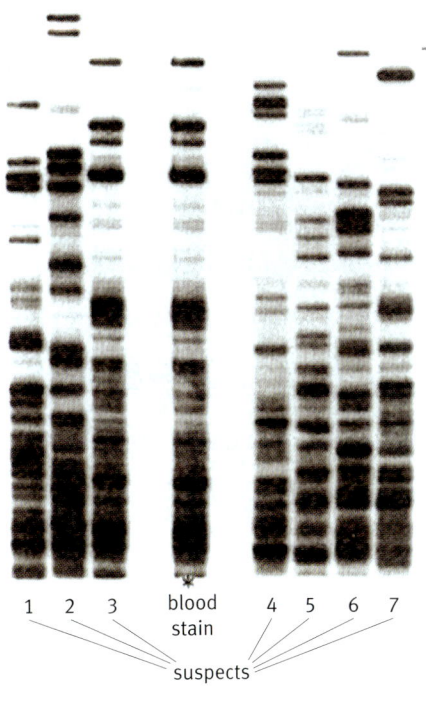

The DNA profile of a blood stain from a crime scene, together with the profiles of seven suspects.

Questions

1 Look at the DNA profiles above. Which suspect left the blood stain?

2 What factors determine how quickly particles move through the gel during electrophoresis?

Bioinformation

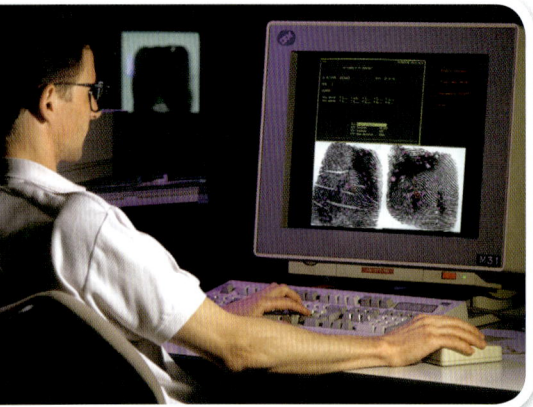

Experts look for distinctive patterns in fingerprints. They study the shapes of the ridges in the skin that make up the print.

Fingerprints and DNA profiles are examples of **bioinformation**. The police use this type of information to investigate crimes.

Fingerprints

Fingerprints develop before birth and remain unchanged throughout life. They are the most commonly used method of identification. No two people have the same fingerprints, not even identical twins.

Fingerprints may be taken from a person when they are arrested or found at a crime scene. They may be taken from a weapon or another item of interest to the police.

A forensics officer uses tape to lift fingerprints from a gun found at the scene of a crime. He is wearing protective clothing to prevent contamination of the scene.

Fingerprints are hard to analyse. This means that trained experts must check them. Identification relies on matching patterns in the prints. It is even harder to analyse partial fingerprints.

When fingerprint evidence is used in court, juries must be aware that when a match is declared, it is never a matter of scientific certainty or conclusive fact. It is the opinion of the expert.

DNA profiles

Investigators collect biological samples such as blood, skin, hair, and semen. These may be found at a crime scene or taken from a weapon or item of interest to the police. Other samples (such as cheek cells) are taken when suspects are arrested. These samples contain all the genetic information about the people involved.

Electrophoresis and other techniques are used to obtain DNA profiles from the samples. After analysis the profile is stored as a sequence of 20 numbers on the National DNA Database.

The chance of two unrelated individuals sharing the same complete DNA profile is around one in a billion. Even so, matches can arise by chance. This is more likely to happen:

* if the crime-scene sample contains only tiny amounts of DNA, giving a partial profile
* between closely related individuals, as the size of the National DNA database expands
* between individuals within an isolated or inbred population.

DNA profiling is generally a very reliable way of identifying a person. However, DNA evidence comes with lots of complicated statistics, which can be difficult for legal professionals and members of a jury to understand. Also, contamination can occur at the crime scene, from police, or from laboratory staff due to poor storage procedures.

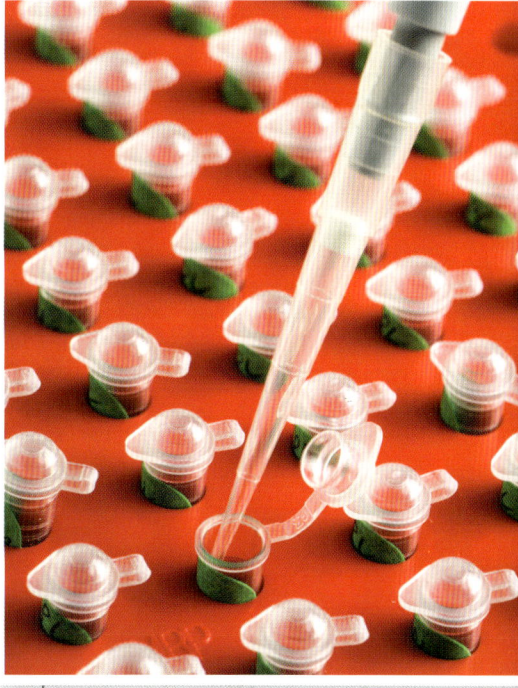

Using a pipette to sample one of a large set of specimens collected at a crime scene.

Ethics and the DNA database

The National DNA database helps the police to solve crimes. However, some people think that it is wrong to keep information about people's DNA. If your DNA is on the database, there is a chance you will be identified as a match to DNA found at a crime scene, even if you are innocent. You may become part of a criminal investigation and be placed under suspicion. Even if you are not charged this can be very distressing.

Sensitive information such as family relationships can be found from DNA samples. Police and forensic science services have access to this information about people without their consent. Some people think that this is an invasion of personal privacy.

The benefits of keeping a database have to be weighed up against the costs.

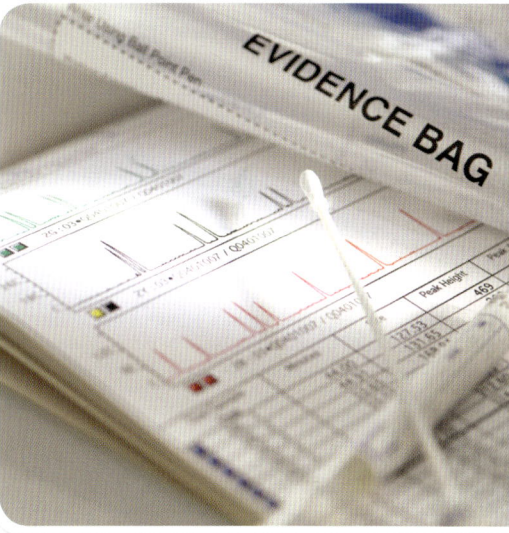

Is this swab containing a DNA sample from a crime scene safe from contamination?

Questions

1 Suggest reasons why fingerprints are more widely used than DNA profiling in crime detection.

2 Suggest why DNA profiles can be used to identify murderers who cannot be detected by fingerprinting.

3 Why is the danger of contamination of a crime scene more serious for DNA profiling than for fingerprinting?

Principles

A sample viewed under a light microscope must be thin enough to allow light to pass through it. Living organisms, therefore, must be very small to be seen under a microscope.

Non-living material can be squashed or sliced thinly. Dyes can be used to stain the sample and show features that cannot otherwise be seen.

Procedure

1 Place the specimen on a microscope slide.

2 Put a drop of water on the specimen.

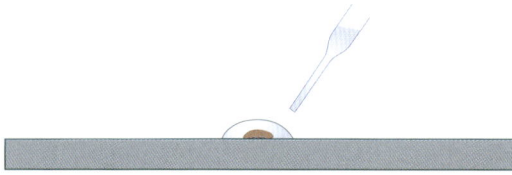

3 Use a mounted needle to lower a coverslip over the drop of water slowly. Avoid air bubbles.

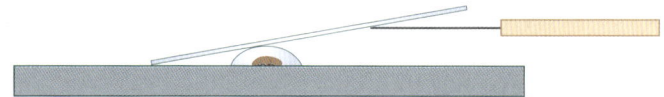

4 Examine the specimen with a light microscope.

5 If required, stain the specimen by placing a drop of stain touching the edge of the coverslip. Use a piece of filter paper to soak up water from the opposite edge of the coverslip. It should draw the stain under the coverslip.

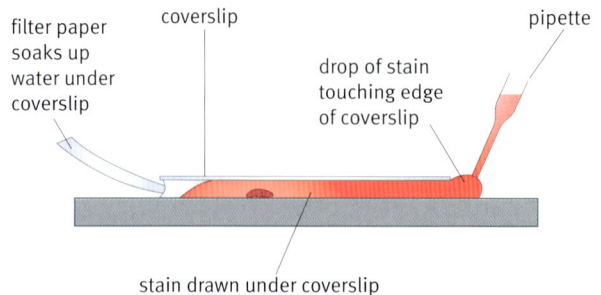

filter paper soaks up water under coverslip   coverslip   drop of stain touching edge of coverslip   pipette

stain drawn under coverslip

6 Re-examine the slide slider the microscope.

PROCEDURES & TECHNIQUES

Estimating concentrations by colour matching

Principles

The intensity of colour of a solution can be used to estimate its concentration. A series of solutions of known concentration are prepared. The concentration of the sample is estimated by comparing its colour intensity with these reference solutions.

Procedure

1 Prepare a suitable reference solution of the coloured compound, for example, at a concentration of 80 ppm.

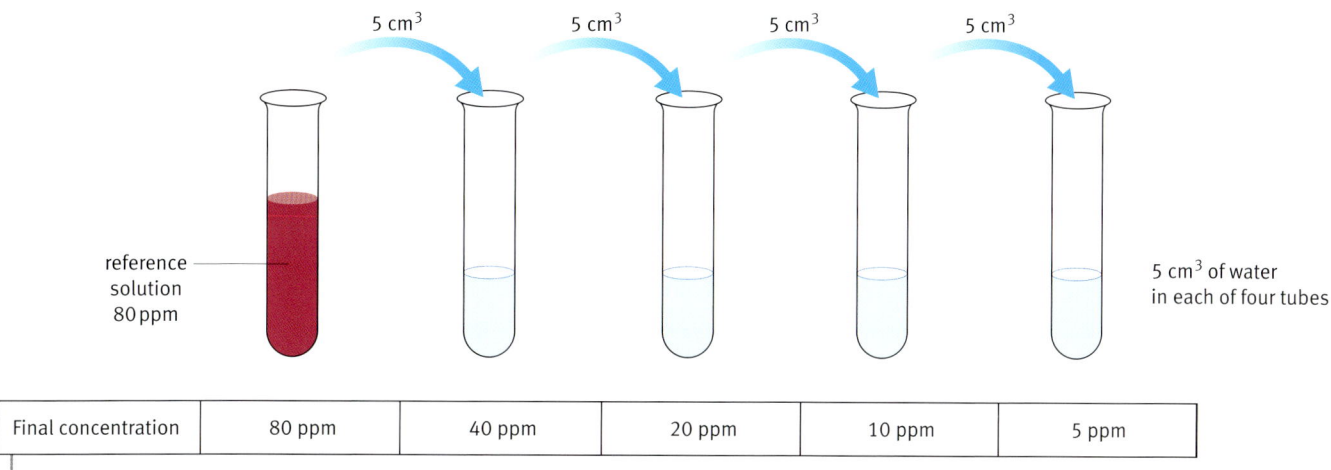

Final concentration	80 ppm	40 ppm	20 ppm	10 ppm	5 ppm

⚬ Reference solutions are prepared by serial dilution.

2 Pipette 5 cm³ of distilled water into each of four test tubes. Label the tubes 40 ppm, 20 ppm, 10 ppm, and 5 ppm.

3 Pipette 5 cm³ of the reference solution into the first tube. Mix by drawing liquid in and out of the pipette a few times.

4 Pipette 5 cm³ of this mixture into the next tube. Mix.

5 Continue in this way to the last tube.

6 Compare the colour intensity of the unknown sample with the prepared solutions. Decide on the best colour match.

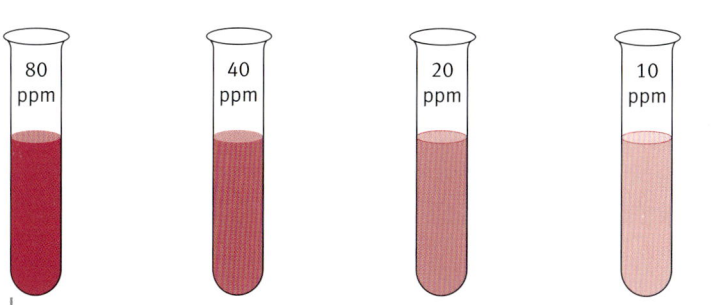
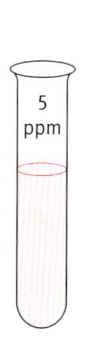
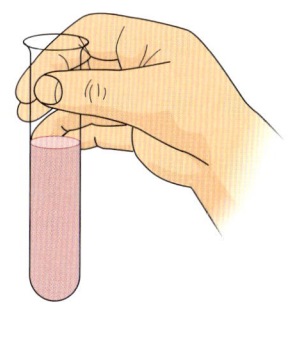

⚬ The concentration of this sample is estimated as 10 ppm.

Principles

Colourimeters are used to measure the exact concentrations of coloured solutions. They compare the amount of light passing through a coloured solution with the light passing through the clear solvent.

The result is a measure of absorbance. A calibration graph is needed to convert absorbance measurements to concentrations.

Procedure

1 Obtain a set of reference standards: solutions of the test chemical in known concentrations.

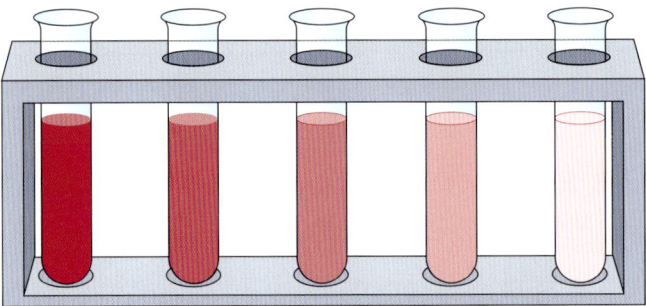

2 Prepare a table for your readings. List the concentrations of the reference solutions in order, starting with the weakest.

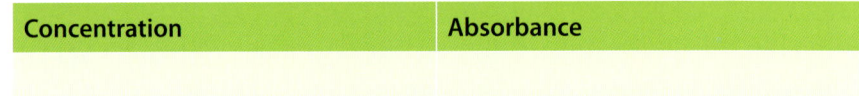

Concentration	Absorbance

3 Put a suitable filter into the colourimeter.

4 Fill a cuvette (the tube for the colourimeter) with pure solvent. Hold it by the top to avoid fingermarks. Make sure the outside is dry.

5 Place the cuvette of solvent into the colourimeter and adjust the reading to zero.

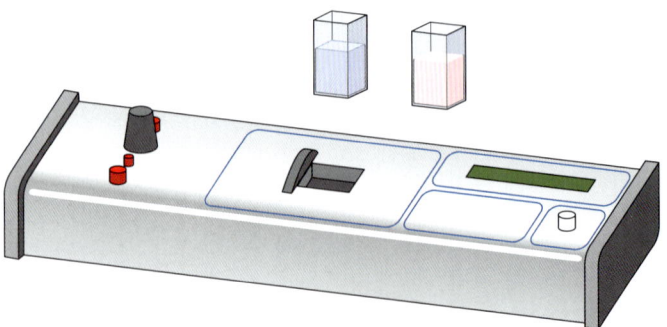

6   In the same way fill another cuvette with the first, least concentrated, reference standard.

7   Place the cuvette of reference solution into the colourimeter and record the absorbance reading.

8   Discard the reference solution and fill the cuvette with the next reference solution in your list.

9   Repeat steps 6–8 until you have a reading for each reference solution.

10 Plot your readings as points on a graph.

11 Draw the best-fit straight line through the points. This can now be used to find the concentrations of solutions of this chemical from their absorbance readings.

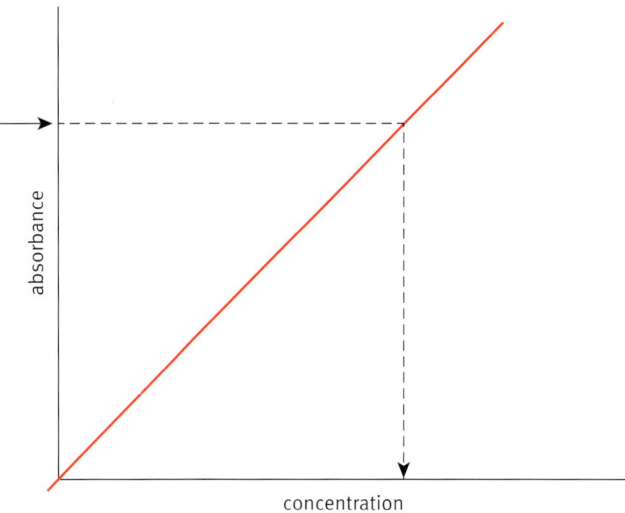

Paper chromatography

Principles

This is a technique for qualitative analysis, to find what compounds are present in a solution. Small, concentrated spots of the sample are made on the paper, which is then placed in a solvent. In paper chromatography the stationary phase is the water trapped in the fibres of the paper; the paper just acts as a support. The solvent is the mobile phase. Reference materials may be run on the same piece of paper, or R_f values may be calculated and compared with known values.

Procedure to run the chromatogram

1 Cut a rectangular piece of chromatography paper to fit a suitable tank.

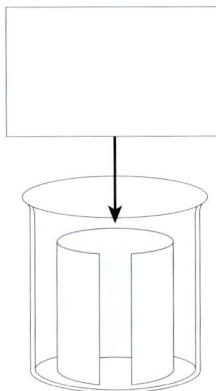

2 Put a 1-cm depth of the chosen solvent into the tank and cover it with a lid.

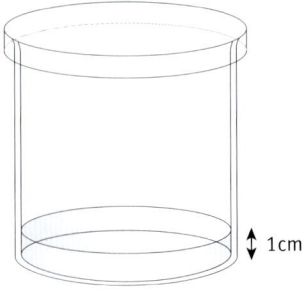

1cm

3 Mark the piece of chromatography paper with a pencil as shown.

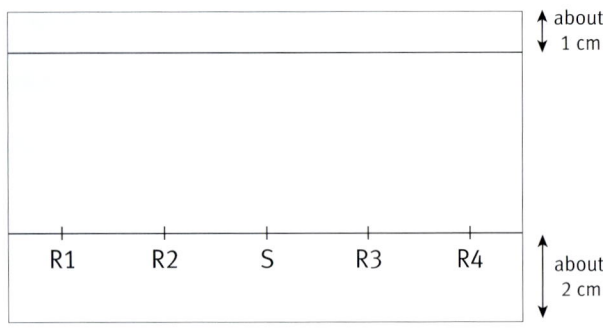

about 1 cm

R1   R2   S   R3   R4   about 2 cm

4 With a fine glass tube, place a small spot of the sample solution at S. Repeat this, using a clean tube, for each reference solution R1, R2, ...

5 Roll the paper into a cylinder and staple it as shown, so that the ends do not touch.

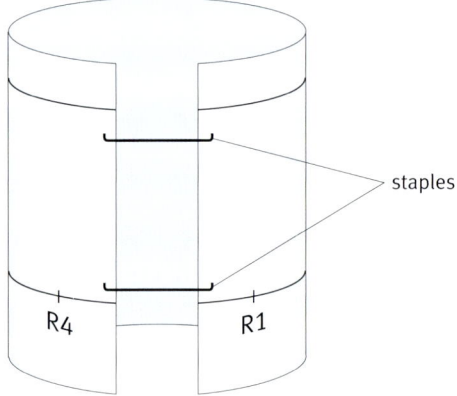

staples

R4   R1

6 Stand the paper in the tank, cover with a lid and leave undisturbed until the solvent front has reached the upper pencil line.

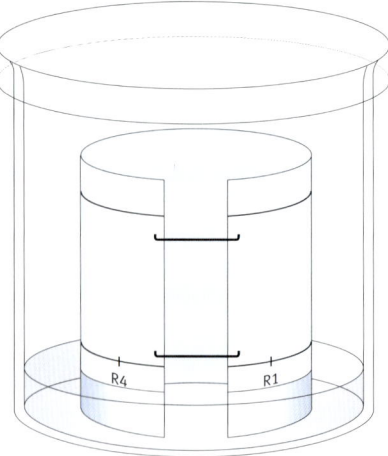

R4   R1

7 Remove the paper from the tank and leave it to dry.

Procedure for calculating the R_f

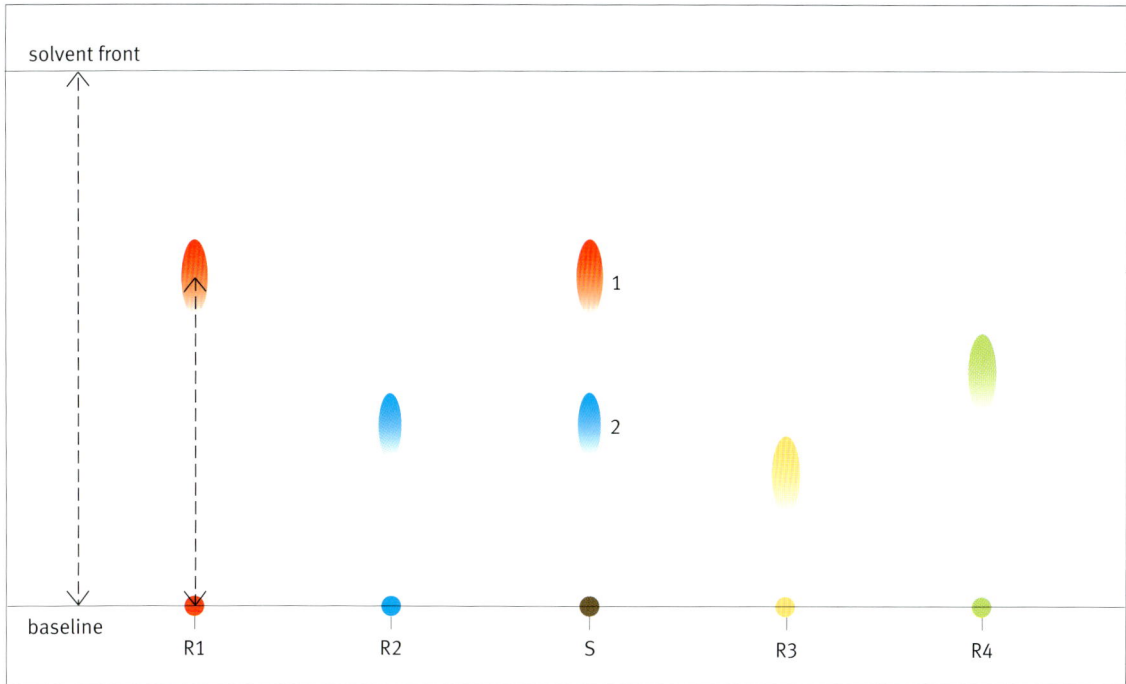

1 If necessary, reveal the positions of colourless spots by developing the chromatogram.

2 Draw around each visible spot with a pencil. If any samples have more than one spot, number the spots.

3 Prepare a table or spreadsheet to record the positions of the spots.

4 Measure and record the distance between the baseline and the solvent front.

5 Measure and record the distance between the centre of each spot and the baseline.

6 Calculate the R_f value for each spot.

Distance between the baseline and the solvent front = cm		
Sample and spot number	Distance from baseline (cm)	$R_f = \dfrac{\text{distance spot moved}}{\text{distance solvent moved}}$

Module Summary

People and organisations

- that scientists need both scientific and technical skills to work effectively
- how public analysts protect people by monitoring standards, giving advice, and doing research
- how the Forensic Science Service collects, analyses, records, and preserves forensic evidence
- why the work of scientists has an impact on the local community
- that the work of scientists is governed by regulations
- that public analysts monitor standards and make sure that the law is obeyed
- why standards are set for food safety and the labelling of products
- how new technology can improve law enforcement such as using DNA profiling, but this raises ethical issues and requires regulation.

The science

- that a colourimeter is used to find the concentrations of solutions by measuring the intensity of a colour
- that the best microscopes have high magnification, good resolving power, and great depth of field; scanning electron microscopes give greater detail but the equipment is complex and can't be used with living samples
- that in chromatography, substances are separated by movement of the solvent (mobile phase) through a medium (stationary phase); substances move between the stationary phase and the mobile phase; different samples have different relative attractions to the solvent and the medium
- that electrophoresis separates mixtures based on their charge and size; charged molecules are drawn to the oppositely charged electrodes
- how a colourimeter calibration curve provides quantitative data
- that microscope images can be interpreted by describing and counting features, making measurements, and comparing with known samples
- that chromatogram samples can be identified by their R_f value $= \dfrac{\text{distance travelled by substance}}{\text{distance travelled by solvent}}$
- how electrophoresis is useful for identifying small samples of biological material such as DNA; DNA profiles can uniquely identify individuals.

Standard procedures

- why a colourimeter calibration curve is plotted using samples of known concentration
- that samples are carefully prepared for the microscope
- magnifying power of a microscope = magnifying power of eyepiece $\times$ objective lens
- that paper and thin-layer chromatography use water or other non-aqueous solvents as the mobile phase
- how biological samples are loaded onto an electrophoresis gel along with reference samples.

Review Questions

1 Describe some of the tasks carried out by people with science qualifications who work for the police force.

2 Jack uses a colourimeter to measure the concentration of a dye in drinks. The maximum safe concentration of the dye is 25 mg/l.

   a Explain how he obtains the calibration graph shown below.

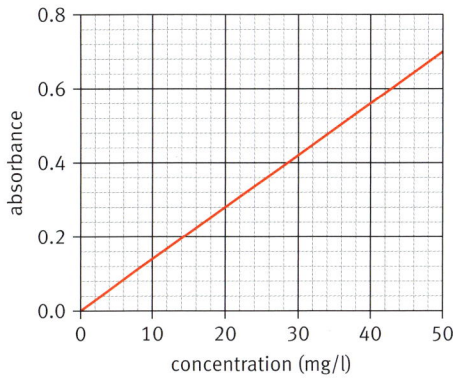

   b One sample has an absorbance of 0.30. Is it safe to drink? Justify your answer.

3 Kerry uses a light microscope to study an organism. Here is the image that she sees.

   a Kerry uses this table to identify the type of organism.

Organism	Number of legs
arachnid	8
insect	6
mammal	4

   What type of organism is she looking at? Justify your answer.

   b The microscope has a magnification of 100 with a × 20 eyepiece. What is the magnification when a × 50 eyepiece is used instead?

   c Explain one advantage and one disadvantage of using a scanning electron microscope to image the organism.

4 Liam uses paper chromatography to separate dyes in a sample of ink taken from a forgery.

   a Explain how the technique separates the dyes.

   b How could Liam use the technique to identify a particular dye in the sample?

   c Explain why dfferent mixtures may need different solvents for the chromatography to be successful.

5 **a** Explain why electropheresis is important in scientific detection.

   b Explain how electropheresis separates the components of a mixture.

6 More and more science is being used to support law enforcement.

   Describe a development in technology that has helped the police or other law-enformcement agencies.

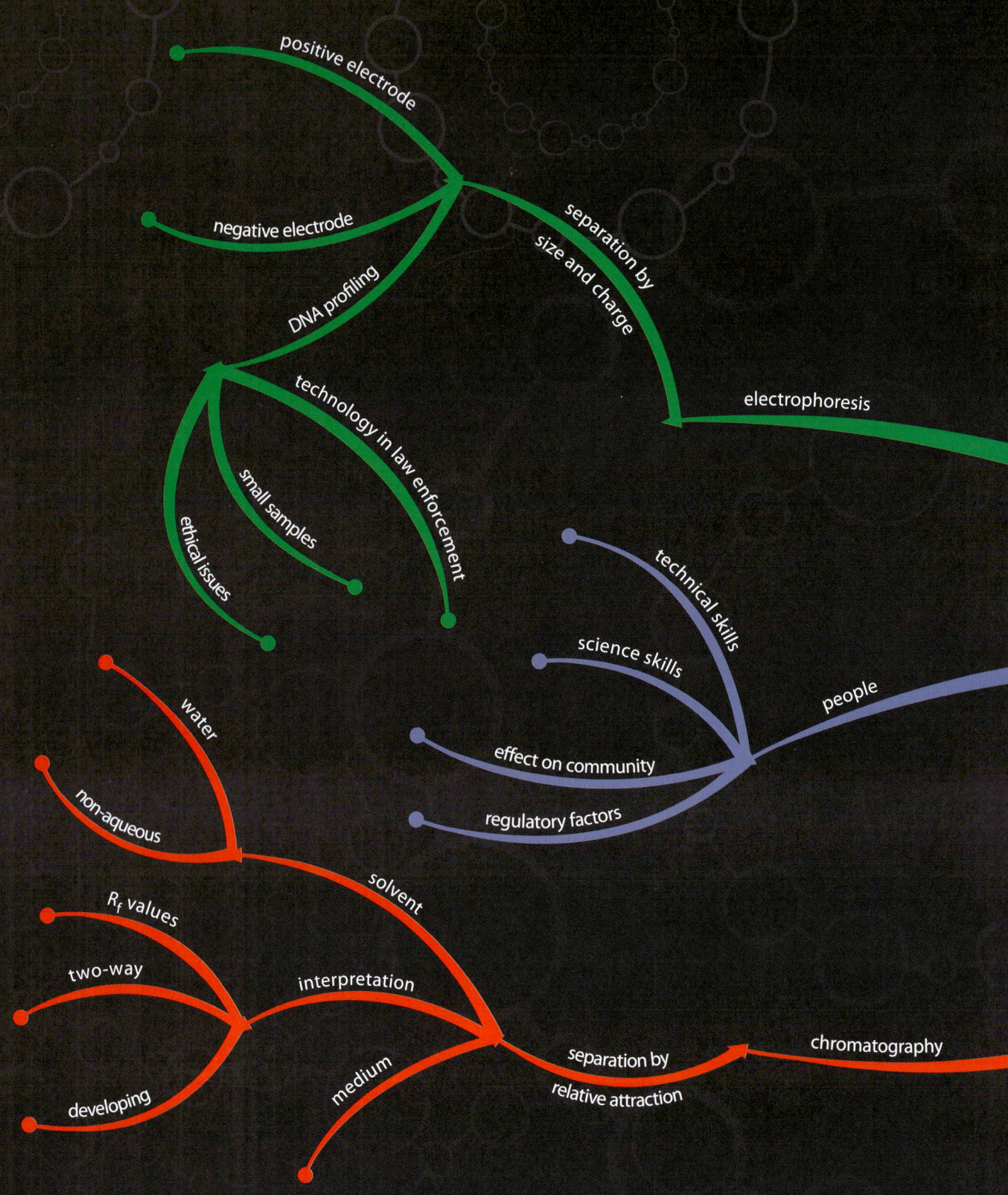

positive electrode

negative electrode

separation by
size and charge

DNA profiling

technology in law enforcement

electrophoresis

small samples

ethical issues

technical skills

science skills

people

effect on community

regulatory factors

water

non-aqueous

solvent

R_f values

two-way

interpretation

chromatography

developing

medium

separation by
relative attraction

SCIENTISTS PROTECTING THE PUBLIC

imaging
- count
- describe
- measure
- depth of field
- resolution
- magnification

microscopes
- advantages and disadvantages
- electron microscope
- light microscope

colourimetry
- coloured chemicals
- reference solutions
- calibration curve
- concentration
- quantitative data

organisations
- law enforcement
- Forensic Science Service
- public analyst
- consumer protection

B1 Sports equipment

Why study sports equipment?

Some people take part in sport to win. Others just do it because it's fun. Whatever your reason for doing sport, you rely on artefacts (such as balls, boots, or bats) to take part. When you select your equipment, it helps if you know what properties are important for the artefact. The correct choice can make all the difference between winning and losing, being comfortable, or in pain. Imperfect equipment can do more than ruin your chances of winning; it can break unexpectedly, perhaps hurting you. You need to know what to look for when making your choice.

What you already know

- how levers do their work
- about the structure of polymers and fibres
- the difference between heat and temperature
- the effect of forces, such as weight and friction, on solid objects
- how forces can change the speed of objects.

Find out about

- the mechanical properties of metals, polymers, and ceramics
- how mixing materials can give improved performance
- how to measure important properties of a material
- how to use shape to make a structure more rigid.

The Science

Scientists use standard procedures to find out if an item of sports equipment meets its specification. Instruments created by scientists, such as accurate rulers and clocks, measure the performance of athletes. Designers of sports equipment use science to match materials to a purpose, often combining materials with different properties to create a composite material with just the right behaviour. The science of materials allows designers to make equipment that is safe, and that won't fail even under the toughest conditions.

Sports performance

Students Amy Cleeton and Blake Raynor are at Birmingham University. After their course in Sports and Materials Science, they plan to help people in different sports to achieve outstanding performances.

They studied the properties of vaulting poles and the structure of the material of which they are made. Blake explains, 'Pole-vault design turned out to be much more detailed than I'd expected. There's the internal structure of the composite material, for example. In a vaulting pole, the fibres are glass and the matrix is an epoxy (polymer) resin. Different composite materials have different amounts of fibres, which can be woven together in different ways.'

International athletes use specially designed vaulting poles made from composite materials.

Want to be an international athlete capable of record-breaking performance? You'd better start training. Want to change what athletes can do, and how they do it? You should become a materials scientist. That's what Claire Davis is, and her work is international too.

In pole vaulting, the highest anybody could jump a hundred years ago was just over 3 metres. Today's athletes can get over 6 metres from the ground. That's almost as high as a house. The difference is mostly due to materials.

To jump as high as possible, a pole vaulter uses the energy of their run-up. They stick one end of the pole in the ground and bend it. This stores the energy from the run-up in the bent pole. The pole is elastic. It springs back and returns to its original shape. When the pole straightens, it transfers the energy back to the athlete and pushes them up. A good pole must be neither too flexible nor too rigid.

Around the year 1900, people used solid wooden poles. Then they switched to bamboo, which is still strong but more flexible. After that, records were smashed at nearly every Olympic Games. By 1956, pole vaulters could reach 4.5 metres, but the rate of improvement was levelling off.

Then, from 1960 onwards, records burst higher and higher again. That was the year when the top athletes started using composite poles. The material has low density, so the poles are light. They are also strong, for safety. If a pole breaks during a jump the athlete might get injured.

You can only win in modern sport if you have the best technology. The knowledge and skills of people like Claire, Amy, and Blake are behind winning performances in the sports stadium.

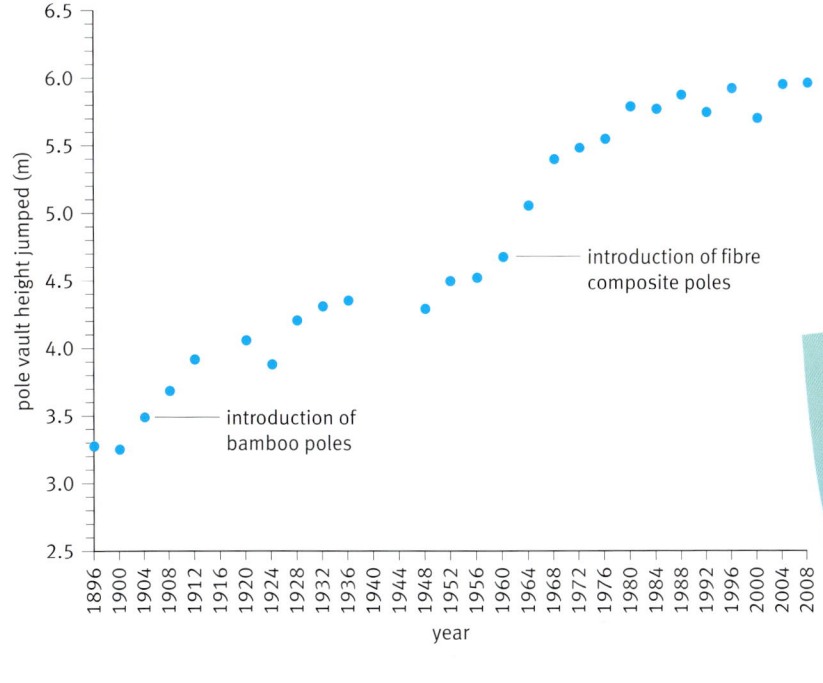

Athletes keep getting better. But when the materials technology improves, they get better faster.

Faster materials

In 1992 a young athlete called Tony Volpentest won the 100-m sprint race in the Olympic stadium at Barcelona. His time of 11.63 seconds is not remarkable until you hear that Tony was born without hands or feet. He was able to cross the finishing line 0.75 seconds ahead of his nearest rival thanks to new technology.

Spring in his step

When Tony runs he wears artificial feet strapped to his stumps. Each foot is L-shaped and springy, so that it can bend and store energy just like a real foot. Tony's feet are made from carbon fibre, a composite material with the following blend of properties:

- high strength, so that they can withstand large forces without breaking
- high elasticity, so that they can store a lot of energy as they deform
- low density, so that they are very light.

Other composites, such as wood or glass fibre, have a similar blend of properties, but nothing like as good as carbon fibre. Tony's feet are so good that they work as well, if not better, than real ones.

Carbon cycle

Carbon fibre made an even more spectacular debut in the 1992 Barcelona Olympics, this time in the cycling events. Up until then, most serious racing cycle frames were made from metal, fashioned into wires and tubes for extra strength. Alloys of aluminium were popular because of their low density, but they were not as strong as steel. Chris Boardman turned up with a bike made from carbon fibre. Not only did it have a mass of just 9 kg but its frame was every bit as strong as steel. It allowed him to lap the world champion in the final heat.

This athlete is running with a carbon-fibre foot.

The cycle is made from 9 kg of carbon fibre, allowing it to be the fastest on the track.

Aerodynamics

Chris Boardman owed his spectacular victory in 1992 to more than just a revolutionary new material. He applied the science of **aerodynamics** to allow air to flow past him more easily. Here are some of the ways in which he did it:

- flat solid wheels instead of wiry spokes, giving less turbulence
- shaped helmet to help the air flow past his head
- low posture and narrow handle bars to intercept less air.

Shape was just as important as material in sweeping Chris to victory.

Swimsuits

The radical new swimsuit worn by Paul Biedermann in 2009 allowed him to surprise everyone and beat the world champion in the World Aquatic Championship in Paris. It used polyurethane to:

- allow water to flow more easily over his skin
- squeeze, stiffen, and slim his body
- trap air to increase his buoyancy.

Polyurethane has been around for years, but the idea of using it in this way was very new.

This swimsuit uses an old material in a revolutionary new way.

Is it fair?

Some say that using advanced technology to get a sporting advantage is cheating. Others point out that unless the athlete has the skill, strength, and stamina to win, the technology is useless. In any case, it doesn't take long before all competitors are able to take advantage of the new devices and the playing field is level once more.

Heat management

What athletes wear can seriously affect their performance, so they try to choose the right materials for the job. Quite often it is a particular combination of properties that makes one material better than another.

Sweat, reflect

Athletes who spend a lot of time in direct sunlight need clothing that will keep them cool. Several properties are important:

- white colour to reflect the sunlight; any other colour will absorb some sunlight, heating up the athlete
- woven material with lots of very small gaps, allowing sweat to evaporate and escape freely; sweating only cools athletes down if the water vapour can leave the body, taking heat energy with it
- water-attracting so that it soaks up sweat; the sweat can then evaporate at the surface of the material.

Natural fibres, such as cotton, have all of these properties.

Fast dressing

Cotton clothes have some disadvantages for athletes who need speed. Artificial fibres (such as Lycra) can perform better than natural ones in a number of different ways:

- lower density, making lighter clothing
- lower drag, reducing friction with passing air
- greater elasticity, allowing clothes to fit closer and catch less air.

Provided moisture can get through the weave, cyclists are better off with artificial fibres.

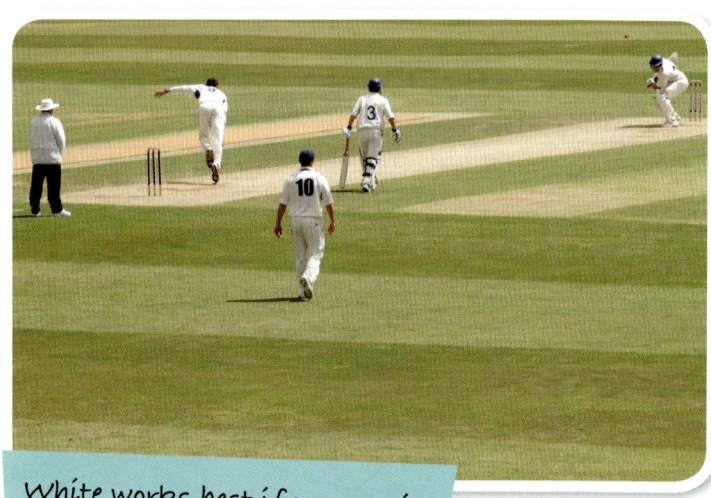

White works best if you are in the sun all day.

Low-friction Lycra makes good clothes for speed cyclists.

Lots of layers

Some sporting activities take place in cold environments. No one material provides all of the properties required, so composite materials are used:

- a moisture-transparent layer next to the skin; this allows evaporated sweat to escape easily
- next comes a layer with low thermal conductivity, to slow down the escape of heat energy from the athlete's body; a material with lots of small air gaps, such as padding, has this property
- the outside layer has to be waterproof and windproof, yet transparent to moisture as well; it therefore has to be woven from a material that repels water and is strong enough to cope with scratching.

The ability to be brightly coloured is often important for clothing worn in the mountains, so that it can be easily seen by the emergency services.

Space blankets

The New York marathon has the same problem every year. It happens in the autumn, when the weather can be cold. Exhausted athletes who are dressed to lose excess heat cool down quickly once they have passed the finishing line and stop running. It may take 20 minutes for them to reach their warm clothing – enough time for hypothermia to set in. In 1979 a new material created for spacecraft came to the rescue. As athletes crossed the finishing line, they were given a sheet of metallised polyester, folded up into a small pack weighing only 90 g. In less than a minute they wrapped themselves in these shiny blankets to keep warm. The high thermal reflectivity of the material, along with its low mass, high strength, and low cost have made it a feature of marathon races ever since.

Mountaineers need clothes that keep them warm and dry.

The high reflectivity of the foil blankets reflects the runners' body heat back to keep them warm.

Setting standards

People working in industry, research, and standards organisations need to know about materials and their properties. This section introduces you to the challenges of their work, including how the use of materials is regulated.

New products have lots of codes and symbols on the packaging. Some of them tell you that samples of the product have been tested against a standard specification. That means it should be safe and reliable. But who sets the standard?

This person's life depends on their parachute. If it breaks, they will die. Worldwide products standards, set and maintained by scientists, mean that they don't have to worry about their equipment failing.

International standards

The International Organization for Standardization (ISO) is a network of standards organisations from 148 countries. The ISO brings people together from all over the world. They include:

- experts from the industries that make the products
- experts from the laboratories that test the products
- people who represent consumers.

They discuss a product and prepare a draft standard. Many people read this and comment on it. Eventually, the ISO creates a new International Standard. So, if you use a plastic bottle that meets standard BS ISO 16929, then you know that it is biodegradable. It will rot away slowly when it is buried and won't clutter the environment forever.

All bank cards are the same size, so a British bank card will fit a French cash machine. It didn't just happen like that. The ISO had to create the standard.

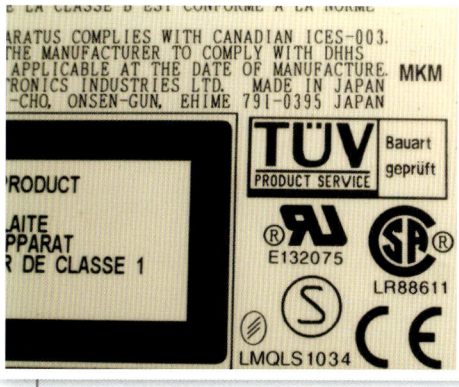

CEN works with standards organisations like BSI from every European country and creates European Standards that are all also British Standards.

European standards

In Europe, including the UK, you might find another mark on the product or packaging – CE marking. That means that the product complies with the directives of the European Union. European standards are set by the European Committee for Standardization (CEN). Their directives cover issues of reliability and safety.

British standards

Most countries have their own national organisation that sets standards for products and services. The British Standards Institution, BSI, was the first in the world. The BSI works with the ISO and many other organisations. It sets standards by consulting experts. It says what should be tested, and how.

The BSI shows that products match up to their standards using the 'Kitemark'. This is a kite-shaped symbol that includes the letters B and S, for 'British Standard'. Each Kitemark has a number, such as BS1877. You'll find this particular number on a pillow for a child's cot. It means air can flow through the pillow, so that the child can still breathe nose-down.

A manufacturer or an importer of a product can go to the British Standards Institution and pay a fee. Then the standard test will be carried out in a laboratory. If the product meets the test specifications, then the Kitemark can be used on the product.

Questions

1   What is a CE mark? What does it tell you?

2   Describe what the ISO does.

3   a   Draw a Kitemark. What does it mean?
    b   If you were a manufacturer, why would you
        want a Kitemark on the goods that you make?
    c   Describe how you would get a Kitemark.

4   Suppose that standards organisations didn't exist. What difference would it make to people buying goods?

The calibration chain

Measurement is important for many sports. Here are some simple examples:

- knowing the height of a hurdle
- timing the duration of a race
- finding the mass used in a weight-lifting contest.

All of these measurements are made by **instruments**, such as tape measures, clocks, and scales. You can usually trust the instruments to give the correct reading because they have been indirectly compared to all other instruments around the world. This **calibration** chain ensures that wherever you are a measurement will always have the same value.

Pressure

An underwater diver depends on the air in her tank. She needs to know how much air there is in it, so that she can work out for how much longer it is safe to stay under water. The pressure gauge on her tank gives her this information.

An electronic handheld timer.

The pressure gauge on a diver's air cylinder.

She has confidence in the readings of her pressure gauge because it has been stamped with ISO and CE marks, guaranteeing that it is accurate to within 5 bar up to its maximum reading of 450 bar. But how is this checked at the factory?

Standard test

The factory has a **standard** pressure gauge. Every day, a random sample of gauges from the production line are tested against this standard. The gauge being tested and the standard gauge are connected to the same air supply. Their readings are taken for a variety of air pressures from 0 bar to 450 bar. If the readings on the two gauges differ by more than 5 bar, the day's output has to be thrown away and the production machinery checked. Something must have gone wrong. This testing procedure ensures that the gauges produced are of consistent quality, an important factor for people buying the equipment.

The standards chain

But how does the factory know that its standard gauge is actually giving the correct reading? Their standard gauge has a certificate from the National Physical Laboratory (NPL) to guarantee its accuracy for five years. At the end of that time, the factory's **secondary standard** gauge has to be tested by NPL against their own **primary standard** gauge. If it passes, it is issued with a certificate for another five years. In its turn, the NPL's primary standard gauge is compared with other primary standard gauges in standards laboratories all over the world.

I'm the customer. I just want to be certain that I can believe what the dial on the pressure gauge is saying.

I do the quality control at the factory, using our secondary standard to check on a random selection of gauges from the production line.

My job at NPL is to check the calibration of secondary standards against our primary standard.

My job at NPL is to check that our primary standards agree with those in other countries around the world.

Questions

1 Describe how six different measuring instruments are used in sporting activities.

2 How does the manufacturer of a clock know that it keeps correct time?

Key words

- ✓ **instruments**
- ✓ **calibration**
- ✓ **secondary standard**
- ✓ **primary standard**

Choosing materials

Why do window designers work with glass, whilst hockey-stick designers do not? Glass allows light to pass through it, but it is brittle. If you used it to hit a fast-moving ball, it would shatter.

What would be a good material for a hockey stick? Steel is stiff and strong. There are over 6000 varieties of steel, but they are all dense. A steel hockey stick would be too heavy. Polymers and wood are less dense. Again, there's a great range to choose from.

These professionals all help to choose the right material for an artefact to do a job:
- A product designer writes a specification for the artefact; this says what it must be able to do.
- A mechanical engineer works out how the artefact twists and deflects when it is used.
- A laboratory technician does tests to check that the artefact meets international standards.
- A materials scientist suggests combinations of materials that allow the artefact to meet its specification.

As well as performance, there are other factors to consider when choosing materials:
- cost – is it affordable?
- durability – how long will it last?
- environmental impact – what damage might the material do, during manufacture, use, or as waste at the end of its lifetime?
- versatility – how easily can the material be shaped or joined to other materials?
- aesthetic appeal – does it look good?

Hockey sticks are made of wood or composites. These materials are stiff, strong, and lightweight.

PEOPLE & ORGANISATIONS

Question

1 Suggest the best material for making each of these items of sports equipment, giving reasons for your choice:

  a a football
  b a snowboard
  c a snooker ball
  d a javelin.

Some golf putters have ceramic heads, making them harder and lighter than steel ones.

In the past, golf balls were made with a smooth surface but golfers found that they flew further when the ball had dents in the surface. Manufacturers began to experiment with different surfaces – these old balls all have different patterns. Modern balls have a dimple pattern, like the ball on the right in this photograph. There are regulations about the maximum speed of the ball, but golfers are always seeking the one that will fly in exactly the direction they want.

Using composites

Tennis was invented hundreds of years ago. At the time, racquets were made of solid wood, and strung with animal gut. The best wood to use was ash as it was strong and elastic, but it had a tendency to change shape in wet weather. This problem was solved by clamping the racquets in special frames when they weren't being used.

Plywood

Manufacturers managed to reduce the problem of warping by developing a plywood racquet. This is built up with six veneers (thin layers) of ash and beech glued together. The wood grain crosses in alternate layers, with a layer of glue between them. The mixture of wood and glue gives a **composite** material with the elasticity of wood and the toughness of glue.

Fibre-reinforced composites

Racquet designers are aiming for a light, strong racquet with exactly the right amount of flexibility. Fibre-reinforced plastics provide the perfect solution. The main **fibres** used are glass, boron, aramid (Kevlar), and carbon (graphite). The fibres are embedded in a plastic **matrix**. This is a thermoplastic that can be injection moulded, or a thermoset that can be moulded and then cured. In all composite materials, the matrix:

- holds the fibres in position
- transfers forces to the fibres
- protects the fibres from surface damage
- stops the fibres buckling under compression.

Composite materials can be light, strong, and tough. They resist breaking because the combination of many fibres and a matrix makes it very difficult for cracks to grow.

Tennis racquets made from solid wood were clamped in special frames to stop them changing shape in the damp.

This table tennis bat is made from plywood. Notice the layers at the edge.

Canoes are often made from woven mats of glass fibres in a solid matrix of resin glue.

Key words
- ✓ composite
- ✓ fibre
- ✓ matrix

Mechanical properties

Key words

- ✔ force
- ✔ deform
- ✔ stiff
- ✔ flexible
- ✔ elastic
- ✔ plastic
- ✔ strong
- ✔ weak
- ✔ brittle
- ✔ tough
- ✔ tension
- ✔ compression
- ✔ extension

Forces and materials

Mechanical properties describe the behaviour of materials when external **forces** act on them. Forces can **deform** (change the shape of) materials by stretching, squashing, or bending them.

- **Stiff** or **flexible**? A material that holds its shape and resists being deformed is **stiff**. A **flexible** material is easily deformed.
- **Elastic** or **plastic**? **Elastic** materials such as rubber spring back to their original shape after the force deforming them is removed. A **plastic** material does not recover its shape like this. It is deformed permanently, like Plasticine.
- **Strong** or **weak**? **Strong** materials such as fishing line require a large force to break them. **Weak** materials such as liquorice lace break easily.
- **Brittle** or **tough**? In an impact, **brittle** materials like glass snap cleanly and sharply. Tiny cracks open up and run through the material. **Tough** materials, like steel in a hammer head, resist the formation and spreading of cracks.

Forces on a beam

Forces act on structures in different directions. Materials respond differently to these forces depending on their mechanical properties.

This beam is in **tension**. The forces are stretching the material. Steel cables are strong in tension – they hold up suspension bridges without breaking.

This beam is in **compression**. A pair of forces is squashing the material. A nail is strong in compression – it doesn't break when hammered.

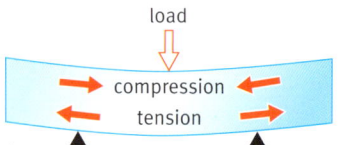

This beam is in compression on the upper surface and tension on the lower surface.

The gymnast's beam is strong in both tension and compression.

Force–extension graphs

To find out how a material behaves in tension, you can hang different weights from it and measure the **extension** (how far it stretches). Plotting a graph of force against extension can tell you a lot about the material's behaviour.

Elastic behaviour

For small forces, the graph is always a straight line. The material is elastic – it returns to its original length when you remove the weights. For an elastic material you can predict the extension produced by a particular force.

$$\text{force constant } k = \frac{\text{force, } F}{\text{extension, } x} \quad \text{and} \quad F = kx$$

Example: A length of fishing line extends by 2.0 mm when it supports a fish of weight 0.5 N. What is the force constant of the line?

$$\text{force constant } k = \frac{0.5\text{ N}}{2.0\text{ mm}} = 0.25\text{ N/mm}$$

A fish of what weight would produce an extension of 5 mm?

$$F = kx$$

$$F = 0.25\text{ N/mm} \times 5\text{ mm}$$

$$= 1.25\text{ N}$$

Plastic behaviour

Many materials become plastic if the force is big enough. They no longer return to their original size when you remove the weight. The force–extension graph becomes curved, as shown for copper.

Stored energy

The area under a force–extension graph tells you how much energy is stored in the stretched material. In the graph opposite:

- each small square is 10 N × 0.001 m = 0.01 J
- the total number of squares under the graph is about 10
- so the amount of energy stored is 10 × 0.01 J = 0.1 J.

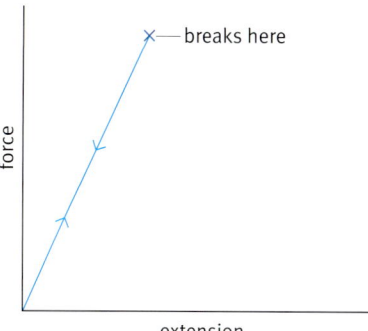

glass

breaks here

force

extension

This force–extension graph is a straight line. The material is elastic. After being stretched it returns to its original size.

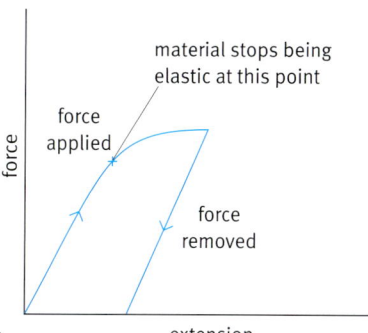

copper

material stops being elastic at this point

force applied

force removed

force

extension

This material is elastic when the force is small. It becomes plastic with larger forces and does not return to its original size.

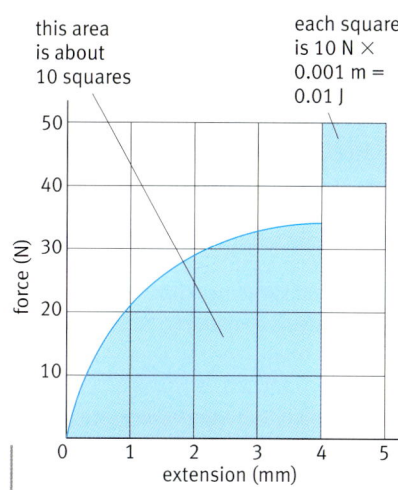

this area is about 10 squares

each square is 10 N × 0.001 m = 0.01 J

A stretched material stores energy. The graph allows you to calculate the energy stored.

Classes of material

THE SCIENCE

Climbing equipment needs to be strong and light. Aluminium makes good stiff karabiner clips. Nylon is an excellent material for making flexible rope.

Key words
- ✔ hard
- ✔ durable
- ✔ density
- ✔ composite

The many different materials used to manufacture sports equipment can be classed into three types. Each has its own range of properties, conferring both advantages and disadvantages, depending on the application.

Metals

Metals commonly used in sports artefacts are iron, steel, and aluminium. Their main useful properties are as follows:

- **strong** in tension and compression; large forces are needed to break them
- **stiff**, so they tend to keep their shape when forces act on them
- **tough**, so they deform rather than break suddenly under stress
- **hard**, so they don't scratch or dent easily.

However, most metals have a high density, making them heavy for their size. Aluminium is popular in sports equipment because it is a low-density metal.

Polymers

Polymers are materials such as cotton, nylon, polythene, polyurethane, and rubber. The following properties make them especially useful in sports equipment:

- **flexible**, so they change shape easily when forces act on them
- **durable**, so they don't rot or rust easily
- **light**, with a low density.

Their high flexibility makes polymers an obvious material to use for sports clothing. Their durability means they do not wear out and need replacing too often. Their main disadvantage is their softness so they dent and scratch easily.

These trainers are made from several different polymers.

Ceramics

Ceramics are materials such as brick, glass, and pottery. Here are their useful properties:

- **strong** in compression, so they don't give way when squeezed
- **weak** in tension, so they break easily when stretched
- **brittle,** so they snap rather than change shape when stressed
- **hard** – even harder than metals.

The brittleness of ceramics means that they are of limited use in sports equipment, much of which requires the ability to withstand sudden impact.

Composites

Sports equipment often uses two different materials acting together to do a job. In particular, **composite materials** have fibres of one material embedded in a matrix of another. For example, glass-reinforced plastic (GRP) has fibres of glass embedded in a polymer matrix. The glass is very strong, but brittle. The polymer keeps the fibres in place, protects them, and makes it very difficult for a break in any fibre spreading to others around it. Composites can have this useful set of properties:

- **strong** in both compression and tension
- **tough**; they don't break easily
- **light**; they can have a low density.

Wood is a natural composite material that is still widely used in sports equipment. It consists of strong fibres of cellulose held together in a tough matrix of lignin. The result is a material that is light, strong, and tough.

Reinforced concrete is a composite material widely used in building. Concrete is strong in compression; inserting steel rods into the concrete increases its strength in tension.

Composites are widely used to make sports facilities. These concrete pillars are part of the 2012 London Olympics stadium.

Questions

1 Give three examples of each class of material: ceramics, polymers, metals, composite.

2 Imagine a cricket bat made from nylon, steel, or glass. Discuss the problems each type of bat is likely to have in use.

Natural materials, such as wood and leather, are still the best for some sports equipment.

Making rigid structures

It's not only the material that an object is made of that determines its mechanical properties. The way the material is shaped also plays an important part. For example, paper and cardboard are folded to make them more rigid in corrugated card packaging, cardboard boxes, and egg boxes.

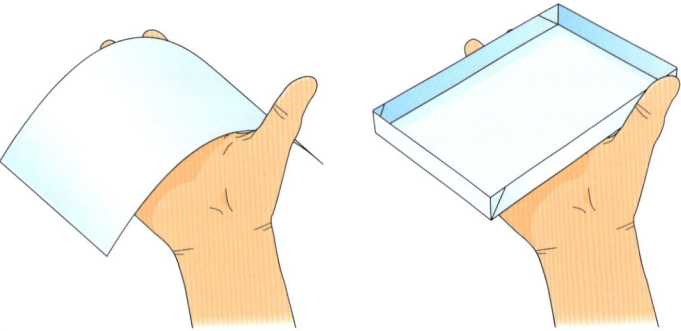

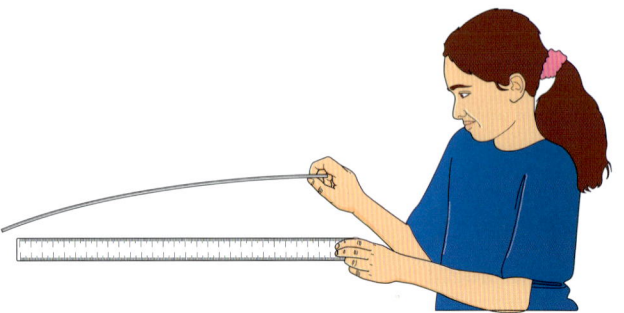

A flat piece of paper is not very stiff. If you fold the edge up as shown, it becomes much more **rigid** – it is stiffer and stronger.

If you hold a metre rule flat, it bends. Hold it edge on and it's much stiffer. It's stronger in compression than tension. The folded edge of the paper is like this.

Imitating bone

Many bones are hollow. This enables them to be lightweight while remaining stiff and strong. This is especially important for birds. They wouldn't be able to fly if their bones were solid to the core.

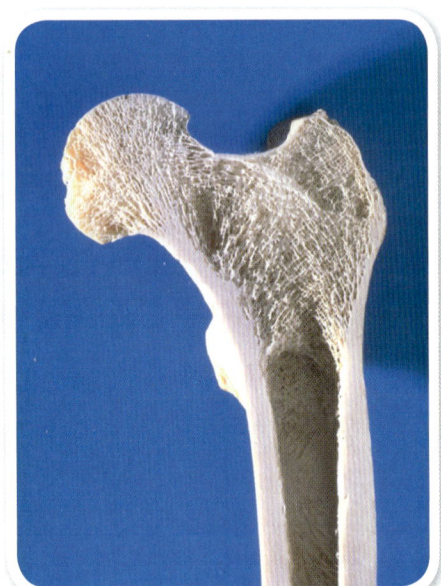

Within a bone, the hollow cavity may have struts that help to keep it rigid without adding much weight. Alternatively, the bone may be filled with a spongy honeycomb of supports. The spaces between are filled with bone marrow.

Racing-car bodies and aircraft wings make use of the same idea. They are made of an outer skin filled with a honeycomb material, giving a composite material that is both light and strong.

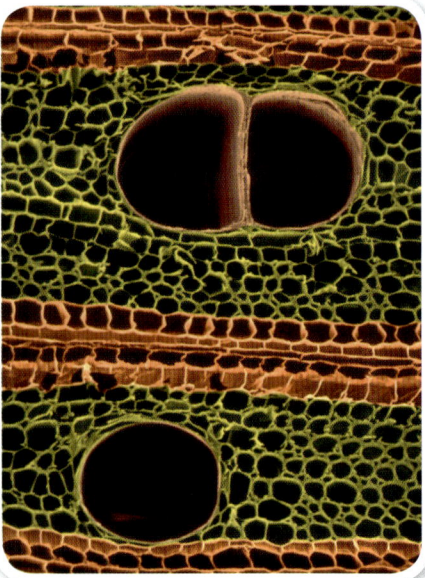

If you look at balsa wood using an electron microscope, you will see that it is made of cells that are long, thin tubes. Its density is about one-fifth that of water, which is not surprising, since it is mostly air. For its weight, balsa wood is a very stiff material. (Magnification ×70.)

Tubes and triangles

Structures made of steel tubes joined as triangles are lighter than solid tubes, but still strong. Tubes of steel are:

*   much stiffer and stronger than flat sheets of steel
*   much lighter and cheaper than solid rods of steel.

Tubes and triangles allow bicycle frames to be very light but very strong.

The tubes that hold the wheels onto the chassis are hollow to save weight without sacrificing strength.

Safety margins

Many sporting artefacts are designed with a generous **safety margin**. They are designed to work with forces beyond the range of those expected to happen. For example, consider a vaulting pole. If it snaps halfway through a lift, the athlete could get hurt. So if the expected maximum mass of the athlete is 75 kg, perhaps the pole should be designed to lift 100 kg before breaking. Here are some other sporting artefacts that should have generous safety margins built into them:

*   cycle frames
*   cricket helmets
*   snowboards and skis.

This athlete doesn't need to worry about the pole snapping in two.

Questions

1   Why is the frame of a bicycle made from hollow tubes rather than solid rods?

2   Why is the outer shell of the racing car above made from curved sheets rather than flat ones?

3   Explain the safety margins built into two different sporting artefacts.

Key words
✔ **rigid**
✔ **safety margin**

Matching the material to its purpose

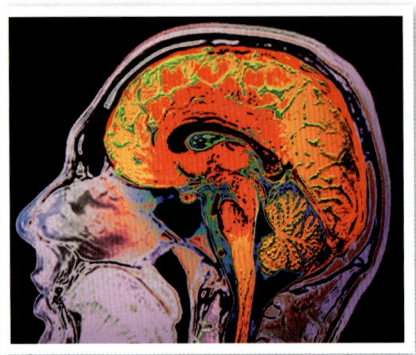

A scan of the brain inside the skull.

Mind your skull

Your brain is soft, complex, and easily damaged. Your skull, which protects the brain, is much stronger, quite hard, and fairly rigid. But if you have a cycling accident, your head may suddenly experience too large a force. Head injuries can kill or cause disabling brain damage.

European standard EN1078, and others like it, guarantee that cycle helmets help protect cyclists in an accident. One lab test simulates what would happen if you fell off a stationary bike and hit your head against a kerbstone or some other hard object. There are also strap and buckle strength tests. Some standards include a test to see if the helmet will stay on when pushed to one side.

How does a cycle helmet work?

In an accident, the helmet protects the head by reducing the forces on the head. The **stiff** outer shell reduces force by spreading the impact force over a large area. It is also **tough** so that the impact doesn't break it. The soft inner foam reduces the force on the head by gradually slowing it down. It is made from **plastic**, so that the impact crushes the bubbles permanently, absorbing energy which might otherwise damage the head.

PVC outer shell – top half
The outer shell is tough and rigid. It spreads the impact force over a larger area. Also, if your head hits the road, a smooth outer shell will make the helmet skid along the surface. Some helmets have a thicker, much stronger polycarbonate outer shell.

tape holding halves of shell together

PVC outer shell – bottom half

polyurethane foam pads held on with Velcro

expanded polystyrene inner helmet
The foam layer makes the head stop moving more slowly. The bubbles in the foam collapse permanently, and this softens the head's impact with the helmet. The foam layer helps to hold the shell together if it cracks.

nylon webbing straps with moulded nylon buckle

The mechanical properties of the shell and foam are **complementary**. They are different but between them they do the job. But it's safer to ride carefully in the first place. You have to use your brain to protect your brain.

Questions

1. Suggest tests that should be used to verify that a cycle helmet meets a standard.

2. A cycle helmet contains two complementary materials. Describe their properties and explain how the materials work together.

THE SCIENCE

Thermal properties

Thermal properties describe the behaviour of materials in response to temperature differences.

Ski clothing is lightweight and flexible, but it keeps you warm.

Clothing for winter sports is designed to keep you warm in temperatures that might be 40 °C below your body temperature. You need to retain the warmth of your body.

A ski jacket feels warm to the touch. It's really at the same temperature as the room, which is colder than you. So why does it feel warm?

Heat from your body escapes into the jacket fabric. However, it doesn't move through the fabric easily, because the fabric is a good thermal insulator. So when you touch the jacket, you are feeling the temperature of your own body.

If you touch a metal object, such as a car, on a cold day, it feels cold. Heat from your finger travels into the metal and then flows away. The metal doesn't warm up. Instead, your finger cools down, so the metal feels cold. It's easy to be fooled by the sense of touch. You think you are finding out about the temperature of something, but you are really finding out how well it conducts heat.

Thermal conductivity

The pattern for **thermal conductivity** is the same as for electrical conductivity:

- Metals are good conductors. Metal objects have high thermal conductivity.
- Polymers (plastics) and ceramics (such as glass) are good thermal insulators and poor conductors. Objects made from polymers and ceramics have low thermal conductivity.

This ice axe is designed for use in very cold places. The head is made from metal, which is stiff and strong. The handle is made from a polymer. This isn't as strong as metal, but has a much lower thermal conductivity. The metal head feels cold to the touch, but the handle doesn't.

CHECK SAFETY
Never work
unsupervised

Measuring stiffness

Principle

The **stiffness** of a material describes how much it changes shape when an external force is applied.

Equipment

- clamp, needle, clamp stand, metre rule, slotted masses, string
- samples of different materials with the same dimensions (same length, width, and thickness)

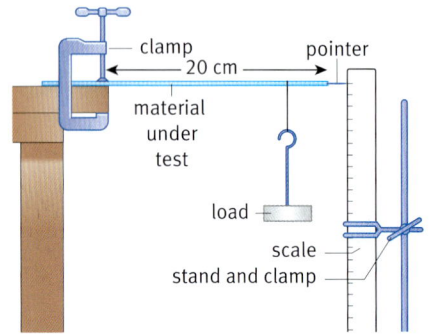

clamp — pointer
20 cm
material under test
load
scale
stand and clamp

Procedure

Method 1 – Material clamped at one end and loaded at the other.

1 Clamp one end of the material to the bench, so that 20 cm sticks out freely.

2 Tie the mass hanger securely to the end of the material using string. Keep feet clear!

3 Fix a needle to the end of the material using sticky tape.

4 Adjust the metre rule so that the needle is lined up with the zero on the scale.

5 Carefully add masses to the hanger until the needle has moved down exactly 10 mm. (Use larger and smaller masses to get the needle as close as you can to 10 mm.) Record the mass.

6 Repeat with samples of other materials.

Method 2 – Material supported at both ends and loaded at the midpoint.

1 Stand the material on two supports, as shown in the diagram, so that 20 cm is unsupported between them.

2 Tie the mass hanger securely to the middle of the material using string. Alternatively, place it on top of the material.

3 Adjust the metre rule so that the underside of the material is lined up with the zero on the scale.

4 Carefully add masses to the hanger until the material has moved down exactly 10 mm. (Use larger and smaller masses to get the needle as close as you can to 10 mm.) Record the mass.

5 Repeat with samples of other materials.

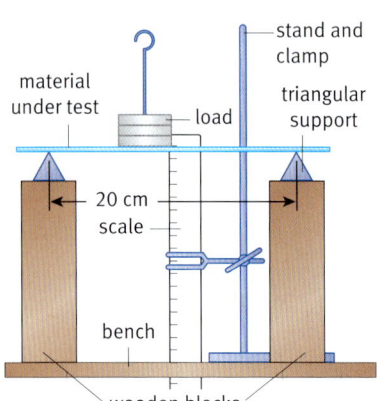

stand and clamp
material under test
triangular support
load
20 cm
scale
bench
wooden blocks

Interpreting the result

The greater the mass needed to bend the material, the greater its stiffness.

Comparing the strength of fibres

Principle

The **tensile strength** of a material describes the stretching force needed to break it.

CHECK SAFETY
Never work
unsupervised

Equipment

- clamp
- clamp stand
- metal slabs
- slotted masses and holder
- samples of different materials shaped as wires with the same length and thickness
- eye protection

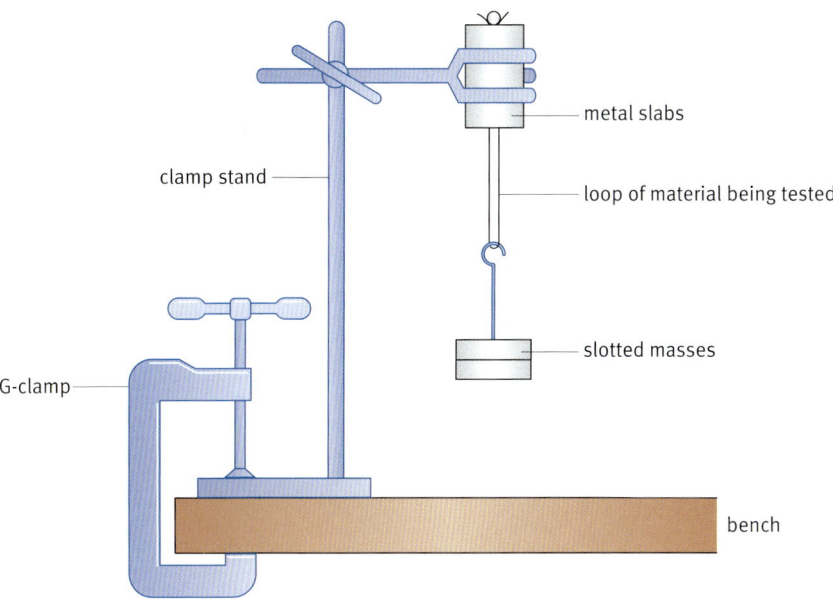

Procedure

1 Clamp the clamp stand to the bench. This stops it falling over.

2 Tie the ends of the material together to make a loop.

3 Tightly clamp the loop of material between the metal slabs.

4 Suspend the first slotted mass from the loop of material.

5 Add slotted masses until the material breaks.

Interpreting the result

The greater the mass needed to break the material, the greater its strength.

Standard procedures – 2

CHECK SAFETY
Never work
unsupervised

Finding the density of a solid

Principle

A dense material has a lot of mass packed into a small volume. If you measure the mass and volume of an object, you can then calculate the **density** of the material from which it is made.

$$\text{density} = \frac{\text{mass}}{\text{volume}}$$

Units: g/cm^3 or kg/m^3

Example: a stone of mass 70 g has a volume of $20\,cm^3$

$$\text{density of stone} = \frac{\text{mass}}{\text{volume}} = \frac{70\ g}{20\ cm^3} = 3.5\ g/cm^3$$

Equipment

- balance
- displacement can
- measuring cylinder
- object being measured

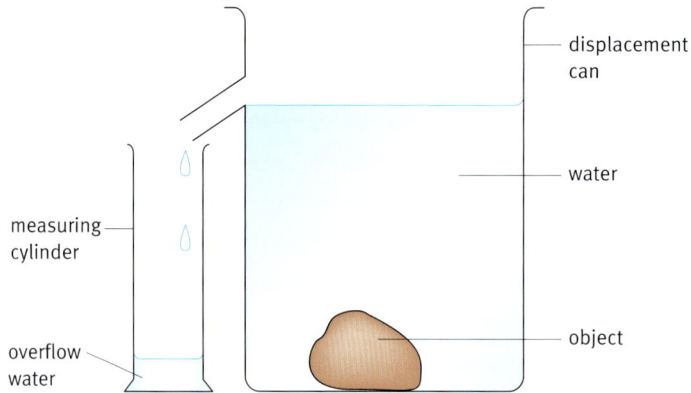

Procedure

An object may have an irregular shape. To find its volume, it is easiest to measure the volume of water it displaces.

1 Use a balance to weigh the object. Record its mass.

2 Fill the displacement can with water to the level of the spout.

3 Immerse the object and collect the water it displaces. Measure and record the volume of water displaced.

4 Calculate the density of the material.

Interpreting the result

The density of water is $1\ g/cm^3$ or $1000\ kg/m^3$. Materials with a density greater than this will sink in water. Those with a density less than this will float.

PROCEDURES & TECHNIQUES

Comparing hardness

Principle

The **hardness** of a material is a measure of how difficult it is to dent or scratch.

Equipment

- plastic tubing
- metal punch
- wooden block guide
- 500-g slotted masses and holder
- string
- samples of materials being tested
- magnifier with scale

Procedure

1 Place the sample on a bench.

2 Clamp the length of clear plastic tubing above the sample. Clamp the ruler beside the tube. Place the metal punch on the sample, directly below the tube.

3 The slotted masses have a mass of 500 g. Tie a length of string to the hook.

4 Hold the string so that the bottom of the hanger is exactly 5 cm above the sample. Let go of the string.

5 Examine the sample to see if the metal punch has dented it. If not, try dropping the masses from a greater height. Record the height needed to produce a visible dent in the material.

6 Measure the diameter of the dent produced. You may need to use a magnifier with a scale in the field of view.

7 Repeat with samples of other materials.

Interpreting the result

The smaller the diameter of the dent (and the greater the height from which the masses are dropped), the harder the material.

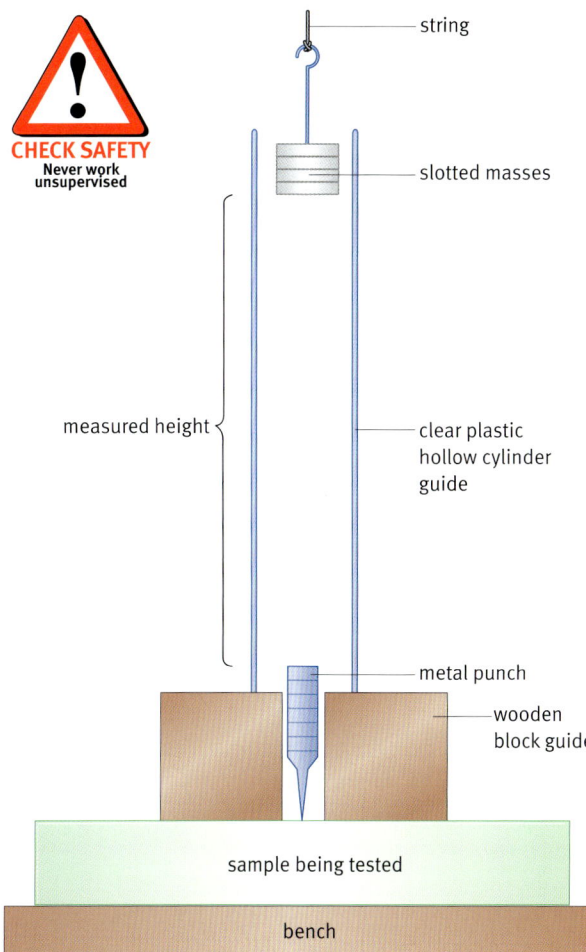

CHECK SAFETY
Never work unsupervised

string

slotted masses

measured height

clear plastic hollow cylinder guide

metal punch

wooden block guide

sample being tested

bench

CHECK SAFETY
Never work
unsupervised

Comparing poor conductors

Principle

An object that conducts heat well has a high **thermal conductivity**. Metal components have high thermal **conductivity**. Polymers, glass, and wood components have low thermal **conductivity**.

The thermal **conductivity** of a poor conductor can be compared with a standard material whose thermal **conductivity** is known.

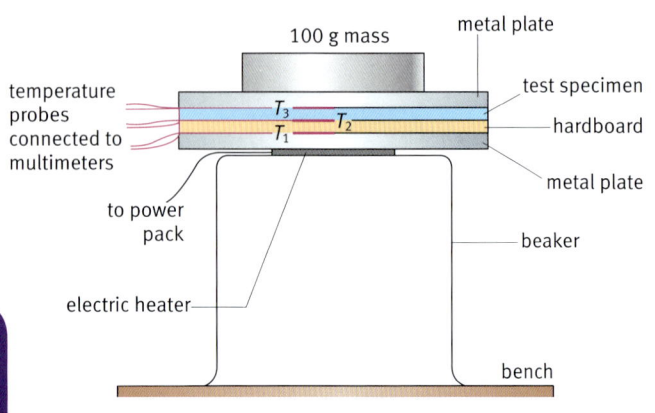

temperature probes connected to multimeters

100 g mass

metal plate

test specimen

hardboard

metal plate

beaker

to power pack

electric heater

bench

Equipment

- metal plates ($\times$ 2)
- multimeters with probes to measure temperature ($\times$ 3)
- 100-g mass
- 250 cm^3 beaker
- 12 V electric heater
- power pack
- hardboard reference plate
- specimens to test, of the same thickness as the reference plate
- eye protection

Procedure

1 Place on top of the upturned beaker, in order:
   - the heater
   - a metal plate
   - the hardboard
   - a test specimen
   - another metal plate
   - the 100-g mass

2 Insert the three temperature probes as shown in the diagram.

3 Connect the heater to the power pack.

4 Set the power pack to 12 V and switch on.

5 Wait until all three temperature readings are steady.

6 Once the readings have been steady for five minutes, record the values of T_1, T_2, and T_3.

7 Switch off the heater and let the whole apparatus cool down before dismantling it.

Interpreting the result

Hardboard has a thermal conductivity, measured in tog values, of 0.3. The thermal conductivity of the specimen is calculated using the formula

$$\text{tog value} = 0.3 \times \frac{T_2 - T_3}{T_1 - T_2}$$

Presenting data from standard procedures

Principles

Suppose you have just performed a standard procedure for a client. How do you present your findings in the most efficient way? You need to do four things:

- State the procedure used.
- Present a table of your measurements, with units clearly shown.
- Draw a graph to present all of your results as a picture.
- State what you can conclude from the procedure.

By presenting data in this way, the client can instantly tell what you did, what you found out, and make a judgement about its reliability.

Report on the tensile strength of Elasmax fibres

Procedure

An Elasmax fibre of length 50 cm and thickness 0.76 mm was clamped at one end and the other end passed over a pulley wheel. The free end was tied to a scale pan. A piece of sticky tape was fastened 200 mm from the clamped end, over the 0-mm mark of a scale. The extension of the fibre was measured as weights were added to the scale pan. Weights were added until the fibre snapped. The procedure was repeated for three fibres of the same length and thickness.

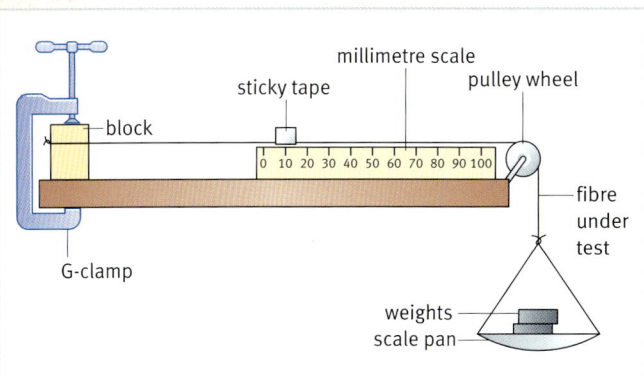

Summary

Although each fibre stretched slightly differently, they all snapped at a loading between 45 N and 50 N. One value, in bold in the table, is probably wrong due to an error in recording the sample.

Table of measurements

Weight (N)	Extension 1 (mm)	Extension 2 (mm)	Extension 3 (mm)	Average extension (mm)
0	0	0	0	0
5	8	12	10	10
10	12	16	20	16
15	30	34	30	33
20	38	44	**75**	41
25	48	51	52	50
30	61	59	60	60
35	63	63	66	64
40	64	63	66	64
45	64	64	66	64
50	snaps	snaps	snaps	

This is an outlier - it is outside the pattern.

Graph to show extension due to loading of fibres.

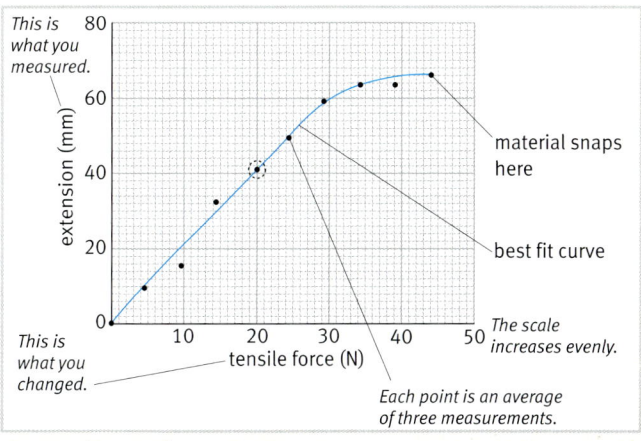

This is what you measured.

material snaps here

best fit curve

This is what you changed.

The scale increases evenly.

Each point is an average of three measurements.

Module Summary

People and organisations

- that people who manufacture sports equipment need to understand the properties of materials
- that athletes who use sports equipment need to know the limits of the materials used in them
- that technicians who test sports equipment need appropriate qualifications to apply standard procedures
- that designers of sports equipment need to know how to select combinations of materials to perform specific tasks
- two examples of the development of new materials that have led to an improvement in sporting performance
- that the British Standards Institution, the European Committee for Standardization, and the International Organization for Standardization set product standards
- why sports equipment is designed with safety margins
- that equipment that carries the Kitemark, CE, or ISO marks meet the product standards set by standards organisations
- that product standards for sports equipment look at its safety, quality, and consistency
- about the role of national standards laboratories in ensuring the traceability of measurement standards by checking the accuracy of measuring instruments.

The science

- how to use these terms in describing mechanical properties of materials: stiffness, flexibility, toughness, brittleness, compressive strength, tensile strength, hardness, density, durability
- how equipment can be made more rigid by changing its shape, materials, or structure
- how to describe materials by their thermal conductivity and thermal reflectivity
- why different materials at the same temperature can feel warm or cold
- that a composite material has one material embedded in another, and be able to give some examples
- how the material properties of metals, polymers, ceramics, and composites are related to their use in sporting equipment
- that it is often a combination of properties that make a material suitable for a particular item of sports equipment, and be able to give some examples
- how to use a force–extension graph to calculate energy stored in a stretched material
- how to use a force–extension graph to predict elastic and plastic behaviour
- how to use the equation $F = kx$
- explain how composite materials combine the useful properties of different materials
- how to interpret information to assess the suitability of a material for use in a piece of sports equipment
- how to match the use of a material for a piece of equipment against required properties, cost, durability, environmental impact, and fashion.

Standard procedures

- how to compare the stiffness of two materials
- how to measure the density of a material.

Review Questions

1 Describe the job of someone who needs to understand the properties of materials used in sports equipment.

2 Mandy finds the CE mark on her new vaulting pole.
  a Name the organisation that awards the CE mark.
  b Explain what the CE mark means and why it is important.

3 Materials can be described in terms of their stiffness.
  a Explain what is meant by stiffness. Use two examples of materials used in sports equipment in your explanation.
  b Describe how you would compare the stiffness of two different materials in your school laboratory.

4 Tom is testing springs for a new toy. He plots this graph.

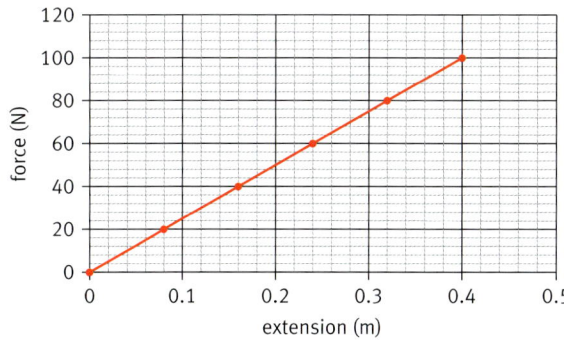

  a How big is the force to stretch the spring by 0.3 m?
  b Use the graph to calculate the energy stored in the spring when it is stretched by 0.4 m.
  c Estimate the amount of stretch when a force of 120 N is applied.

  d Explain why you can only make an estimate in your answer to part c.
  e Use the graph to calculate the spring constant k of the spring in N/m.

5 Nicky uses an ice axe to help her climb. Explain why the wooden handle feels warm when she touches it, but the metal blade feels cold.

6 In some applications of materials the thermal reflectivity is important. Use two different examples of materials used for sport to explain what is meant by thermal reflectivity.

7 Making vaulting poles from fibreglass instead of wood allows athletes to jump higher.
  a Fibreglass is a composite material. Explain what this means.
  b Suggest why athletes could jump higher with vaulting poles made from fibreglass.
  c Give another example of how a new material has changed sporting achievement.

8 Mark designs a new tennis racket. He consults this table of material properties.

Material	Density (kg/m^3)	Strength (MN/m^2)
wood	800	50
aluminium	2700	110
carbon fibre	1800	5600

  a Explain why density is important when choosing material for a tennis racket.
  b Explain why strength is important when choosing material for a tennis racket.
  c What would be the best material for Mark to use? Justify your answer.

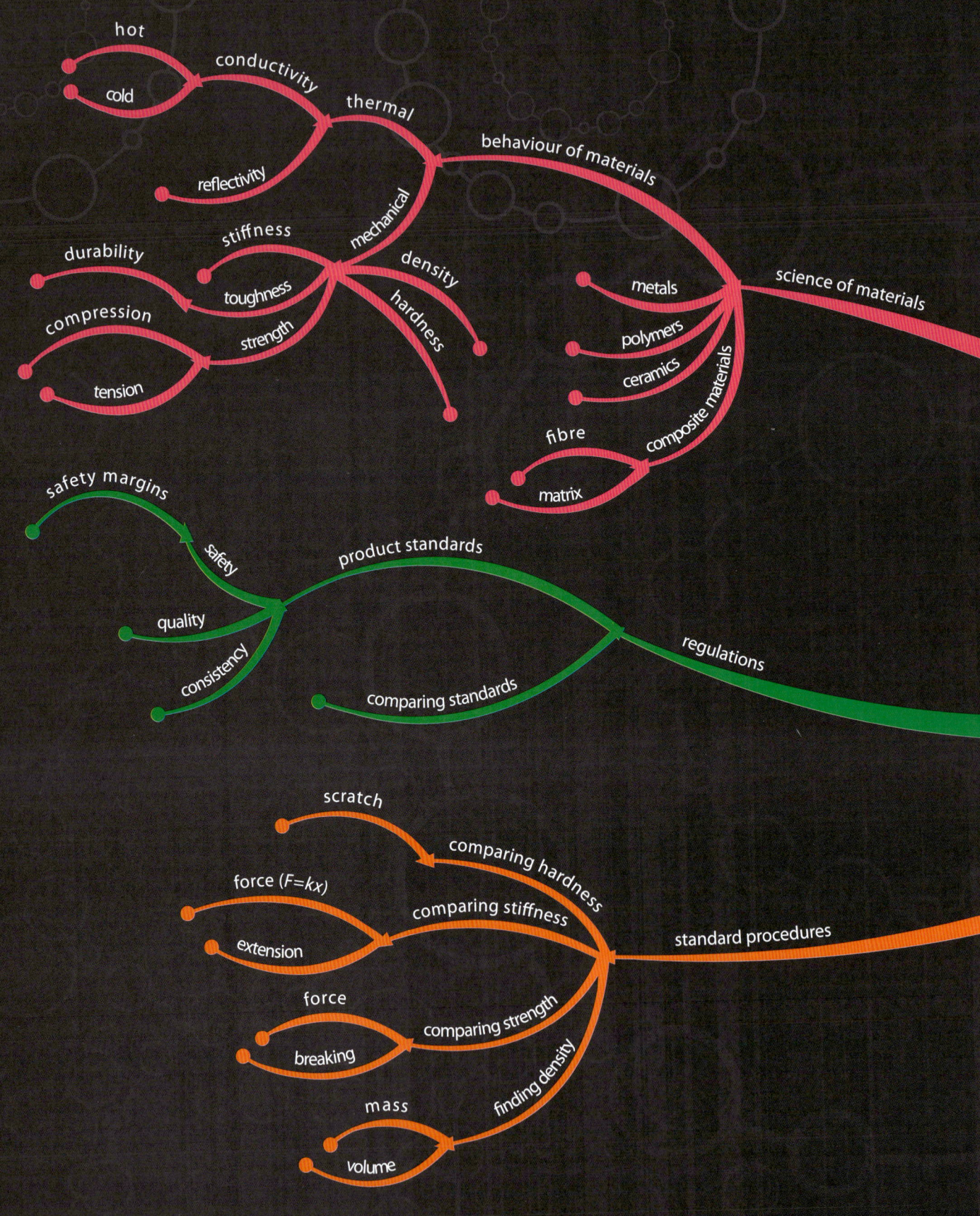

hot

cold

conductivity

thermal

reflectivity

behaviour of materials

stiffness

mechanical

durability

toughness

density

compression

hardness

strength

tension

science of materials

metals

polymers

ceramics

composite materials

fibre

matrix

safety margins

safety

product standards

quality

consistency

regulations

comparing standards

scratch

comparing hardness

force (F=kx)

comparing stiffness

extension

standard procedures

force

comparing strength

breaking

finding density

mass

volume

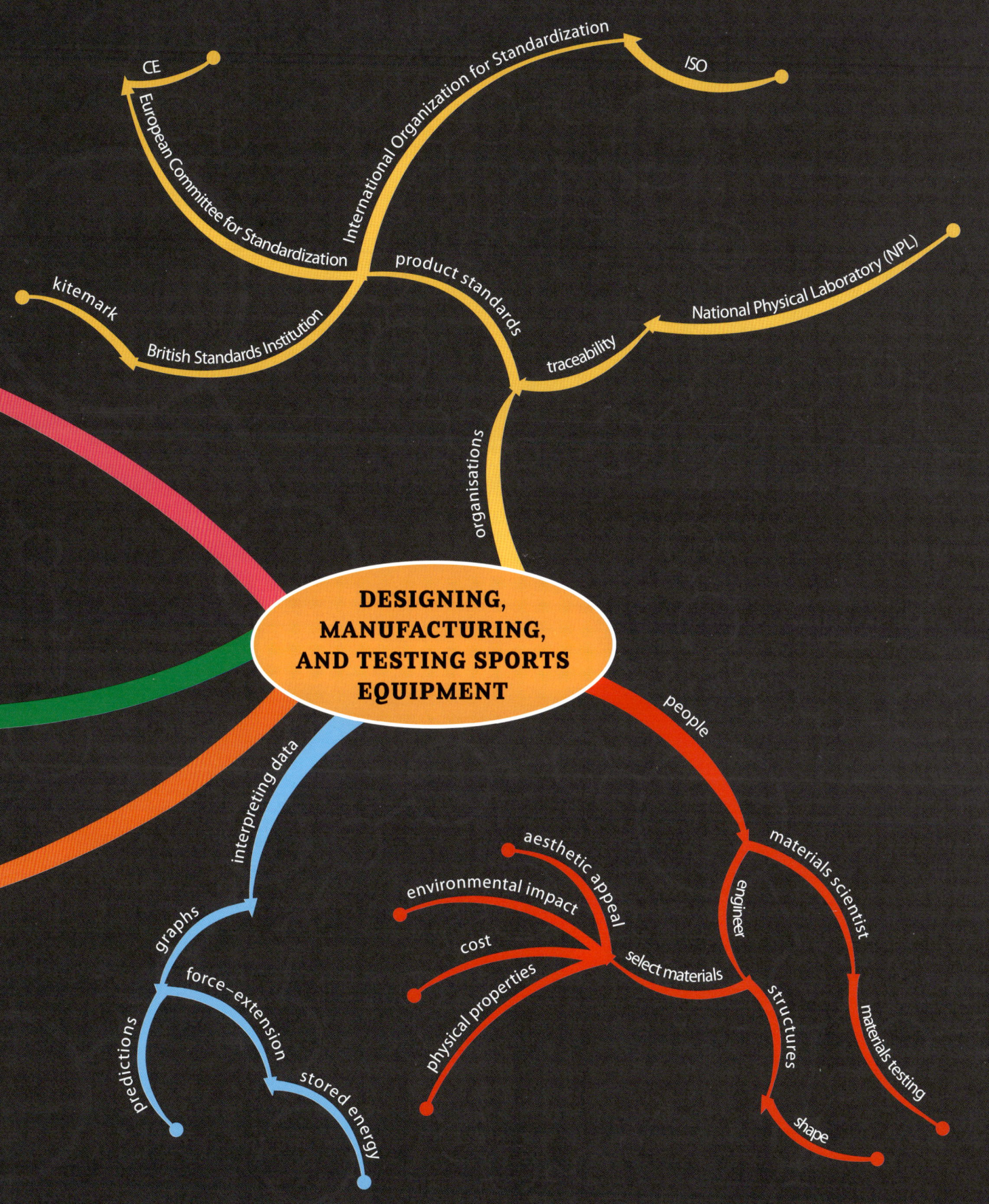

CE

European Committee for Standardization

International Organization for Standardization

ISO

product standards

National Physical Laboratory (NPL)

kitemark

British Standards Institution

traceability

organisations

DESIGNING, MANUFACTURING, AND TESTING SPORTS EQUIPMENT

people

materials scientist

engineer

aesthetic appeal

environmental impact

cost

physical properties

select materials

structures

materials testing

interpreting data

graphs

force–extension

predictions

stored energy

shape

B2 Stage and screen

Why study stage and screen?

It isn't just the actors who make it seem so real. There's a whole army of people behind the camera or in the wings. Lighting directors, makeup artists, sound technicians...they all play their part in telling the story. Each applies their expertise to fool your senses into seeing or hearing something that isn't there at all. Between them, they feed your senses. Your imagination does the rest.

What you already know

- about different types of electromagnetic radiation
- how to use mirrors to make images
- how sound is made and how it travels
- about using electricity to make light.

Find out about

- how cameras use lenses to fix images
- how to amplify sound to fill a concert hall
- the use of materials to control the flow of sound in a building
- the use of filters to create moods through lighting

The Science

Good lighting is more than just plugging in a lamp and switching it on. Science is required to dispose of harmful radiations that could damage the actors. Technicians need to understand how combining filters can alter colour and mood. Lots of things can go wrong with a sound system. Too much echo and the sounds blur; too little and it sounds far away. Unwanted feedback can ruin the most sensitive performance, and sound getting into the wrong part of a building can make others' lives a misery. The science of sound gives technicians the understanding to get it right on the night.

The lighting director

Mike Le Fevre is a lighting director for BBC Resources in West London.

I work in the studios at Television Centre and on location – anywhere from the Cannes Film Festival to the Caribbean!

Sounds like a great job! What does a lighting director do?

Well, there are two sides to the job, really. First, I use my technical knowledge to get enough light from the actors and set to the TV or movie camera. I have to understand a lot about electrics and light to do that. The second part is artistic. I have to create the mood and use light to support what the performers and producers are trying to say.

What kind of lights do you use?

I could talk all night about different kinds of light. But there are two basic kinds of light: soft and hard.

Hard light?

Hard light is a single source of light. It's a bit like the rays from the Sun. I'll do you a sketch. See, with hard light the rays are all coming at the subject in parallel lines, and the shadows are clear and hard.

Sun

hard light (deep shadows)

soft light (no shadows)

Mike's sketch

. . . and soft light?

Soft light is like the light coming from a cloudy sky. The rays are scattered and come at the subject from all directions. This makes a range of pale, soft shadows.

These lights must get hot . . .

Yes, this unit has a tungsten filament that can reach around 3000 °C. The unit is made of aluminium with vents on top so that hot air can escape.

It feels hot at the front!

Some units have special reflectors made of materials that absorb the infrared and only reflect the visible range. These ones are much cooler. The glass on the front of the units is vital because it absorbs ultraviolet. It stops actors getting sunburnt!

Do you use coloured lights?

Yes, all the time. For example, we might put a pale blue filter on these tungsten lights to make it look like daylight. Otherwise the camera picks it up as orange.

Do you use the filters for effects?

Yes, we use light to show emotion. There's a pale greeny–blue we call 'dismal blue'. And I might use a warm red or pink for a happy event.

So there's a whole range of colours?

The filters come in all sorts of colours, including the basic primary colours of red, green, and blue. I can mix any colour you want from these. But the position of the lights is just as important. It's not so much the light as where shade is that makes a scene.

Do you get to work with famous people?

Oh yes, all the time. And it's my job to make them feel comfortable with the look the lights are giving them. The most successful stars are often the nicest to work with!

Safe performance

Thanks to comprehensive Government regulations, you are more likely to burn to death in your own home than in a theatre or concert hall. Sadly, those regulations only came into being because of a few spectacular disasters. The Exeter Theatre fire of 1887 was the first and worst. It still holds the UK record for the most number of people killed in a single building fire.

Naked flames

'Twas in the year of 1887, which many people will long remember,
The burning of the Theatre at Exeter on the 5th of September,
Alas! that ever-to-be-remembered and unlucky night,
When one hundred and fifty lost their lives, a most agonising sight.

> from *Burning of the Exeter Theatre* by McGonagall

Like all theatres at the time, the Exeter Theatre was lit by gas. Halfway through a performance, a naked gas flame in the wings set fire to some curtains. Within minutes, the entire stage was on fire. Most of the people sitting downstairs escaped the building safely, but people upstairs in the balcony were not so lucky. There was only one exit, partially blocked by the ticket office. In the panic, 186 people were either crushed or burnt to death.

Safety curtain

By the time the Exeter Theatre was rebuilt, new government regulations required it to have a **safety curtain**. It was made of asbestos, a fire-resistant material, so that it could be used to separate the audience from any fire backstage. Modern fire curtains use materials other than asbestos, which is now known to be a cause of cancer. They are designed to slow down a fire for long enough to allow an audience to leave safely.

The death toll of this fire shocked Parliament into drafting the first health-and-safety legislation for theatres.

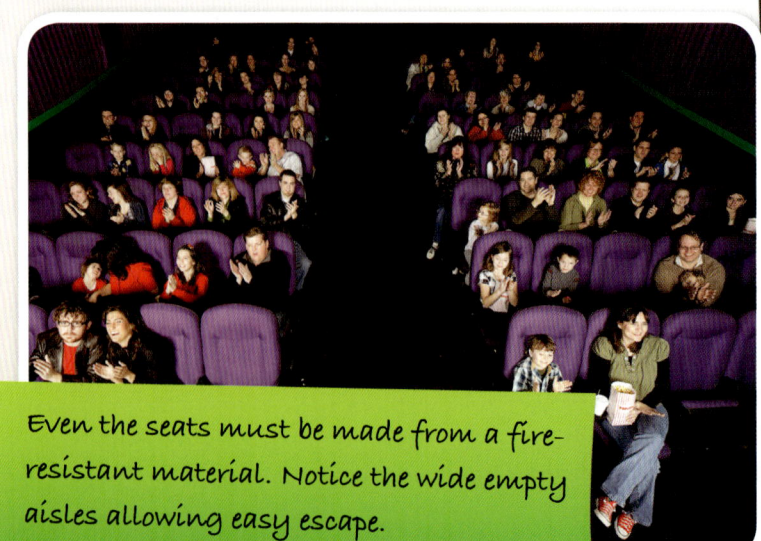

Even the seats must be made from a fire-resistant material. Notice the wide empty aisles allowing easy escape.

No time to think

In an emergency, most people act automatically. When trapped, their first instinct is to leave by the way they came in. Since most people enter a hall by the same entrance, this easily leads to a dangerous crush. Fire regulations require that:

- many exits are provided, spread around the building
- routes to those exits are clearly indicated by signs that stay lit in a power failure
- exit routes are free of obstructions that could cause bottlenecks
- emergency exit doors open outwards when pushed, with no key required to open them.

Planning for disaster

Advance planning for disaster saves lives. So does training. In an emergency, staff must know what they are supposed to do and calmly get on with it. Of course, members of the paying public won't put up with regular fire drills, so the planners have to think ahead for them. They hope that their ideas will never have to be put into action, but also know that they will save lives in the unlucky event of a fire.

An emergency exit door should always be obvious and open outwards when pushed.

This self-illuminated sign shows two ways out.

Advance planning for emergencies saves lives.

Managing performance

Although the band makes the music, someone else has to control the delivery of sound and light to the audience.

PEOPLE & ORGANISATIONS

The loudspeakers are arranged to send sound towards the audience and away from the band.

Steve and Jane are getting ready for a rock concert. Their job is to get the sound and lighting systems set up before the band arrives at a venue. Between them, they manage what the audience will see and hear that evening. Steve takes charge of the sound, and Jane supervises the light effects.

Loudspeakers

Steve spends most of the afternoon getting the loudspeakers sorted. He has two columns of them, one on each side of the stage and slightly in front. The shape of each column directs the sound evenly over the whole audience, so that wherever they sit they hear the same level of music. As part of the sound check with the band during the afternoon, Steve uses a decibel meter to check that the music won't be too loud to be safe.

Microphones

Tonight's band have decided to use wireless microphones. Steve likes this – it's much safer without all the wires trailing all over the stage to trip people up. However, he has to make sure that all of the transmitters are operating on different channels. He does this by tagging each transmitter with a different colour, and making a note of which singer or instrument has which colour.

The microphone clips onto the performer's shirt and the transmitter tucks into their waistband.

Mixing

Steve has to spend the whole concert in front of the mixing desk. Although the sound check in the afternoon sorts out most of the balance problems, performers often produce different levels of sound on the night. So Steve watches the levels coming in from each microphone, and adjusts them to keep the balance right. He also listens for early signs of instability in the sound system and lowers the sound level if necessary. The audience don't like howl; they've come to hear music.

Computer control

Jane spends her afternoon sorting out the lights and special effects. She always worries about the laser – it mustn't shine into anyone's eyes, so she has to watch for stray reflections off shiny surfaces. All of the lights are controlled by computer, and most of the sequences are programmed into it beforehand. Many of the changes of lighting required are either too rapid or too complicated for one person to do on their own. During the concert Jane cues the computer when the next lighting sequence is needed – and enjoys the music for the rest of the time!

Steve uses this mixing desk to control the flow of sound from instruments and microphones to the loudspeakers.

Jane tells the computer when to start each lighting sequence during the concert.

Jane needs a good knowledge of electrical safety for her job.

Making it sound right

The design of music rooms

The acoustic environment is very important in music rooms. Sound quality inside the music room should be optimised. However, sound from a music room should not disturb neighbouring classrooms, and the music department should not suffer outside noise from a road or sports ground, for example. The two main objectives are good sound quality and good acoustic insulation.

School classrooms are individually designed for their purpose. For example, the government and local councils publish guidelines for the design of music rooms in schools. These have to be followed by architects and builders of schools.

Acoustically soft ceiling tiles help reduce reverberation time.

Acoustic insulation

To keep music in music rooms, guidelines say schools should ideally:
- put music rooms in their own block, separate from the rest of school
- build music rooms with materials that have the right acoustic properties
- think carefully about the shape and size of the rooms
- make sure that sound can't leak from one room to another.

Acoustically hard materials should be used in the floors, walls, ceilings, and roofs of a music block. These materials reflect sound. Cavity walls and double glazing should be used for the outside of the block, to prevent noise getting in or the music getting out.

The quality of sound depends on a balance between how clear it is and fullness of the tone. This balance depends on reverberation time – the time taken in seconds for a loud sound to decay and no longer be heard.

Panels and a moulded rail on the wall help prevent flutter echo.

Sound quality

Acoustically hard materials on the walls, ceiling, and floor keep the sound in the room. But this allows sound to bounce around the room for a long time before it dies away. The music sounds muddy, because new notes start before the previous ones have died away.

To solve this problem architects put acoustically soft materials inside music rooms. These materials absorb sounds. For example, each room may have a false ceiling made from foam plastic. The foam is fragile, but on the ceiling it will not be damaged.

Rooms the right shape

Square classrooms should be avoided. With parallel walls, sound bounces repeatedly between the walls, giving a ringing sound called flutter echo. Walls at right angles can also produce annoying reflections. If the room has to be square, then a curtain on one wall can solve the problem.

Noisy neighbours

Sounds may leak from one music room to the one next door. Here are some possible solutions:

- no keyholes in the doors
- acoustic breaks built into any pipes and joists that go between rooms
- any gaps in walls for services such as water and electricity filled with foam.

Designing a good suite of music rooms is a skilled job. It requires a lot of experience and a good understanding of the acoustic properties of building materials.

A steeply pitched ceiling prevents flutter echoes developing between the ceiling and the floor.

Questions

1 Explain, in your own words, what the two main design issues are for a school music room.

2 Why do the government and local councils publish design guidelines for school music rooms?

3 A school hall will be used for speeches and drama as well as music performances. Suggest how the acoustic design of a school hall may differ from the design of a music room.

Light sources

Filming outdoors in the sunlight.

The light of an incandescent lamp comes from a white-hot strip of metal.

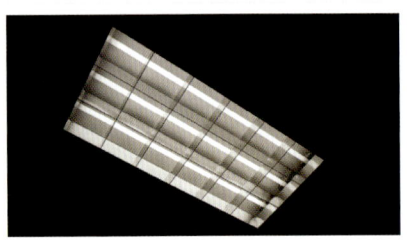
The reflectors behind these three fluorescent lamps project their light downwards.

THE SCIENCE

Not everything that emits light is useful for illuminating stage or screen. Some **light sources** are too feeble for normal use; these include candle flames and light-emitting diodes (LEDs). You can tell that they are light sources because you can see them in the dark, when there are no other light sources around.

Sunlight

The Sun is our largest light source. However, unless you live in the right place, it isn't very reliable. It also changes colour and strength during the day, being bright and white at midday but dim and orange at dawn and dusk. It produces strong shadows and bright colours.

Incandescent lamps

The most common source of light for stage and screen indoors relies on filaments of metal heated to 3000 °C by electricity. Like sunlight, their glow contains all the wavelengths that we can see, as well as some that we can't. An incandescent lamp gives out more red and less blue light than sunlight. The light comes out of the lamp in all directions, so lenses and mirrors are usually used to direct it to where it is needed.

Fluorescent lamps

Incandescent lamps are very power hungry – at least 95% of the electrical energy put into them comes out as heat instead of light. A fluorescent lamp is much more efficient, so doesn't get as hot as an incandescent lamp. However, it does have some disadvantages for stage and film:

- It doesn't emit all wavelengths of visible light, so can make things look the wrong colour.
- It flickers on and off 100 times a second. This is a problem for film cameras, which take 24 shots per second.

Lasers

Although **lasers** emit far less light than lamps, it comes out as a bright narrow beam. This can be moved around rapidly by mirrors to create special effects, such as moving shapes and writing. Only a few colours are available – white-light lasers do not exist.

Key words

- light source
- incandescent lamp
- fluorescent lamp
- laser

Questions

1 Describe three different light sources used for stage or screen.

2 List the important properties of a light source for stage work.

3 Compare fluorescent and incandescent lamps.

Filters

A **filter** is a transparent sheet of glass or plastic that alters the colour balance of the light passing through it. Filters not only change the colour of light but also remove harmful components.

Primary colours

High-power light sources emit **white light**, but they also emit other radiation that we cannot see. At one end of the light spectrum is **ultraviolet** or **UV**. You can't see it, but it can give you skin cancer. At the other end is **infrared** or **IR**. It is also invisible, but you can sense it through its heating effect on your skin. Our eyes detect **visible light**. The colour sensors in our eyes respond to red, green, and blue light and we call these **primary colours**.

ultraviolet   spectrum of visible white light   infrared

The spectrum of light from a lamp contains both visible and invisible radiation.

Filter action

All filters work in the same way:
- They let through, or **transmit,** some regions of the white-light spectrum.
- They block, or **absorb,** light in all the other regions of the spectrum.

Filters that let through two of the primary colours produce the **secondary colours.** The secondary colours are yellow, cyan, and magenta.

Filter	Transmits	Absorbs
yellow	red and green	blue
cyan	blue and green	red
magenta	blue and red	green

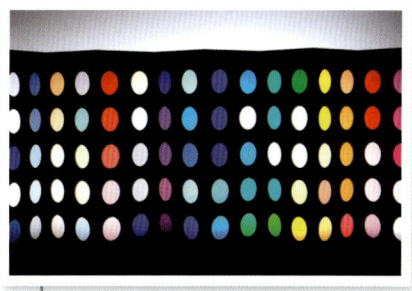

Each disc filter lets through only some of the colours from the white-light source.

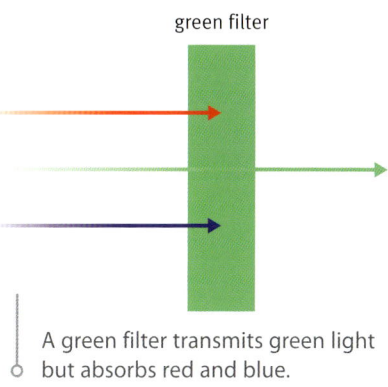

green filter

A green filter transmits green light but absorbs red and blue.

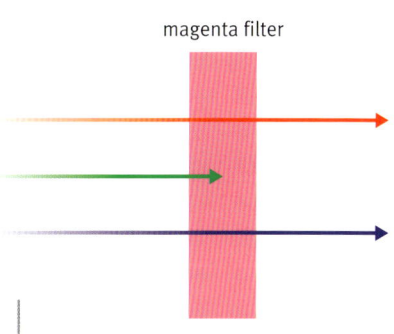

magenta filter

A magenta filter transmits red and blue light but absorbs green.

Questions

1 Name three different regions of the spectrum of a white-light source.

2 Explain why the invisible radiation from a lamp should be filtered.

3 Explain how a filter changes the colour of white light from a lamp to yellow.

Room lighting

A carefully placed mirror makes a room or a product seem larger.

An office or home needs lighting that suits a variety of activities. Lighting types include:

*   natural light – from windows or light pipes
*   background lighting – for example, uplighters, or ceiling or wall lights
*   accent and task lighting – for example, a table lamp or spotlights.

A mirror will reflect light and can make a room seem larger.

Windows

A window is a sheet of **transparent** glass that will let light into a room. But did you know that ordinary window glass is not 100% transparent? It only transmits about 80% of the light that falls on it. Of the remaining 20%, most is reflected, and some is absorbed by the glass.

Frosted glass has an irregular surface. The light rays are jumbled as they pass through, and you don't see a clear image. It's described as **translucent**. Walls are **opaque** (no light passes through).

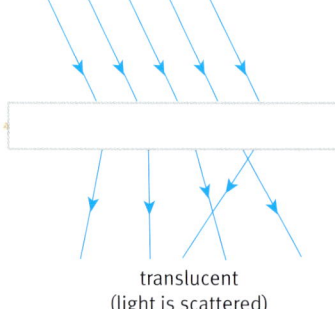

transparent
(light is not scattered)

translucent
(light is scattered)

opaque
(light is absorbed)

This Dutch water tower was converted into smart offices with clear views, thanks to self-cleaning glass.

Designing windows

Sunlight contains both visible light and infrared (heat) radiation. You want these to get into your home, but you don't want the heat to get out again.

Metals are shiny materials that produce mirror reflections. The metal silver forms a **coating** on the back of most glass mirrors. Similarly, double-glazing glass has a thin coating on one surface. This lets heat and light through from the Sun, but reflects the heat back inside.

Self-cleaning glass is the latest technology. On the outer surface, it has a very thin coating of titanium dioxide. When dirt lands on the window, this coating helps to break it down, using the energy of ultraviolet light from the Sun. Any rain then washes the dirt off. Self-cleaning glass is expensive but ideal for buildings with inaccessible windows.

Light pipes

Here is a new way of lighting a room: use a light pipe.

- On the roof, a glass dome collects the daylight that falls on it.
- A pipe with shiny metal walls reflects the light, guiding it towards a room below.
- A translucent diffuser in the ceiling spreads the light around the room.

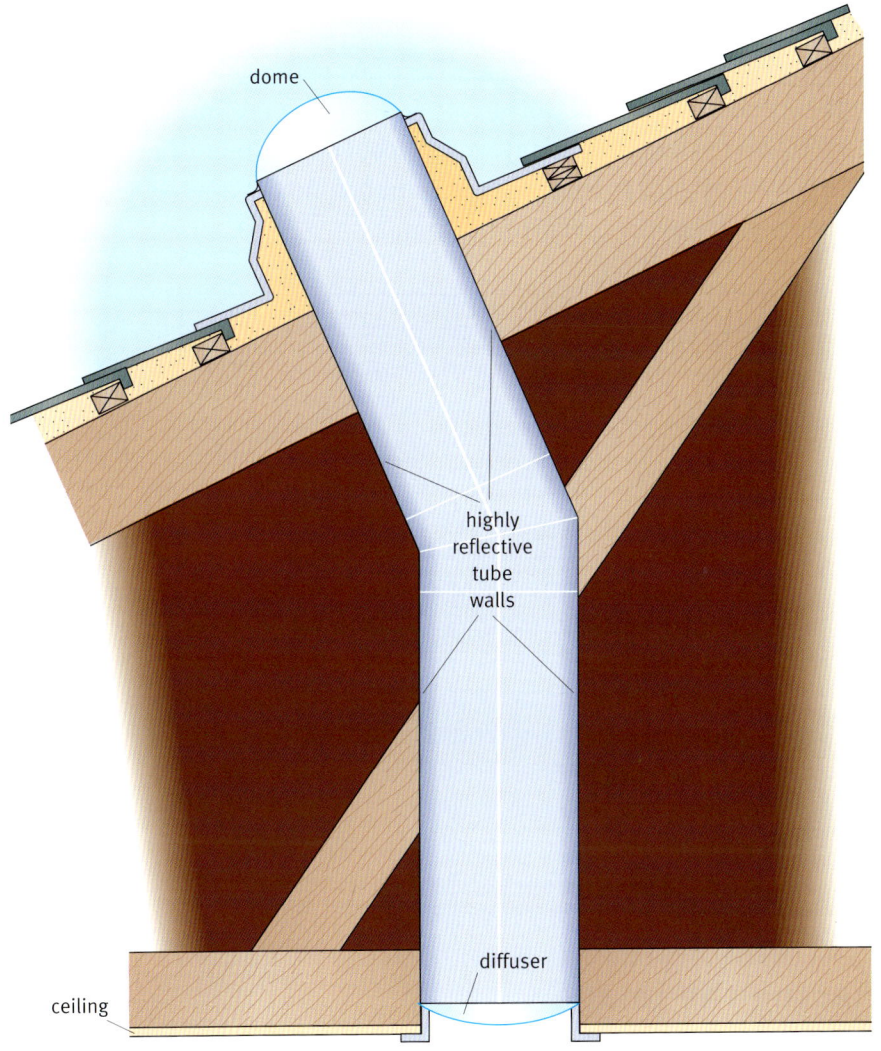

dome

highly reflective tube walls

diffuser

ceiling

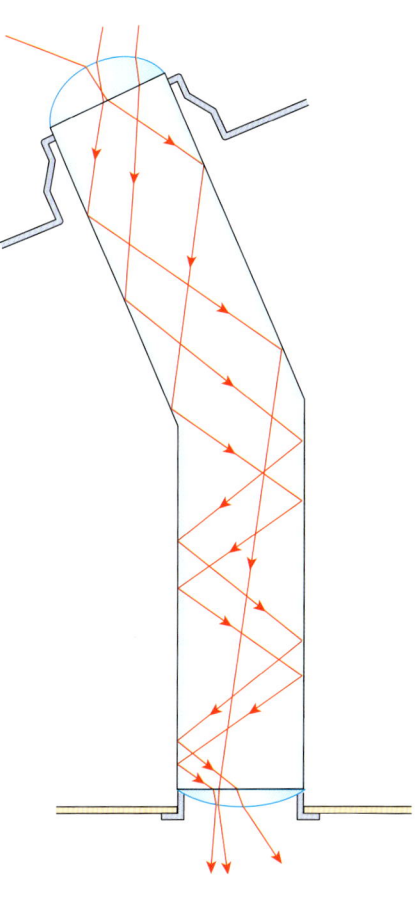

Questions

1 Describe some advantages of putting lots of mirrors in shops.

2 Give examples to explain the meaning of these terms: opaque, transparent, translucent, reflective.

3 Describe the properties needed for the ideal window glass. Justify each property.

Key words

- ✓ **transparent**
- ✓ **translucent**
- ✓ **opaque**
- ✓ **coating**

This actress uses her image in the mirror to adjust her stage makeup.

Key words
- ✓ **image**
- ✓ **reflect**

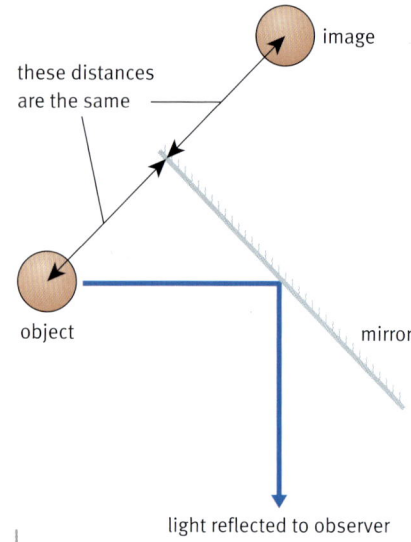

these distances are the same

image

object   mirror

light reflected to observer

The image is as far behind the mirror as the object is in front.

Each time you use a mirror, you see something that isn't really there. The **image** in the mirror is an illusion – your brain is fooled. What the mirror does is change the direction of the light from the **object** and make it appear to come from somewhere else. Lenses do the same thing to create images for cameras.

Reflection

Light rays obey a very simple rule when they **reflect** off a mirror. The angle between the ray and the mirror doesn't change. Rays coming from one point on the object (a lamp) are reflected off the mirror into the girl's eye. The rays appear to come from a point behind the mirror. This is where the girl sees the image of the lamp.

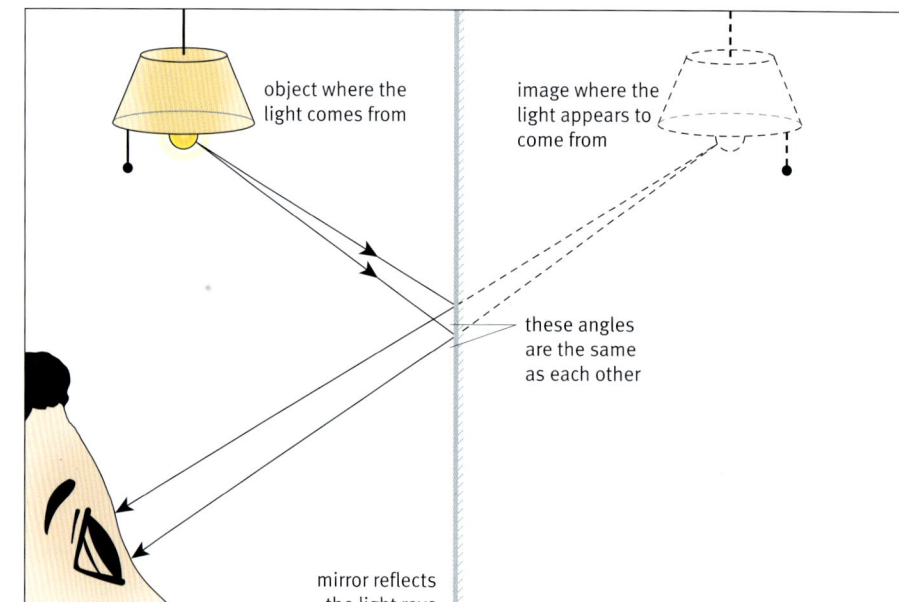

object where the light comes from

image where the light appears to come from

these angles are the same as each other

mirror reflects the light rays

Placing the image

Images in mirrors are easy to place.
- Start from the object and go straight towards the mirror.
- Then go the same distance again behind the mirror.
- That's where a viewer will think the light rays are coming from.
- So that's where the image is.

Ghosts

Mirrors are sometimes used for special effects in the theatre. They can make ghosts appear on stage, and even move through solid objects. They rely on the fact that a mirror can transmit light as well as reflect it. Of course, it can only do this if the silvering is very thin.

ghost actor   lightly silvered mirror

They do it with mirrors – a ghostly figure appears to be sitting on the bench.

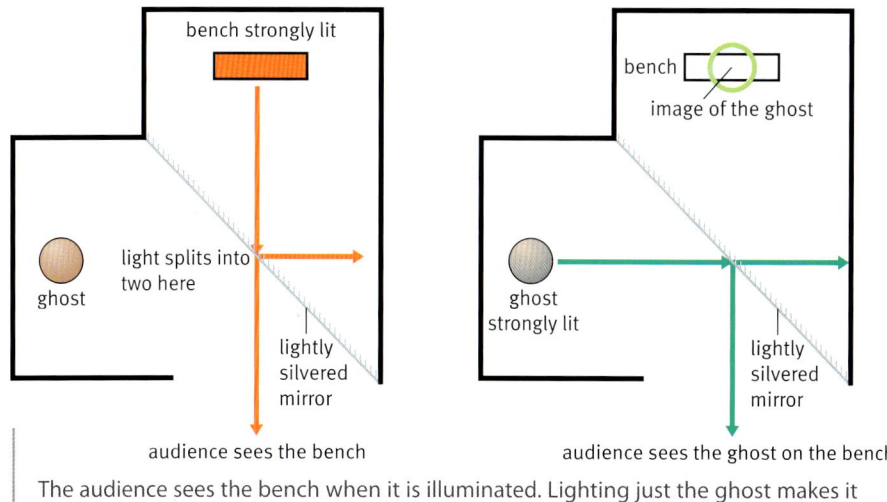

bench strongly lit

light splits into two here

ghost

lightly silvered mirror

audience sees the bench

bench

image of the ghost

ghost strongly lit

lightly silvered mirror

audience sees the ghost on the bench

The audience sees the bench when it is illuminated. Lighting just the ghost makes it appear to be where the bench was. Illuminate both at once and the ghost and bench appear to occupy the same space.

Questions

1 Draw a diagram to show how light from an object reflects off a mirror.

2 What is the difference between an object and an image?

3 Explain how a half-silvered mirror can create illusions on stage.

All digital cameras rely on a **converging** lens to place an image on a **light-sensitive surface**. Each point on that surface senses the colour and intensity of light at that point and converts that information into an electronic signal that can be stored in the camera's memory.

The lens uses light from the object to create an image on the light-sensitive surface.

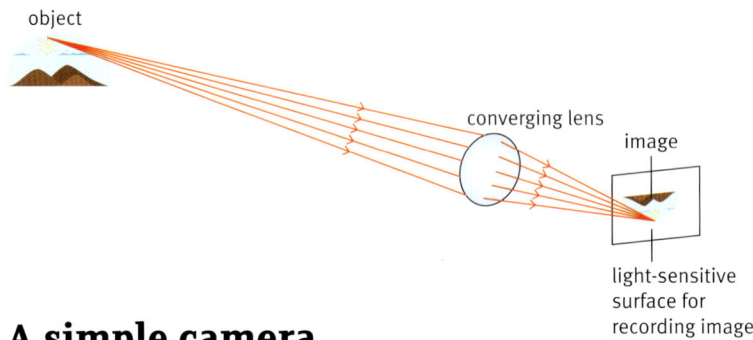

A simple camera

Each part of the camera has a part to play in recording the image.

- The **shutter** normally blocks light from passing through the lens, except for the instant that the photo is taken.
- The size of the gap in the **aperture** can be adjusted to allow just the right amount of light through the lens.
- The light-sensitive surface is usually placed in the **focal plane**, so that the image of a distant object is clearly in focus.
- The **viewfinder** allows the photographer to see what the image will be like before it is recorded.

Focusing a camera

Some cameras can be adjusted to make a clear image on the light-sensitive surface, whether the object is near or far away.

- For distant objects you place the lens so that the light-sensitive surface is in the focal plane.
- For an object closer to the camera, the image will be further away from the lens (beyond its focal length). To refocus, you move the lens towards the object, away from the back of the camera. This increases the size of the image.

Other cameras, like those found in mobile phones, have a fixed focal length. If the object is too near the camera, the image is blurred.

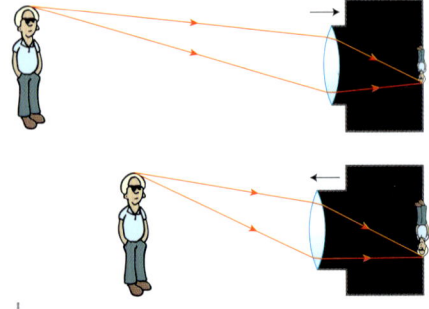

Focusing a camera by moving the lens.

Converge, diverge

A camera contains two types of **lens**.

Lenses are usually made of glass. Lenses are shaped to change the direction of the rays of light that pass through.

A **diverging lens** is thinnest in the middle. The diverging lens in the viewfinder creates a small image that is upright and in front of the lens – just what the photographer needs to view the image.

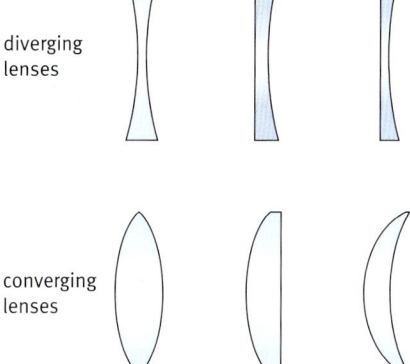

diverging lenses

converging lenses

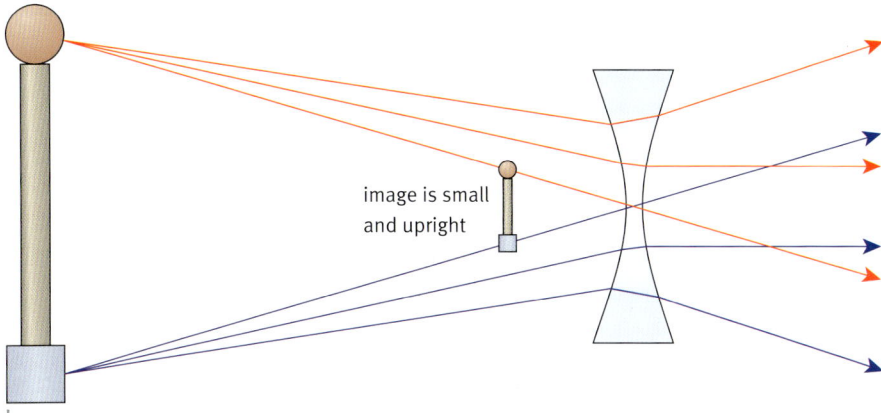

image is small and upright

A diverging lens makes the light rays spread out.

A converging lens is thickest in the middle. The converging lens in a camera also creates a small image, but it is upside down and beyond the lens. The light from a distant object that passes through the lens is brought to a focus in the focal plane. This is the place to put the light-sensitive surface, which detects the image. The **focal length** of the lens is the distance from its centre to the focal plane.

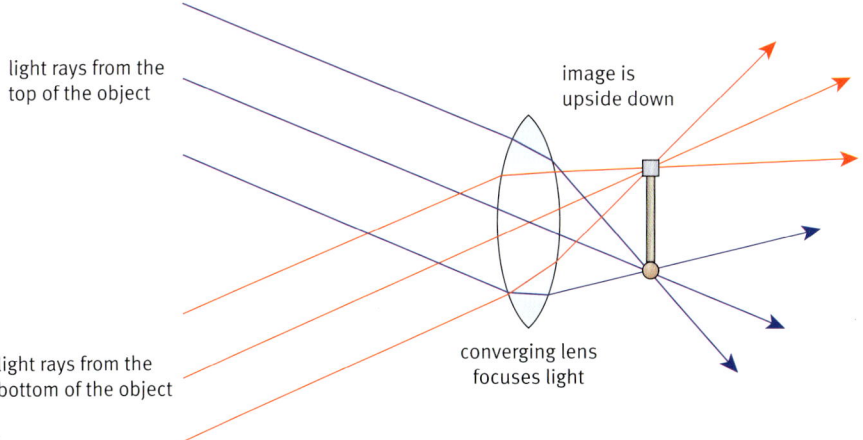

light rays from the top of the object

image is upside down

light rays from the bottom of the object

converging lens focuses light

A converging lens makes the light rays come together.

Key words

- ✓ **converging lens**
- ✓ **shutter**
- ✓ **aperture**
- ✓ **focal plane**
- ✓ **viewfinder**
- ✓ **lens**
- ✓ **diverging lens**
- ✓ **focal length**

Questions

1 Draw a diagram to show how a converging lens forms an image of a distant object.

2 Explain the function of the lens, aperture, and shutter in a camera.

Stage lighting

Theatre-lighting control desk. Each lamp has its own switch and dimmer control.

The lighting director needs to control the brightness of each lamp separately from all the others. This means that each lamp has to be wired in series with a **switch** (to turn it on and off) and a **dimmer** (for fine control of the power delivered to the lamp).

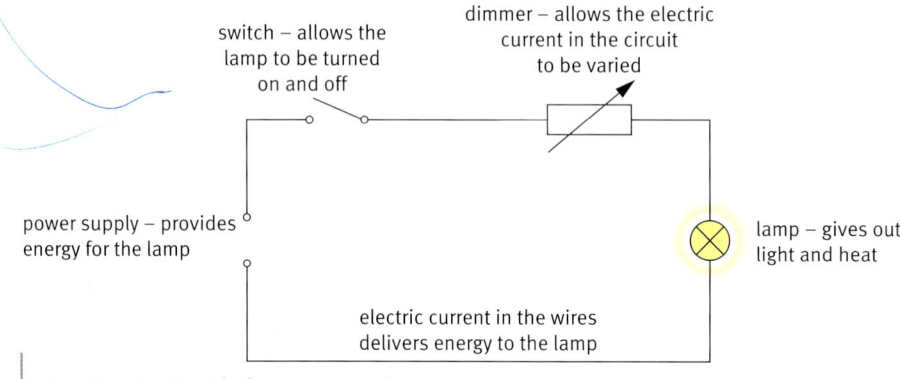

switch – allows the lamp to be turned on and off

dimmer – allows the electric current in the circuit to be varied

power supply – provides energy for the lamp

lamp – gives out light and heat

electric current in the wires delivers energy to the lamp

A series circuit with dimmer control.

Energy flow

When the lights are on, a lot of heat energy gets pumped into the theatre. This needs to be removed, otherwise the audience can get unpleasantly hot. **Ventilation** fans remove stale hot air from the auditorium, replacing it with fresh cool air. The rate at which fresh air is circulated needs to be carefully controlled to keep the audience at a comfortable temperature.

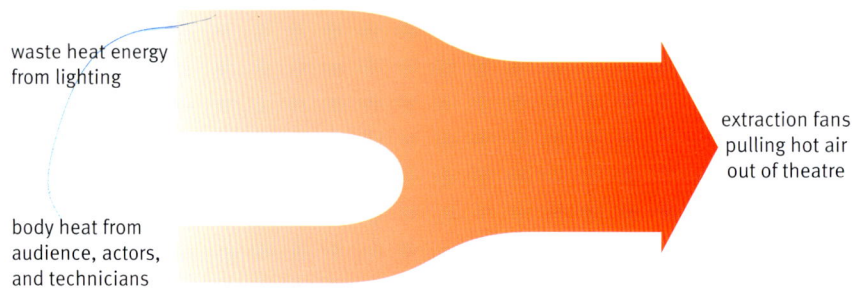

waste heat energy from lighting

body heat from audience, actors, and technicians

extraction fans pulling hot air out of theatre

This energy flow diagram shows that about one third of the energy extracted from the theatre by the fans comes from the people in it!

Luminaires

One of the most common lighting units (luminaires) used in a film studio uses a Fresnel lens and curved reflector to control light from a hot tungsten filament. The design of the luminaire has to solve two problems:
- The lamp emits light in all directions. Most of the light needs directing onto the stage.
- The waste heat must be carefully managed. If the lamp overheats it will damage its housing; it may even start a fire.

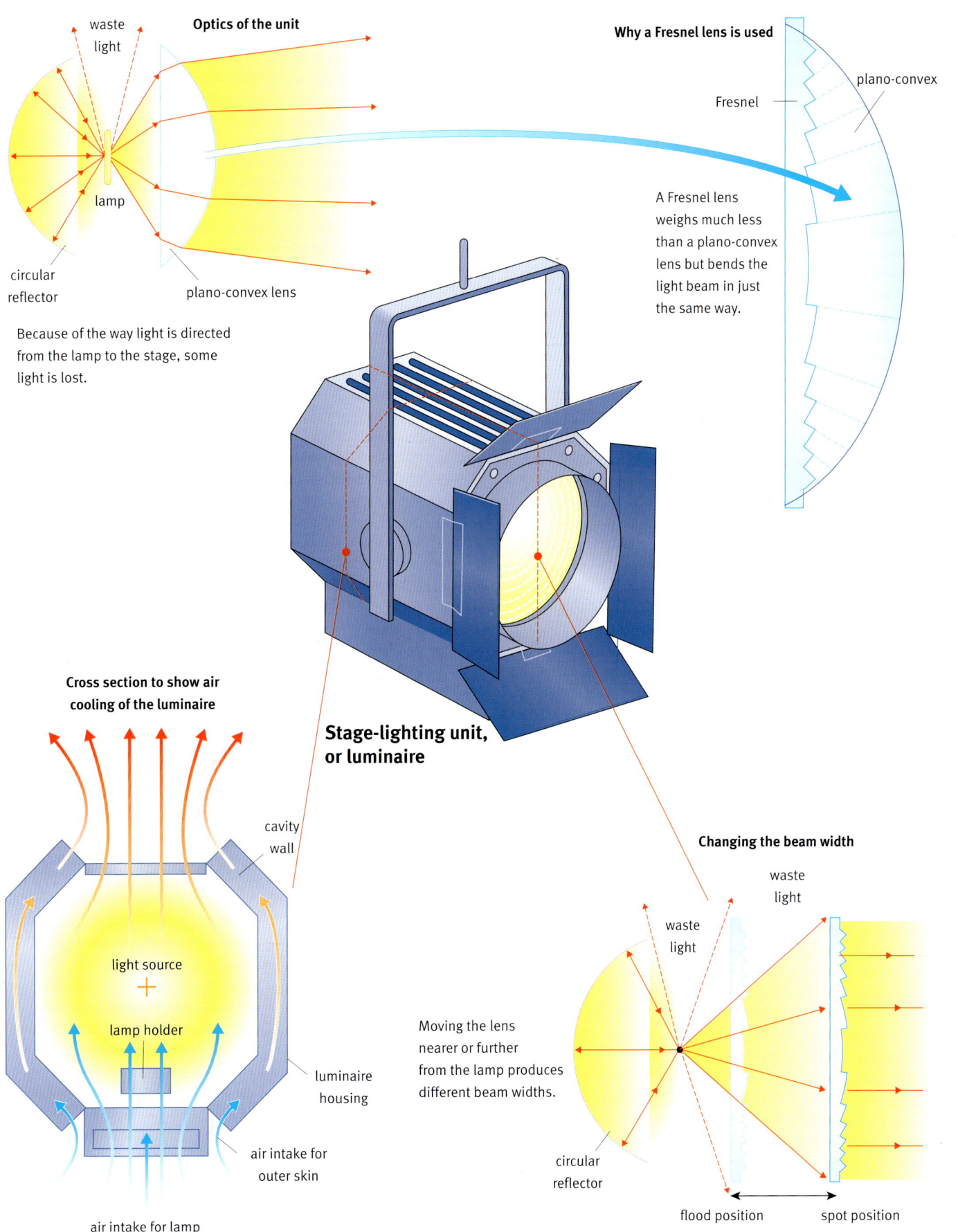

Optics of the unit

waste light

lamp

circular reflector

plano-convex lens

Because of the way light is directed from the lamp to the stage, some light is lost.

Why a Fresnel lens is used

Fresnel

plano-convex

A Fresnel lens weighs much less than a plano-convex lens but bends the light beam in just the same way.

Stage-lighting unit, or luminaire

cavity wall

Cross section to show air cooling of the luminaire

light source

lamp holder

luminaire housing

air intake for outer skin

air intake for lamp

Changing the beam width

waste light

waste light

Moving the lens nearer or further from the lamp produces different beam widths.

circular reflector

flood position

spot position

Refractive index

The broken-pencil illusion.

Material	Refractive index
air	1.00
water	1.33
glass	1.4–1.6
acrylic (eg, Perspex)	1.49
diamond	2.4

Some materials and their refractive indices.

This pencil in water looks broken. This is because light changes direction as it enters and leaves the water. The bending effect is called **refraction**. Glass and other transparent materials also refract light, though by different amounts.

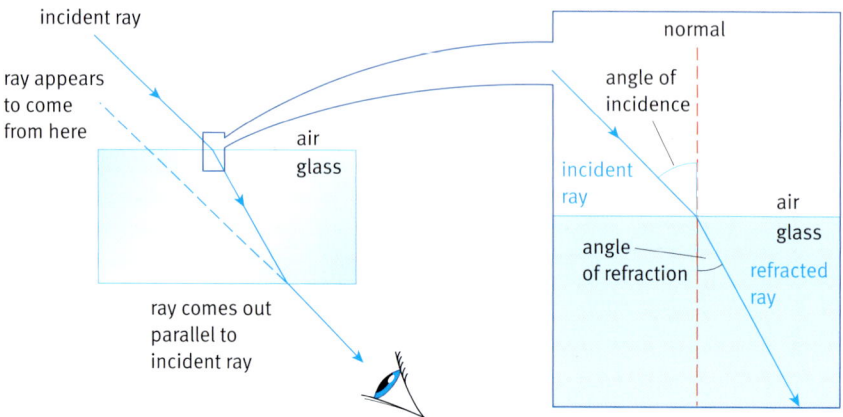

The diagram above shows how a ray passes through a glass block. The line at right-angles to the surface of the block is called the normal. The ray is refracted towards the normal when it enters the block, and away from the normal when it leaves it.

With a rectangular block like this one, the ray comes out parallel to its original direction. If the ray strikes the block at right-angles (square on), it goes straight through without being refracted.

The light ray is refracted when it enters glass because it slows down. A material that slows down light a lot has a high **refractive index**.

The greater the refractive index of a material, the more it can bend rays of light. So making a lens from glass with a high refractive index allows the lens to be thinner. This also makes it lighter, which is important for portable devices like binoculars or spectacles.

The refractive index of glass depends on what it is made of. Glass with a lot of lead has a higher refractive index than window glass, making it ideal for use in decorative cut glass.

Key words
- refraction
- refractive index

Questions

1 Explain why the best spectacle lenses are made from glass with a high refractive index.

2 Your eyes refract light that enters the eyeball to create a focused image. Use ideas about refraction to suggest why you need to wear goggles to see clearly under water.

Amplifying sound

A singer's sound is often a delicate thing. It is the sound engineer's job to ensure that everyone in the audience has an equal chance of enjoying it. The problem sounds quite simple. Take a low-energy sound wave, make a high-energy copy of it, and direct it into the auditorium. In practice, it isn't that simple.

Connecting blocks

A useful sound system has three separate parts:

- A **microphone** – this picks up the sound and converts it into an electrical signal. The mounting of the microphone is important to prevent unwanted vibrations getting through from the stage floor.
- An **amplifier** – this makes a high-power copy of the microphone's electrical signal. It must not change the signal in any other way. In practice, hum and noise often get added at this stage. A good sound engineer can usually prevent this happening. They also use tone controls to adjust the frequency balance of the signal.
- A **loudspeaker** – this is supposed to create an exact copy of the sound that entered the microphone, but much louder. In practice, the job needs sharing between several loudspeakers, each dealing with a different range of sound frequencies.

This microphone must be plugged into ....

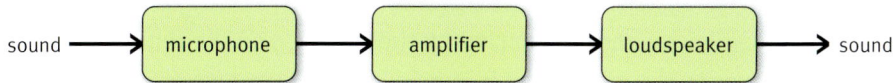

A block diagram shows the flow of signals between each of the units in a simple sound system.

Howl

A badly installed sound system can easily become unstable and **'howl'** loudly. This happens when a certain amount of sound from the loudspeaker leaks back to the microphone. It is passed on to the amplifier, producing a feedback loop in which the sound just gets louder and louder. There are two ways to avoid this:

- Keep the sound level low enough for this not to happen.
- Arrange the loudspeakers so that they send the sound out to the audience and not back to the microphone.

... an amplifier and loudspeaker to make a complete sound system.

THE SCIENCE

Questions

1 Draw a block diagram of a sound system. Explain the function of each block.

2 What is howl? Explain how it can be avoided.

Key words
- ✓ **microphone**
- ✓ **amplifier**
- ✓ **loudspeaker**
- ✓ **howl**

Acoustic properties

Key words

- ✓ amplitude
- ✓ absorb
- ✓ reflect
- ✓ transmit
- ✓ acoustic

Noise is more than just a nuisance. It can also seriously damage your health. It is important to know how the materials in a building affect sound waves:

- Will they stop sound getting in from outside?
- Will they stop echoes interfering with people's conversations?
- How much sound will leak from one place to another?

Careful choice of building materials keeps unwanted sound away from people.

Sound is caused by vibrations. The greater the **amplitude** of the vibrations, the more energy carried by the sound, and the louder it will be. The energy of the vibrations is carried by pressure waves through solids, liquids, or gases in contact with the sound source. How far that energy travels depends on the **acoustic** properties of the material:

- Materials with lots of tiny cracks in them are acoustically soft. They **absorb** sound, because vibrations warm the material.
- Solid materials with no cracks are acoustically hard. They either **reflect** sound or **transmit** it.

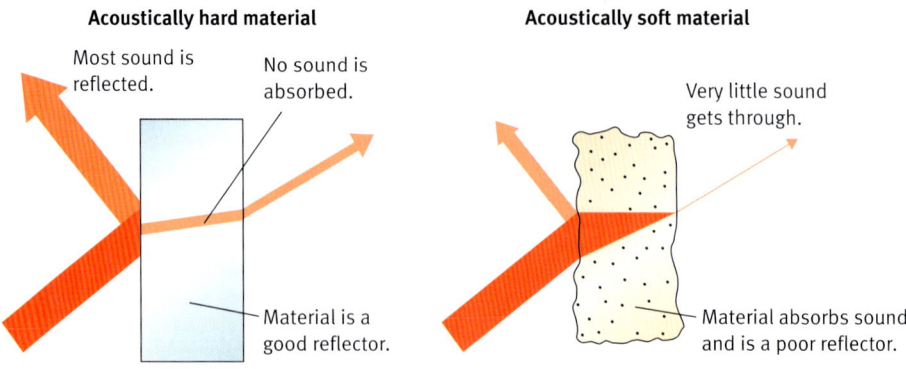

Acoustically hard material

Most sound is reflected.

No sound is absorbed.

Material is a good reflector.

Acoustically soft material

Very little sound gets through.

Material absorbs sound and is a poor reflector.

Unwanted sound can be dealt with in two ways. It can be reflected by hard surfaces, changing its direction. Or sound can be absorbed by soft surfaces, transferring its energy into heat.

Measuring sound

The pitch and intensity of a sound determine how much of a nuisance it is. The **pitch** depends on how rapidly the source is vibrating. It is usually given as the **frequency** of the sound. This is the number of vibrations of the source per second, measured in **hertz** (Hz).

The **intensity** of a sound is a measure of how much energy reaches your ear every second. The **loudness** of a sound is how you perceive it. Loudness depends not only on the intensity of a sound, but also its frequency and how long it lasts.

Humans can hear sounds only in the frequency range of 20 Hz to 20 000 Hz (20 kHz), and the ear is most sensitive to sounds at about 2 kHz. In other words, if the intensity of a sound stays the same while its frequency changes, it will sound loudest at about 2 kHz.

Intensity is measured in **decibels** (**dB**). A sound that increases by 10 decibels doubles its loudness.

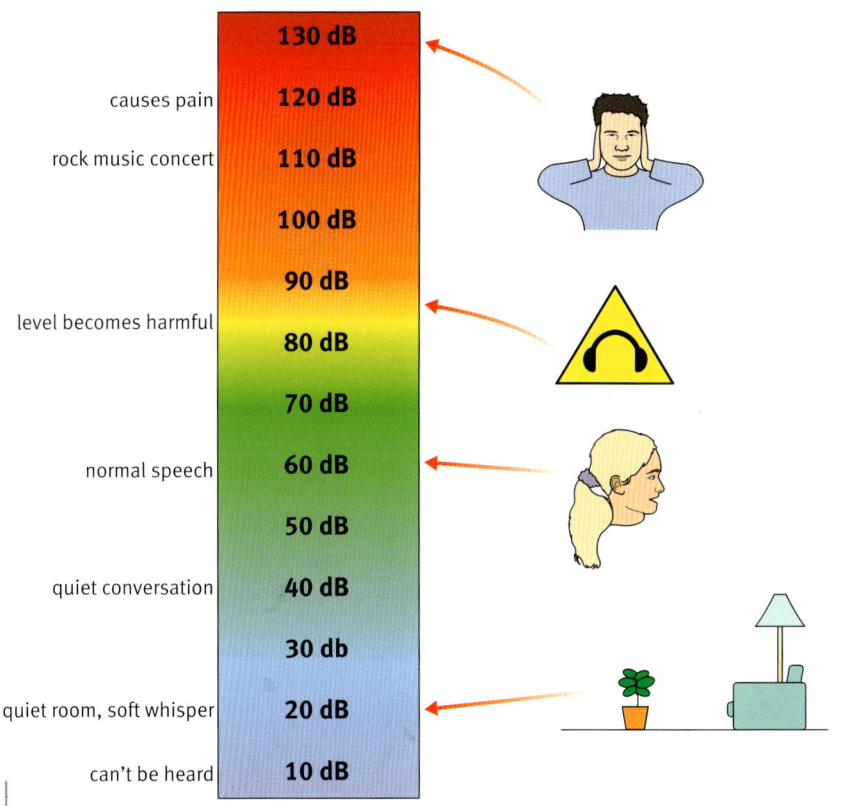

The decibel scale for sound intensity.

Questions

1 A guitar string vibrates 256 times a second. What is the frequency of the sound it produces?

2 Describe the difference in structure between an acoustically soft material and an acoustically hard material.

3 Why are sound levels above 90 dB dangerous?

4 Describe and explain two ways of reducing sound levels.

Sound control

Pete Townshend of The Who is now partially deaf after many years of playing very loud music.

Ringing in the ears

Have you ever had a ringing sound in your ears after listening to music, perhaps at a rock concert? This was probably the result of exposure to sound levels above 85 dB. That level of sound often results in temporary loss of hearing too.

The damage may become permanent if you listen to sound at 90 dB or more for long periods of time, perhaps at work. You gradually lose the ability to hear high frequencies, eventually becoming deaf.

In some cases, the ringing sound doesn't go away but carries on and on. This distressing condition is called **tinnitus**.

There are three ways of reducing the level of unwanted sound in your ears:
- Move away from its source.
- Absorb the sound before it gets to you.
- Reflect it somewhere else.

Double glazing

Window glass is acoustically hard. Any sound reaching one side of it goes straight through, with almost no absorption. So single-pane windows transmit outside noise into buildings.

Using two panes of glass, as in double glazing, cuts out noise very effectively. Most of the sound that gets into the cavity is trapped there. It bounces off the inner surfaces of the glass until it has been absorbed by the air. Triple glazing is even more effective.

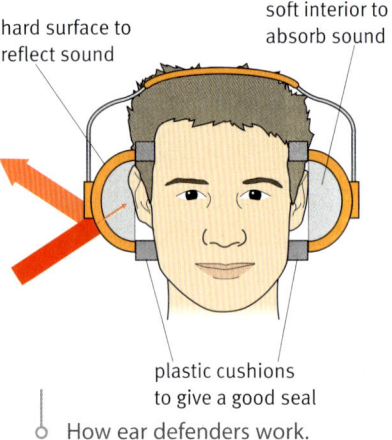

hard surface to reflect sound

soft interior to absorb sound

plastic cushions to give a good seal

How ear defenders work.

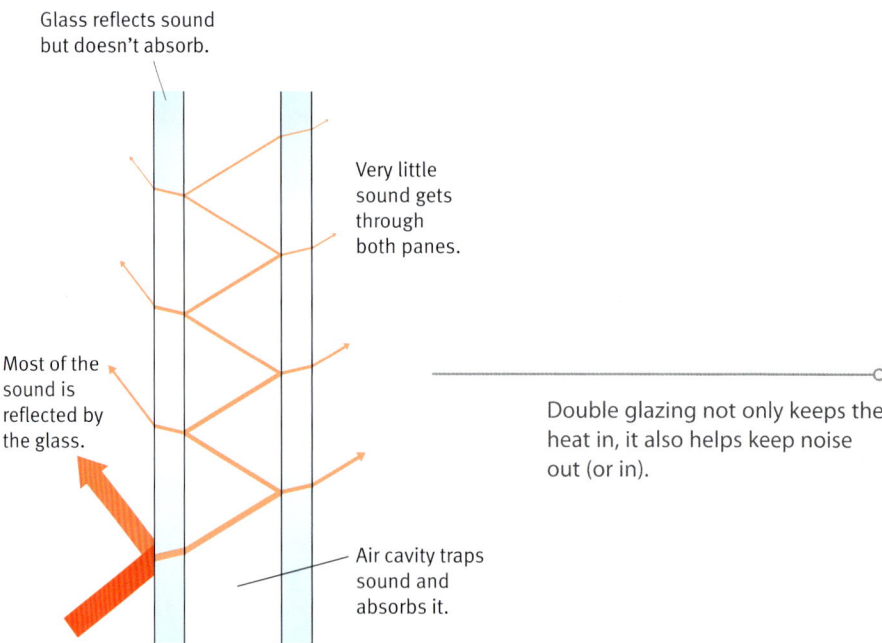

Glass reflects sound but doesn't absorb.

Very little sound gets through both panes.

Most of the sound is reflected by the glass.

Air cavity traps sound and absorbs it.

Double glazing not only keeps the heat in, it also helps keep noise out (or in).

Questions

1 Describe the consequences of prolonged exposure to high sound levels.

2 Explain why motorways near housing estates are often edged with high walls.

3 Explain how foam-filled cavity walls in buildings can provide sound insulation.

Bad vibrations

Low-frequency sound is difficult to manage. This is because acoustically soft materials absorb high-frequency sound much better than low-frequency sound. The graph opposite shows this. Frequencies below the audible range (under 20 Hz) are particularly troublesome. You feel them, instead of hearing them.

There are two ways of dealing with low-frequency vibrations:
* stop them happening in the first place
* isolate the source by mounting it on special absorbing material.

Sympathetic vibration

Annoying vibrations happen if the panels of a car vibrate in sympathy with the engine at a particular speed. Stiffening each panel by the right amount can suppress the vibration, making the car interior quieter.

Damping vibration

A car engine works by using rapidly repeating explosions, so you cannot stop it vibrating. However, it is possible to reduce the vibrations transmitted to the rest of the car by mounting the engine on pieces of rubber. This is because rubber is a good **damping** material. Rubber pads transfer the energy of vibrations into heat, as shown in the graph on the right.

Rubber is also good for supporting floors to prevent sound being transmitted. Larger buildings use **fluid-filled dampers**. The vibrations force pistons to move inside cylinders of oil, transferring the energy of the vibration into heat.

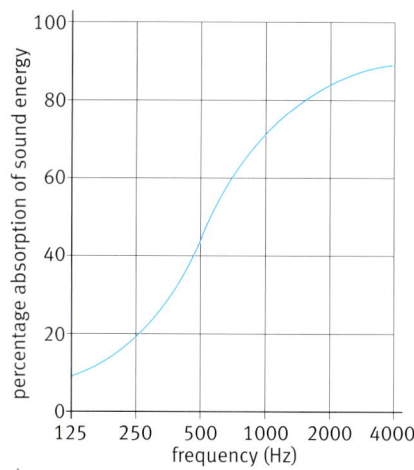

Foam tiles absorb high-frequency sound much better than low-frequency sound.

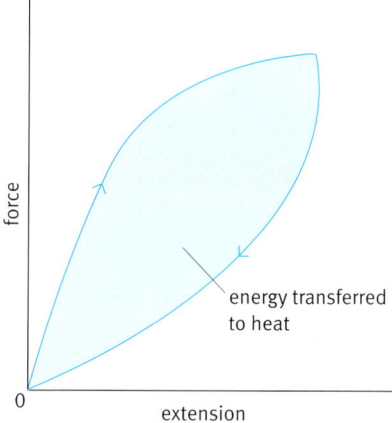

Rubber extends and contracts differently, heating up in the process.

Questions

4 Describe two ways of damping vibrations of machinery.

5 Why are low-frequency sounds troublesome?

Key words
* tinnitus
* damping
* fluid-filled dampers

Focal length of a converging lens

Principle

The **focal length** of a lens is the distance from the lens to its focal point. A fat lens has a short focal length. A thin lens has a longer focal length.

There are several ways to measure the focal length of a converging lens.

CHECK SAFETY
Never work
unsupervised

Equipment

*   converging lens
*   metre rule
*   light box with cross wires and screen
*   plane mirror

Procedure

1   Stand near the wall at the opposite side of the room to the window.

2   Hold up the lens and use it to focus an image of the window on the wall. Hold the lens so that the image is as clear as possible.

3   Measure the distance from the lens to the wall. This is the focal length f of the lens.

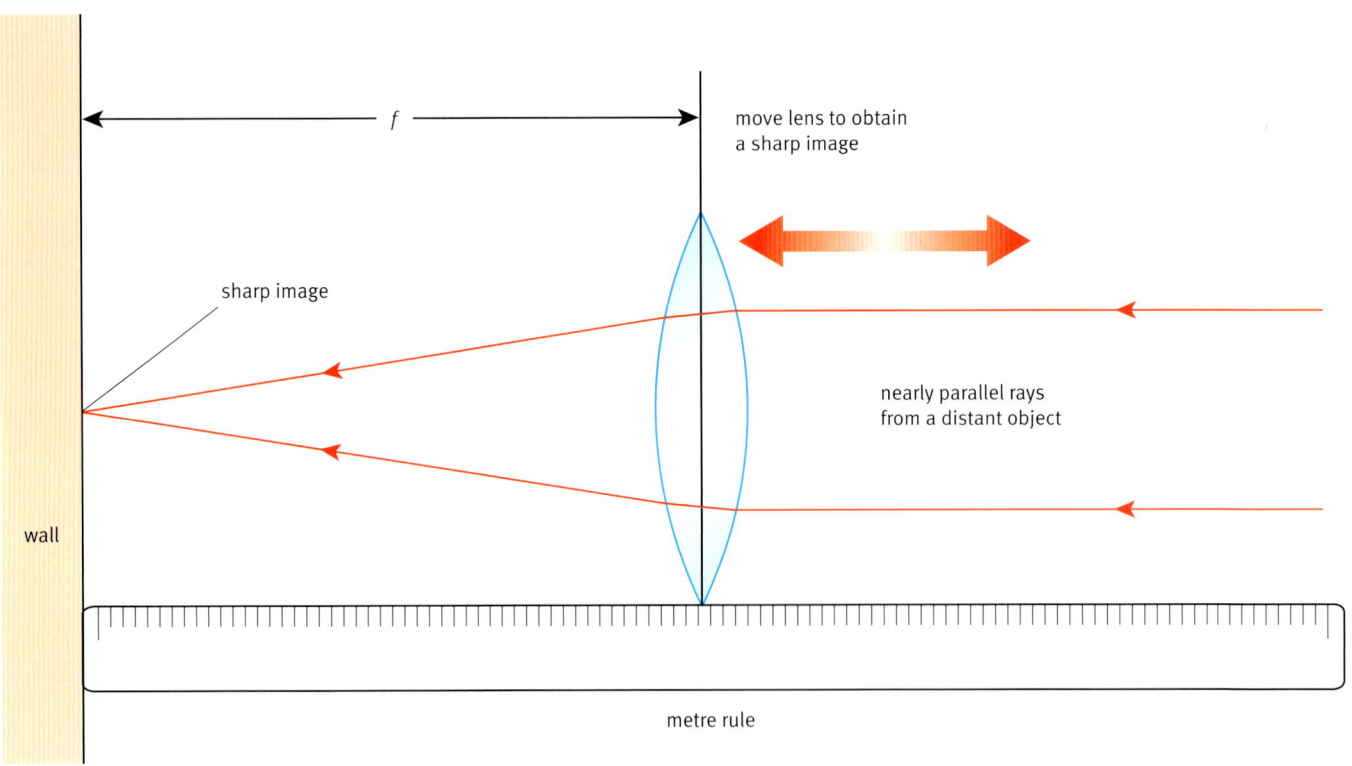

f

move lens to obtain
a sharp image

sharp image

nearly parallel rays
from a distant object

wall

metre rule

Measuring sound levels

Principle

The sound level or **intensity** of a sound in **decibels** measures how much energy it delivers to surfaces that absorb it.

Equipment

* decibel meter
* sound sources

Procedure

1 Use the rotary switch to set the meter to 'BATT' (battery test). If the needle is not in the red region, replace the battery.

2 Set the weighting switch to match the type of sound you are measuring. Use A for noise and C for music.

3 Set the response switch to SLOW if you are measuring noise or FAST to follow rapid changes.

4 Point the microphone at right angles to the source of sound, as shown in the diagram.

5 Rotate the switch to 120 dB and reduce it until the needle stays in the middle of the range.

6 Add the reading of the needle to the setting of the switch to obtain the sound level in dB. For example, if the needle reads 9 and the switch setting is 110, the sound level is 119 dB.

Interpreting the result

The intensity of a sound reduces as you get further away from its source, so to compare sound sources you must measure them at a fixed distance.

It is illegal to have sounds above 80 dB in the workplace without ear protection.

CHECK SAFETY
Never work
unsupervised

microphone — to source of sound

rotary switch

SOUND LEVEL METER

CHECK SAFETY
Loud sounds (85
dB and over) can
damage hearing.

PROCEDURES & TECHNIQUES

Module Summary

People and organisations

- that sound engineers use their understanding of acoustics to enhance stage and screen performances
- that lighting directors use their understanding of light to enhance stage and screen performances
- that theatre managers are required to adopt health-and-safety procedures for public entertainments
- that health-and-safety regulations require emergency lighting, emergency exits, signage, fireproof curtains, and fireproof doors in public-entertainment venues
- that health-and-safety regulations require planning for person flows and evacuation times in public venues.

The science

- that sunlight, incandescent lamps, fluorescent lamps, and lasers are light sources for film and theatre
- that some light sources give off unwanted radiation such as ultraviolet and infrared radiation
- how filters work by absorbing some colours and transmitting others
- that removing blue from white light makes it yellow, removing green makes it magenta, and removing red makes it cyan
- how to use these words to describe optical properties of materials: transparent, reflective, translucent, opaque, and refraction
- that the amount of refraction at a material's surface depends on its refractive index
- how the action of a lens depends on its shape and what it is made from
- that moving a distant object towards a lens increases the size of the image and moves it beyond the focal plane
- that increasing the frequency of a sound raises its pitch and increasing the amplitude of its vibrating source makes it louder
- that the human ear is most sensitive to sounds at a frequency of 2000 Hz
- that when sound doubles in loudness its intensity increases by 10 dB
- the sound intensities of some situations: normal conversation is 60 dB, temporary hearing loss occurs above 85 dB, and pain is felt above 130 dB
- that acoustic tiles, double glazing, carpeting, and curtains are used to reflect or absorb sound
- how to arrange sound systems to avoid howl
- that rubber pads and wire suspension can be used to isolate vibrations in solid structures
- why low-frequency sounds cause more problems in buildings than high-frequency sounds
- how to draw and interpret simple images formed by a plane mirror
- how to identify converging and diverging lenses
- how to draw and interpret ray diagrams
- how to identify the lens, shutter, aperture, focal plane, and viewfinder of a simple camera
- how to design and interpret simple sound systems made from microphones, amplifiers, and loudspeakers
- how to design and interpret simple lighting circuits including switches and dimmers.

Standard procedures

- how to measure the sound intensity at a venue
- how to measure the focal length of a lens.

Review Questions

1 The work of actors is essential to stage productions. Describe the role of another qualified practitioner whose work is essential for stage productions.

2 Spotlights are used to direct high-intensity light in a particular direction.

   a Explain how a filter can change the colour of light from white to yellow.

   b Describe a use of filters that does not change the colour of the light.

   c Suggest why a spotlight often contains a mirror and lens.

   d Describe another kind of light source and how it is used in stage or film productions.

3 The diagram shows three rays from an object being focussed by a lens.

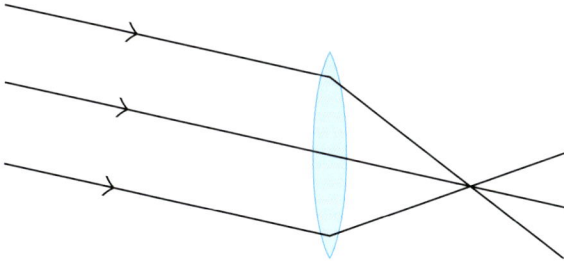

   a How can you tell that the object is a long way away?

   b What happens to the image as the object moves towards the lens?

   c What is the name of the process that makes the light change direction as it passes through the lens?

4 Light behaves in different ways when it strikes the surface of a material. Use examples of different materials to explain the meaning of:

   a transparent **b** translucent **c** opaque

5 Draw a diagram to explain why the image formed by a mirror appears to be behind the mirror.

6 A simple stage sound system in a large hall contains these components:

   amplifier   loudspeaker   microphone

   a Explain the function of each component in the system.

   b The system makes the sound eight times louder. What is this increase in dB?

   c Suggest why sound engineers often place the microphones behind the loudspeakers.

   d Explain why the venue has sound-asborbing materials on the walls of the hall.

7 Nicky builds this model of a stage lighting system.

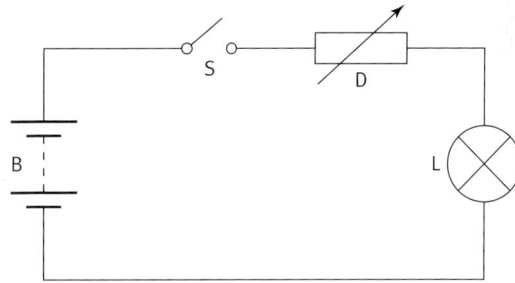

Name the four different components **B, D, L,** and **S** and explain their functions in the system.

8 Sam is helping to organise a charity concert. She makes a list of some of the safety features needed in the hall. For each feature, explain why it is needed:

   a emergency lighting

   b exit signs

   c stage fire-safety curtain

   d fire doors

STAGE AND SCREEN PRODUCTIONS

managing light
- optical properties of materials
  - reflection
    - mirror
      - ray diagram
        - image
        - object
  - refraction
    - refractive index
      - lens
        - image
        - object
        - focal plane
        - shutter
        - viewfinder
        - converging
        - diverging
    - transparent
    - opaque
    - translucent
    - filter
      - cancer
      - UV
      - blue
      - green
      - red
      - IR
        - heat
- light sources
  - incandescent lamp
  - laser
  - fluorescent lamp
  - sunlight

managing temperature
- heat sources
  - audience
  - lights
  - extraction
    - ventilation
    - air conditioning

managing sound
- acoustic propeties of material
  - acoustic tiles
  - double glazing
  - curtain
  - amplification
    - microphone
    - amplifier
    - loudspeaker
  - loudness
    - amplification
      - decibels
        - temporary hearing loss >85dB
  - pitch
    - frequency
      - ear sensitivity

interpreting data
- mirror images
- ray diagrams
- schematics
  - sound systems
  - circuits
  - heating

standard procedures
- measuring brightness
- measuring focal length of a lens
- measuring sound intensity
- decibel meter

regulation

216

feedback

amplification

balance

soundproofing

vibration isolation

absorbing sound

sound engineers

lighting engineer

computer control

lamps

switches

dimmers

heating engineer

ventilation

people

venue management

air conditioning

organisations

performance venues

music studies

Health and Safety Inspectorate

Fire Service

Town Council Planning

health and safety

fire

evacuation

emergency lighting

temperature

sound

fire curtain

emergency signage

auditorium

decibel limits

emergency exits

emergency lighting

stage

people flows

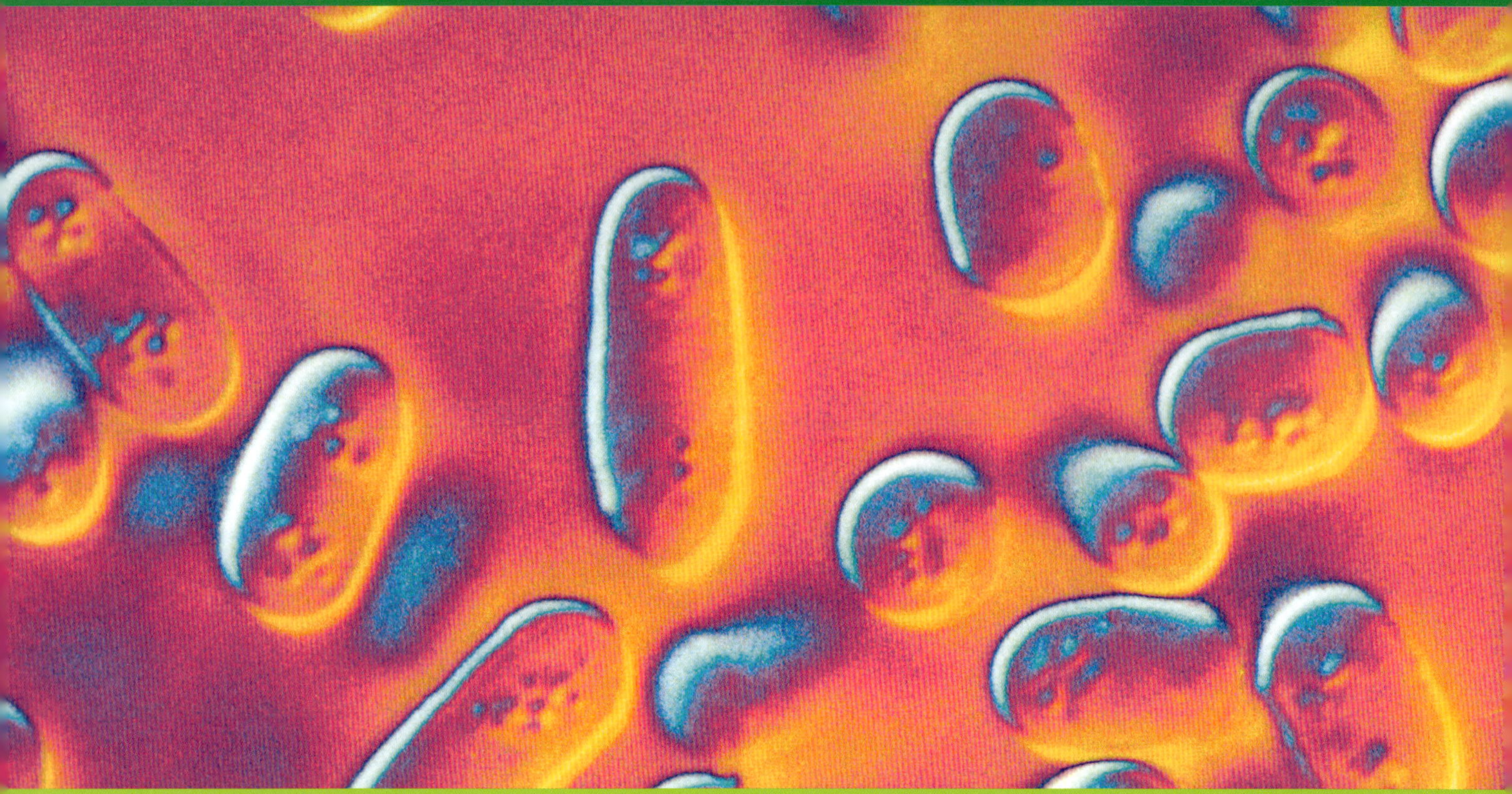

B3 Agriculture, biotechnology, and food

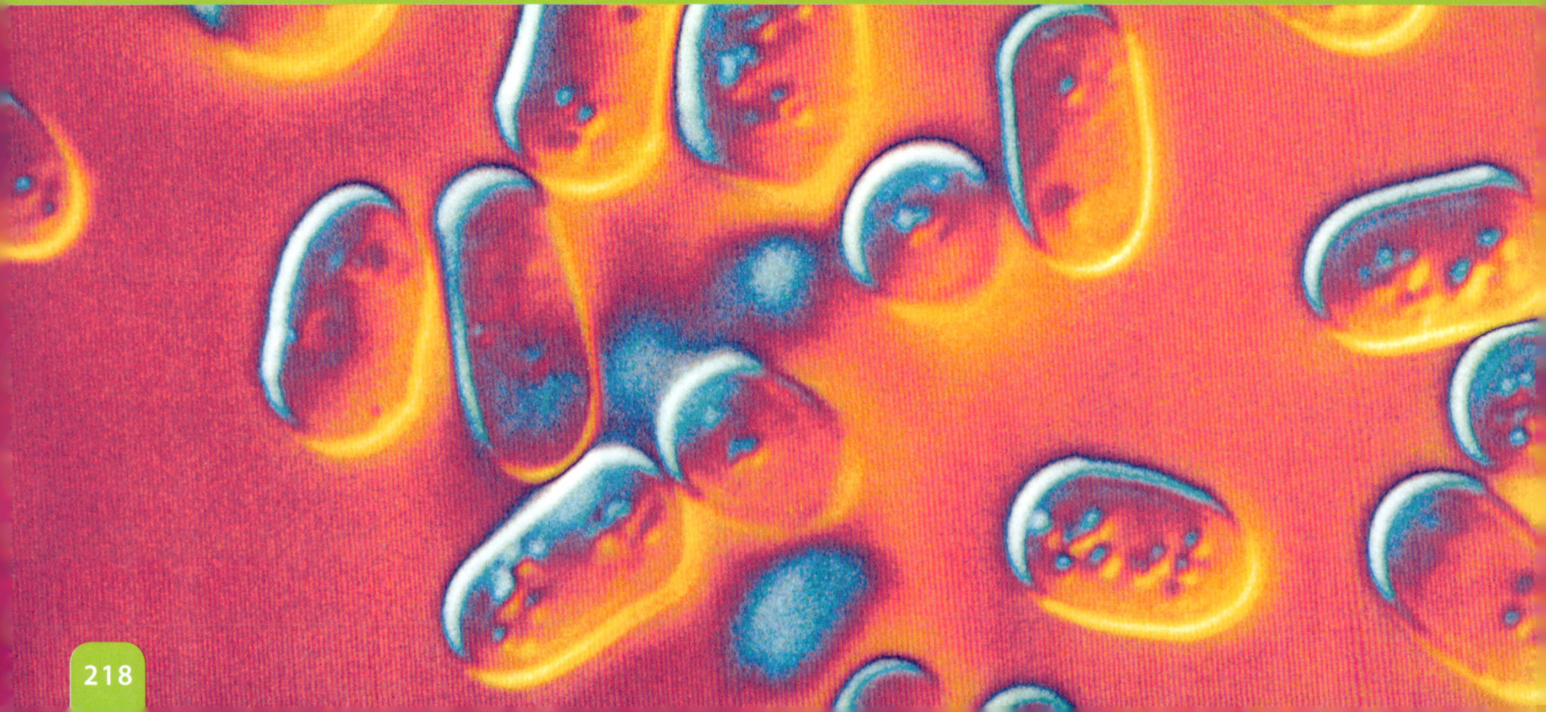

Why study agriculture, biotechnology, and food?

British farmers produce nearly 70% of our food. In this topic you will learn about food production from 'farm to plate' in dairy farming and growing wheat for bread-making. We use microorganisms to make bread, yoghurt, cheese, and alcoholic drinks by fermentation. This natural process has been harnessed by humans for thousands of years. Microorganisms can also lead to food spoilage and food poisoning. Understanding the conditions in which microorganisms grow and multiply helps us to keep our food safe to eat and prevent it going 'off'. Regulation and testing of food and drink products is important for public health and safety and to protect the environment.

What you already know

- that crop varieties are influenced by genetic and environmental factors

- in the right conditions populations of bacteria and viruses can increase very quickly

- genes are sections of DNA molecules that determine which proteins are made by cells

- that enzymes are proteins

- industries are regulated to maintain standards and protect the public.

Find out about

- the process of wheat and dairy cow production

- the factors that affect crop yield and milk production

- how some microorganisms can be used to make useful food products while others can lead to outbreaks of food poisoning

- how sensors and systems are used to control bioreactors

- the organisations that regulate the food industry.

The Science

Dairy cattle are bred to produce large quantities of milk. This is used to produce a variety of milk products. Farmers can apply science to encourage growth and high yields.

The UK bread market is one of the largest sectors in the food industry. Farmers work to maximise yields of wheat to provide enough good-quality bread-making flour at a competitive price.

Scientists control fermentation processes to get the right products and maximise yield. Some processes use genetically modified microorganisms to produce food ingredients and enzymes used in food processing.

Brewing

What's in beer?

British brewers produce over 1200 different beers using their skill, scientific knowledge, and four basic ingredients: malt, water, hops, and yeast (a type of microorganism).

Malt

Malt is barley grains that have begun to germinate. Timing is tricky. A brewer at St Austell Brewery explains: 'If you can see a green shoot, it's already too late.' At exactly the right point the malt is roasted. The roasted grains are roughly milled before being used for brewing.

The malting process converts some of the starch in the barley seeds into sugars. These can be used by the yeast to make alcohol. The roasting gives flavour to the beer.

Water

St Austell Brewery has its own water supply – a spring of mysterious origin with a water level that hardly drops, even in times of drought. The water affects the quality and taste of the beer produced. It is important that it is free of contaminants that could affect the brewing process.

Hops

Hops are the dried flowers of female hop vines. The brewers at St Austell select a blend of different hops. Some varieties add bitterness. Others add flavour and aroma to the beer.

The flowers of female hop vines are picked and then dried. They are used to add flavour or bitterness to a beer.

Barley grains that have begun to germinate and become malt.

Brewer's yeast

Brewer's yeast is a strain of yeast that is particularly good for making beer.

When malt, water, and hops are heated together they make a dark liquid, known as the hopped wort. The wort is cooled and transferred to large fermentation tanks, where yeast is added. Yeast is a living microorganism that uses sugars in the wort as food and produces alcohol. This is an example of fermentation.

During fermentation microbes break down complex molecules in food to provide the energy for their growth. As by-products they sometimes produce the fermented foods and drinks we enjoy.

The yeast cells multiply by budding during the fermentation process and are saved at the end, so that some can be used for the next fermentation.

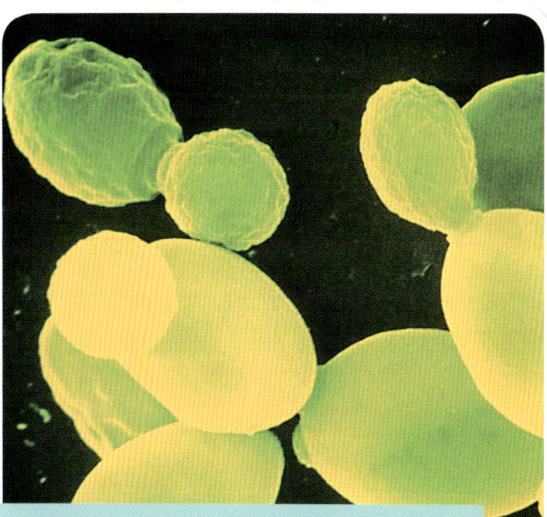

False-colour electron microscope image of yeast cells budding, magnified 10000 times.

Avoiding problems

St Austell Brewery uses a mixture of traditional and modern techniques to make sure each fermentation is good. Large containers called mash tuns are used to heat the malt and water. At the end of mashing somebody has to climb into the mash tun to clean it before another batch of malt is added. Other equipment, such as the fermentation tanks and casks, have a computer-controlled cycle of cleaning and sterilisation between each fermentation.

Brewer's yeast can become contaminated with wild yeasts. Analysts check the yeast regularly using laboratory tests and microscopic examination. It is also important to watch every fermentation carefully, as the laboratory manager explains: 'You can look down a microscope all you like, but if fermentation slows down, that's a sure sign the yeast is no good.'

The laboratory manager at the brewery tests the beer for correct colour, pH level, bitterness, and yeast count. He has computers and equipment to help him. His first job every day is to taste samples of the previous day's brew.

Baking bread

Daily bread

Every day in the UK, millions of loaves of bread are made. Eighty per cent of this bread is made by large industrial bakeries and sold sliced and wrapped. In-store supermarket bakeries supply most of the rest. Only a very small proportion of bread is now made by master bakers working in local bakeries.

High-protein, hard-milling varieties of wheat have been developed for bread-making.

Ingredients

Whatever size the bakery, the same basic ingredients are needed: flour, the microorganism yeast, and water.

Yeast is a type of fungus and like all living things needs food, water, and warmth to grow. Baker's yeast is a strain of yeast that is particularly suitable for making bread. It comes in two forms:

- compressed fresh yeast – the yeast is alive and active
- dried into granules – the yeast is alive but dormant.

The yeast uses sugars present in the flour as its food. It breaks them down to provide itself with energy for growth. As the yeast grows it produces bubbles of carbon dioxide. These bubbles cause the dough to rise. When the dough is baked the heat kills the yeast.

The process of making bread has three stages: mixing the ingredients, allowing time for the yeast to act on the dough mixture, and baking the loaf.

How good that loaf is depends on the quality of the starting ingredients and the skill of the baker controlling each stage of the process.

New wheat for better bread

Not all wheat produces flour that is suitable for bread-making. Plant breeders have recently developed new varieties of wheat that produce good bread-making flour and grow well in the UK climate. This means less wheat needs to be imported from North America.

Outbreak

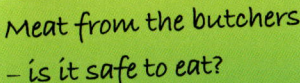
Meat from the butchers – is it safe to eat?

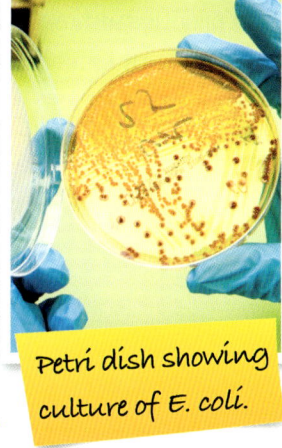
Petri dish showing culture of E. coli.

The incident

In 2010 a newsflash announced that:

'Seven more people have been taken ill in an outbreak of the *E. coli* bug in Leeds, bringing the total affected to 21, health officials have confirmed.'

At least 12 of those affected needed hospital treatment, including a five-year-old girl and an 82-year-old woman. None of the cases was believed to be life-threatening.

A spokesman from the Health Protection Agency (HPA) said:

"*E. coli* has an incubation period of 2 to 14 days and we're now 13 days into this outbreak. We expect some more cases but hope it will start tailing off soon."

Investigating the cause

Environmental health officers (EHOs) sent samples from patients to the HPA's specialist laboratory in London for further testing.

Investigations traced the source of this outbreak of *E. coli* O157:H7 to a butchers in the Armley area. The EHOs reported their findings to the government.

The public was warned not to eat or feed to pets any meat products known to have come from the butchers.

The shop, which supplied 18 other local food outlets, was closed while investigations continued.

The science behind the story

E. coli O157:H7 is a harmful strain of *E. coli* bacteria. It is a pathogen. It is often found in undercooked minced beef and unpasteurised milk. Outbreaks have also involved sprouted seeds, unpasteurised fruit juices, leafy greens, and cheese. It particularly affects the very young and the very old and is potentially fatal.

Food poisoning is caused by consuming contaminated food or water. Poor hygiene, cross contamination, and inadequate heat treatment are the most common causes. In 2007, there were an estimated 850 000 UK cases of food poisoning, over 19 500 hospitalisations, and over 500 deaths. Restaurants (42%), non-residential caterers (21%), and retail (7%) sectors were the major sources of outbreaks.

Prevention and regulation

In the UK, Food Hygiene Regulation (EC) 852/2004 sets out rules for food businesses handling and processing foods. It requires food business operators to:

- put in place and maintain food-safety management procedures
- supervise, instruct, and train food handlers on food hygiene and provide training on food-safety procedures
- monitor the colour and temperature of meat
- take corrective action when needed
- evaluate and document their procedures.

Food industries

Many plants don't grow all year round in this country. If we want fresh fruit and vegetables these have to be imported.

Farming is a high-tech industry that often uses the latest machinery for the greatest efficiency.

Cows are milked in a milking parlour. A computer records the yield of milk from each cow.

Food production is a high-tech, multibillion-pound business. Households in the UK spend over £90 billion every year on food and drink. Over 10 million tonnes of milk is produced and over 15 million tonnes of wheat is grown in the UK each year. British farmers look after more than 75% of the UK land area and produce almost 70% of the food we eat. Some foods are imported because we want to have fruit and vegetables all year round.

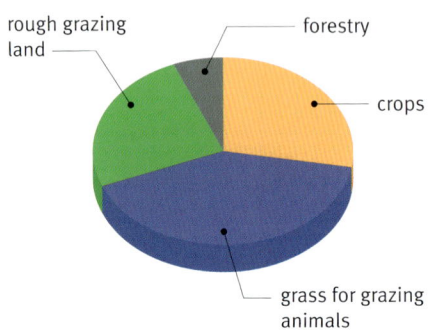

Agricultural land use in the UK: June 2008 (adapted from DEFRA Agricultural Statistics in your Pocket 2008).

There are many steps and many people involved in putting food on our plates. Each stage must be carefully coordinated so that there is enough food, it arrives fresh, and we can afford it.

WHO'S INVOLVED IN MAKING BREAD?

farmers	agronomists
farm-machinery suppliers	dock workers
seed growers	flour mill workers
pesticide dealers	bakery employees
fertiliser dealers	food technologists
grain inspectors	lorry drivers

WHO'S INVOLVED IN SUPPLYING MILK?

dairy farmers	dairy workers
vets	microbiologists
cattle-feed suppliers	food technologists
milking-parlour workers	lorry drivers
tanker drivers	milk purchasers

Key word
- ✓ chain of food production

The chain of food production

Wheat is grown on the farm.

Milk is produced on the farm.

Production

It is transported to the mill.

Transportation

It is transported to the dairy.

Wheat is ground into flour.

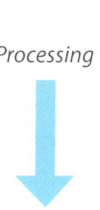

Processing

Milk is pasteurised and homogenised.

Flour is stored until needed for baking bread.

Storage

Milk can be heat-treated or dried for storage.

Freshly baked bread can be delivered to our door.

Delivery

Milk can be delivered fresh to our door or supermarket, or made into butter, cheese, yoghurt, cream or powdered milk.

Food products are sent to the shops from large regional food distribution warehouses. The process is highly automated to ensure that food stays fresh.

Questions

1 Which type of agriculture uses least land in the UK?

2 Dairy cattle need good-quality grass to produce the most milk. What proportion of UK agricultural land suits cattle grazing?

3 Not all of the land used for cultivated crops is sown for wheat. What other crops are grown in the UK?

4 Think about foods that you like. Draw a flow chart to show all the steps involved before it arrives on our plate. Try to name the different people that are involved.

Key words
- yield
- skimmed milk
- UHT
- pathogen
- cheese
- yoghurt
- fermentation

From the stockperson looking after the cows to the shopkeeper selling the products, people with a wide range of skills are needed for the production, processing, and selling of dairy products.

The farm

The milk **yield** from each cow is recorded. Road tankers collect the milk from a holding tank on the farm and take it to the processing dairy. Samples of milk are taken from each farm so that the composition (quality) and standard of hygiene (safety) can be checked.

The milk-processing plant

The milk first goes through a separation unit to remove some of the cream, leaving standard, semi-skimmed, or **skimmed milk**. This is then homogenised to mix the fat evenly through the milk.

Most milk is pasteurised before being sold as fresh milk. A heat exchanger is used to bring the milk to a specific temperature (71.7 °C) for 15 seconds and then quickly cool it. This kills most but not all of the **pathogenic** and spoilage microbes present. **Pasteurisation** allows fresh milk to keep in the fridge for longer, but does not affect its flavour. More tests are carried out on the milk to check the effects of pasteurisation. It is then put into bottles or cartons, labelled, date stamped, and taken in refrigerated lorries to the dairy depot or supermarket.

Some milk is processed by ultrahigh-temperature **sterilisation** (**UHT**). This treatment kills all the microbes and heat-resistant bacterial spores in the milk by heating it to a very high temperature (135 °C) for a very short time (1 second). UHT milk will keep in sealed packs for several months, without refrigeration.

A heat exchanger pasteurises the milk in a continuous process before the bottling machine fills and seals the bottles.

Milk that has not been heat treated is called raw milk. It may contain microbes such as *Campylobacter* and *Salmonella*, which are **pathogens**.

In Britain, consumers can only get raw milk directly from dairy farms, for example, at farmers' markets. Some people like to drink unpasteurised milk because they feel it is healthier than pasteurised milk and also tastes better.

Yoghurt and cheese

Milk is delivered from the farm to plants that make **cheese** and **yoghurt**. On arrival the quality of the milk is checked. It is usually pasteurised first. Then a starter culture is added to begin the fermentation process together with other ingredients.

A technician taking a sample of yoghurt from an incubation tank. The sample is tested as part of the quality-control program.

Keeping the industry competitive

Dairyco is funded by a charge on all UK milk sold. Dairyco carries out research into the supply and marketing of British milk and dairy products. It supports dairy farmers with practical advice, and runs marketing campaigns to inform the public about the value of dairy products.

Keeping milk safe

Milk hygiene and safety is watched over by an independent agency. Its job is to protect public health and safety. An inspectorate registers and inspects dairy premises for the agency. They make sure that good standards of hygiene are followed all along the chain of milk production and supply.

Questions

1 Draw a flow diagram to show what happens to milk from cow to refrigerator.

2 Explain why UHT milk keeps longer than pasteurised milk.

3 Explain why more people buy pasteurised milk than UHT milk.

4 What organisation promotes the dairy industry?

5 What organisation looks after your interests when you buy milk?

Harnessing microorganisms

Food produced using microorganisms.

Microorganisms such as **bacteria**, **viruses**, and **fungi** (including **yeasts** and **moulds**) are found all over the globe. They are too small to be seen without a microscope but their activities are very important to us. Some cause decay and are essential for the recycling of materials. Some cause disease. But some can be used to make useful chemicals and even to process food.

Scientists and technologists work in industry, universities, and government institutions to create or improve food and drink products and monitor their quality. They are needed at all stages of manufacture including:

- initial product development
- maintenance of microbial starter cultures
- monitoring the production process
- managing food safety and quality control.

Brewing industry

Microbiologists work in the brewing industry to ensure the best-quality beer is produced. They also produce improved strains of yeast for new types of beer. Yeast is used to produce **alcohol** in alcoholic drinks or to use as a fuel.

Microbiologists supervise the fermentation process. During fermentation beer may become contaminated with microbes from the surrounding air, from the barley malt, or from the fermentation vessels and associated pipe work. They develop tests to detect the microbial contaminants in the beer and develop ways of preserving the product.

They test the culture yeast (inoculum) to make sure that it is not contaminated with beer-spoiling bacteria and wild yeasts. They also carry out tests to make sure the yeast is still viable (alive).

PEOPLE & ORGANISATIONS

Key words

- ✓ microorganisms
- ✓ bacteria
- ✓ viruses
- ✓ yeast
- ✓ alcohol
- ✓ food poisoning
- ✓ environmental health officers

Question

1 Why do microbiologists need to monitor and control the quality of yeast cultures in the brewing industry?

Food poisoning

Food-safety regulations

Food poisoning is any illness due to eating food contaminated by chemicals (which is rare) or by pathogenic microbes or the **toxins** they produce. Poison, whatever the source, makes us ill.

The most common symptoms of food poisoning are diarrhoea and vomiting. The main food-poisoning microbes are bacteria, fungi, and viruses.

Enforcing food-safety regulations

Local environmental health officers (EHOs) visit food manufacturers, shops, and restaurants to advise them on food **regulations** and safety. They have the power to withdraw foods or close premises if they are a danger to public health.

In April 2004 the EHOs at two city councils, Nottingham and Derby, found that a brand of chilli powder was contaminated with high levels of aflatoxins. They reported their findings to the Food Standards Agency (FSA), which immediately contacted all other district councils. EHOs all over the country had to make sure the product was removed from sale in local shops. The EHOs also visited local restaurants to chase up batches of the contaminated powder.

For certain types of food production, the premises have to be registered and are inspected centrally by an independent government agency.

Government agencies monitor a wide range of food-safety issues, from safe production, storage, and transport to the right sort of packaging materials to use for different foods. They make sure that food is labelled correctly and give advice about safe food handling and healthy eating.

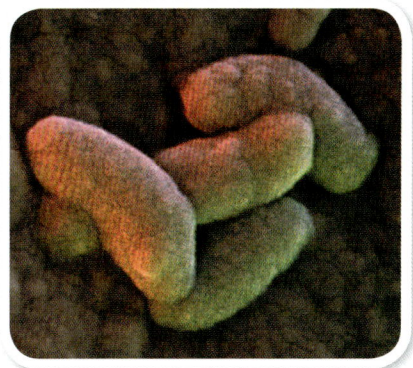

The bacterium *Campylobacter* magnified about 36000 times. This bacterium is the most common cause of food poisoning. It has been found mainly in poultry, red meat, unpasteurised milk, and untreated water. Even a few bacteria in a piece of undercooked chicken can make someone ill.

PEOPLE & ORGANISATIONS

An inspector checks the temperature of carcasses in an abattoir.

Questions

1 Explain why each of the following can help to avoid food poisoning:
   a washing hands before preparing food
   b preparing cooked and uncooked food on different surfaces
   c storing cooked meat in a refrigerator.

2 Why is it often impossible to tell just by looking at food whether or not it is safe to eat?

Growing wheat for bread production

There are many commercial varieties of wheat. This variety is called *Triticum aestivum*. Different varieties grow best in certain conditions and produce high-quality flour.

THE SCIENCE

Key words
- ✓ **bread wheat**
- ✓ **spring wheat**
- ✓ **winter wheat**
- ✓ **durum wheat**
- ✓ **insecticide**
- ✓ **herbicide**
- ✓ **fungicide**
- ✓ **fertilisers**

The UK bread market is worth almost £2.9 billion each year. Farmers try to grow the greatest amount of wheat – but the wheat must also be high quality and a competitive price. Each choice that a farmer makes affects the costs and the yield, including how to prepare the soil, which seed to sow and when, which chemicals to use, when to harvest, and how to store the crop.

Bread-making wheat

Wheat has been selectively bred over thousands of years to give a large amount of high-quality grain. 'Domesticated' wheat is different from wild grasses because the grains don't fall off easily and the seed heads are far larger. **Bread wheat** *Triticum aestivum* is best for baking because it has enough of the protein gluten to make bread dough elastic, trapping gas bubbles. This gives bread a light texture. Plant scientists have tested different varieties and cross bred those with the best characteristics. Some varieties of wheat have different growing seasons, like **spring wheat** and **winter wheat**. Winter wheat is sown in autumn and unlike many plants is not killed by frost in winter. Other varieties have high or low gluten content. **Durum wheat** has a very high gluten content and is good for making bread and pasta. It is not good for making cakes.

Scientists have genetically modified wheat to resist pests. In the past every farmer used a different variety of wheat. These days a more limited range of special seeds is available from the suppliers or is specified by the supermarkets. This monoculture can be vulnerable if a disease or pest takes hold.

Nitrogen fertiliser is added to crops to boost their yield. The amount of fertiliser required depends on the soil type and how regularly the field is cultivated. The table below shows the amount of nitrogen fertiliser recommended for winter wheat (kg N per hectare).

Questions

Soil type	Previous cultivation and fertiliser application	
	Little	Regular
Sandy soil	160	0-40
Clay soil	250	60
Silty soil	220	40

(Based on the Defra fertiliser manual 2010.)

1. What are the advantages of adding extra fertiliser to the field?

2. Which soil type would need the most nitrogen fertiliser to grow winter wheat?

3. Which soil type would need the least amount of added fertiliser?

4. For clay soil, how many times less fertiliser is needed after regular cultivation?

Growing wheat

A tractor ploughing a field.

Soil preparation

Plants need water but can get mouldy if it is too wet, so wheat grows best in well-drained soil. Plants are also sensitive to pH. Farmed soils tend to become acidic. This can release toxic metals that stunt the plants' growth. Adding lime neutralises the acidity, changing the pH. It is easier to use machinery in large, level fields. The previous year's stubble can be ploughed in or left to bind the soil and reduce erosion.

Sowing

Wheat is sown in spring or autumn when the ground is firmer for machinery. A seed drill plants the grains under the surface – the distance between each seed is chosen to suit the conditions. Commercial seed varieties have been bred to give high yields in particular conditions and resist common plant diseases. Short plants don't fall over. Some countries allow genetically modified seeds to be used.

A modern seed drill sowing wheat.

Pesticides being sprayed onto a growing crop of wheat.

Chemicals

Modern farming uses chemical pesticides to get high yields:
- **Insecticide** kills insects that might eat the plant as it grows.
- **Herbicide** kills weeds that compete for space and nutrients.
- **Fungicide** kills fungi that can infect the crop and make it mouldy.

These chemicals are expensive and can harm helpful insects and other wildlife. Some people worry about traces of pesticides left in food.

Adding **fertilsers** to soil helps the plants to grow but fertilisers can pollute waterways when they are carried off fields in rainwater. Organic farms use naturally occurring fertilisers such as manure. Inorganic chemical fertilisers are made in factories. They are readily available and easy to spread in a measured way. However, they can be expensive and use large amounts of energy in their production.

A combine harvester harvesting wheat in a British field.

Harvesting

Farmers analyse their crops carefully to harvest them at the best time. Wheat is ready when it is about 1m tall and turns from green to gold. A combine harvester cuts and threshes the wheat. Threshing separates the grains of wheat from the chaff and stalks. Straw can be used for animal bedding. The grain collects in a hopper until it can be unloaded into a trailer. On-board electronics allow the operator to monitor and maximise the amount of wheat harvested.

Drying and storage

Wheat grains are separated from any dirt, straw, and chaff. Air is blown over the grains to dry them. This prevents mould growing and destroying the crop. Wheat is stored in a tall building called a grain silo, protecting it from vermin and damp weather. Wheat is sent to the mill to be ground into flour or is used to make animal feed.

Wheat is stored in grain silos.

The stages of wheat production.

Rearing dairy cows

The Holstein is the most popular breed of dairy cow in the UK. It gives the greatest milk yield.

Milk from the Jersey cow is rich in butterfat and protein.

Bull semen is stored in thin tubes in refrigerated tanks until it is needed. The semen is chosen from a pedigree bull whose offspring are known to give high yields of good-quality milk.

There are over two million adult dairy cows in the UK producing over 30 million litres of milk each day. The quality of milk is just as important as the quantity. Milk contains butterfat and protein as well as important sugars and minerals. It is also important that the milk has a low bacterial count.

Dairy cows only produce milk once they have had a calf, and then only for one year. The cows' milk output is measured twice a day, when the milk is collected. The farmer makes sure that each dairy cow continues to produce milk by ensuring that they have a calf once a year. The calves are used to replace dairy stock or for meat.

Animal breeding

Ever since the first domesticated animals, people have improved stock by **selective breeding**. Good-quality animals with some of the desired characteristics are selected and allowed to breed. Gradually breeds develop with the right combinations of characteristics. Dairy cows are different to those used to produce beef. They have large udders and little muscle.

If the farmer wants calves that will produce a high yield of milk, a bull that has come from a good dairy breed will be allowed to mate with the cows. A bull from a good beef breed would be used to give calves that are better for meat.

Artificial insemination

Many farms do not keep a bull. **Artificial insemination** (AI) is now frequently used for cattle instead of natural mating.

Semen from a good-quality bull is collected into a device that provides a stimulus to the bull and keeps the collecting vessel at the right temperature. The **sperm** is checked for quality and measured into disposable plastic straws. The straws are frozen and stored in liquid nitrogen.

The farmer will call the vet if she is worried about a cow's health. A veterinary service monitors the health of all farm animals in the UK and reports problems to government departments. All cattle have a permanent tag fixed to their ear to identify them. Farmers must inform the government database of all births, deaths, and movements. This helps enforcement officers to trace the sources of problems or illnesses.

Dairy cow breeds

Different breeds of cow are suited to different conditions. In the UK Holstein cattle are by far the most popular for milk production.

The cow must be inseminated at the correct time in its reproduction cycle. As the developing egg matures in the ovary, **hormones** are released that make the cow come into oestrus (on heat). At this time a long pipette is used to deposit the contents of the straw beyond the cervix. Implants or injections of reproductive hormones can be used to adjust the natural cycle. This means that many cows can be inseminated at the same time. The advantages of AI include:

- sperm from high-quality, tested bulls is widely available
- there are fewer problems of sexually transmitted diseases
- it is usually cheaper than keeping a bull
- sperm from different bulls can be chosen to suit different breeding needs.

Animal welfare

Like all animals, cows need access to food and water, warmth and shelter, and to be protected from disease. In the UK there are regulations that apply to farm animals about inspections, record keeping, freedom of movement, buildings and equipment, and feeding and watering. These regulations are to protect the welfare of farm animals.

Female calves (heifers) are separated from their mothers at birth and fed on a milk substitute. Later, they are given clean water and they eat grass. They regurgitate the grass with digestive enzymes from their stomachs to chew again before finally swallowing. This is called 'chewing the cud'. In winter when there is less grass available they eat silage and food supplements.

Healthy cattle can tolerate a wide range of temperatures. However, they do need shelter at times of above or below average temperatures, or adverse weather. This not only improves the welfare of the animals, but also helps to maintain productivity.

The farmer will call the vet if she is worried about a cow's health. A veterinary service monitors the health of all farm animals in the UK and reports problems to government departments. All cattle have a permanent tag fixed to their ear to identify them. Farmers must inform the government database of all births, deaths, and movements. This helps enforcement officers to trace the sources of problems or illnesses.

Although a cow could live for 20 years they are often only kept for five years to maximise milk production. Their meat is of a lower quality than beef cattle, so it is used to make processed meats like burgers.

Cows can be kept indoors in winter when there is little fresh grass available. They are fed silage – grass that has been fermented during storage in plastic-wrapped bales, or in a plastic-covered air-free pit.

THE SCIENCE

Dairy cows need shelter, food, and clean water. An automatic cow back-scratcher makes life more comfortable too.

Questions

1 Name two food products that are made from milk.
2 What factors influence how much milk a farmer can get from her dairy cows?
3 Why is the quality of milk as important as the quantity for farmers?
4 Describe the conditions that cattle need to stay healthy and productive.

Key words
- ✓ **selective breeding**
- ✓ **artifical insemination**
- ✓ **sperm**
- ✓ **hormone**

Yoghurt production

Small-scale yoghurt production.

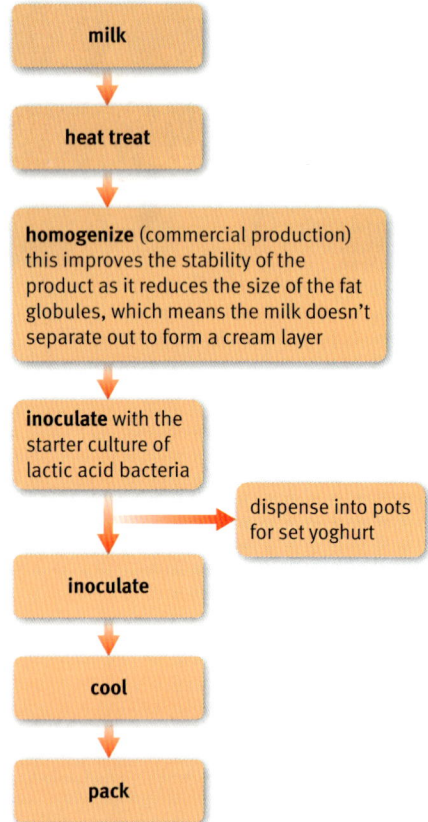

milk

↓

heat treat

↓

homogenize (commercial production) this improves the stability of the product as it reduces the size of the fat globules, which means the milk doesn't separate out to form a cream layer

↓

inoculate with the starter culture of lactic acid bacteria

→ dispense into pots for set yoghurt

↓

inoculate

↓

cool

↓

pack

Flow diagram showing the stages in the large-scale production of yoghurt.

Small-scale yoghurt production

Yoghurt has been around for thousands of years. It was first made in Mesopotamia, the present-day country of Iraq. Stored goats' and sheep's milk fermented naturally in the warm summer climate. The mixture was then hung in animal skins to cool. A soft curd called 'jugurt' was formed. This is where the word yoghurt comes from.

In modern commercial yoghurt-making, milk is inoculated with a starter culture containing two different types of 'lactic acid bacteria' called *Streptococcus thermophilus* and *Lactobacillus bulgaricus*.

First the milk is heated to a very high temperature of 85–95 °C for 15–30 minutes. This kills off any unwanted microbes that may be present. Then the milk is cooled and the mixture of lactic acid bacteria is added.

The mixture is incubated at 38–44 °C for 12 hours to allow the two microbes to grow. As they grow they use lactose (the milk sugar) as an energy source and produce lactic acid. Initially *Streptococcus thermophilus* ferments the lactose; as the level of acid accumulates it is suppressed. *Lactobacillus bulgaricus* is more acid tolerant so it continues to ferment the remaining lactose.

During this process the pH drops from 6.5 to around 4.5. This slows the growth of spoilage microbes so the yoghurt keeps for several days in the fridge.

Lactic acid causes the structure of the milk protein to change, giving yoghurt its special thickened texture. The lactic acid also gives the yoghurt its sharp taste. Other fermentation products such as acetaldehyde give the yoghurt its characteristic smell.

Large-scale yoghurt production

When yoghurt is produced on a commercial scale the solid content of the milk is increased to improve the final texture of the product. Adding milk powder usually does this. Small amounts of natural or modified gums, which bind to water and thicken the product, may also be added. For set yoghurt the inoculated milk is poured into retail pots before incubation.

Questions

1 Why is a starter culture used in yoghurt production?

2 Name two conditions that need to be controlled to keep yoghurt fresh. What happens if these conditions aren't controlled?

Choosing the right organisms

Food producers

For a long time people have taken advantage of natural fermentation processes to make drinks, bread, dairy products, and pickles. Microorganisms are added to food and kept in the right conditions. As the microorganisms grow, their by-products create the desired product. It is very important that the right microorganisms are used and that the process is carefully controlled. The two main types of microorganism used to make fermented foods are lactic acid bacteria and yeast.

Bacteria and fungi (predominantly moulds) are also used in industrial fermentation systems to make enzymes used in processing food. Large amounts of enzymes are produced by a mould called *Aspergillus*.

Aspergillus mould on some nectarines.

Vitamins

Vitamin B12 is produced by *Pseudomonas* bacterium. Vitamin C is produced by *Acetobacter* bacterium.

Organic acids

Acetic acid is produced by *Acetobacter* bacterium. Citric acid is produced by *Aspergillus* mould.

Coffee

In coffee production, pulp surrounding the coffee beans can be removed by fermentation using cellulose-digesting enzymes.

Cheese and yoghurt

The manufacture of dairy products such as yoghurt and cheese involves the addition of a starter culture of lactic acid bacteria to milk. This is an example of anaerobic fermentation.

Soy sauce

Soy sauce fermentation is a two-stage process. The first stage involves growing the mould *Aspergillus* on soya beans. The mould releases enzymes to break down the beans to provide it with food. This stage of fermentation is aerobic. The mash that results is rich in amino acids and sugars that can now be used by lactic acid bacteria in further fermentations. The mash is mixed with brine to provide anaerobic conditions, which stops the growth of the mould but allows the growth of the lactic acid bacteria.

Soy sauce is made from fermentation of soya beans.

THE SCIENCE

The fermentation process

Microorganisms are grown commercially in very large vessels called **fermenters**. The process of growing the organism is fermentation.

A fermenter for growing microorganisms. Water can flow through the water jacket to adjust the temperature. Excessive foam can be controlled by adding an anti-foaming agent. Nutrients, pH, aeration, and stirring paddles can be adjusted to suit the organisms being grown.

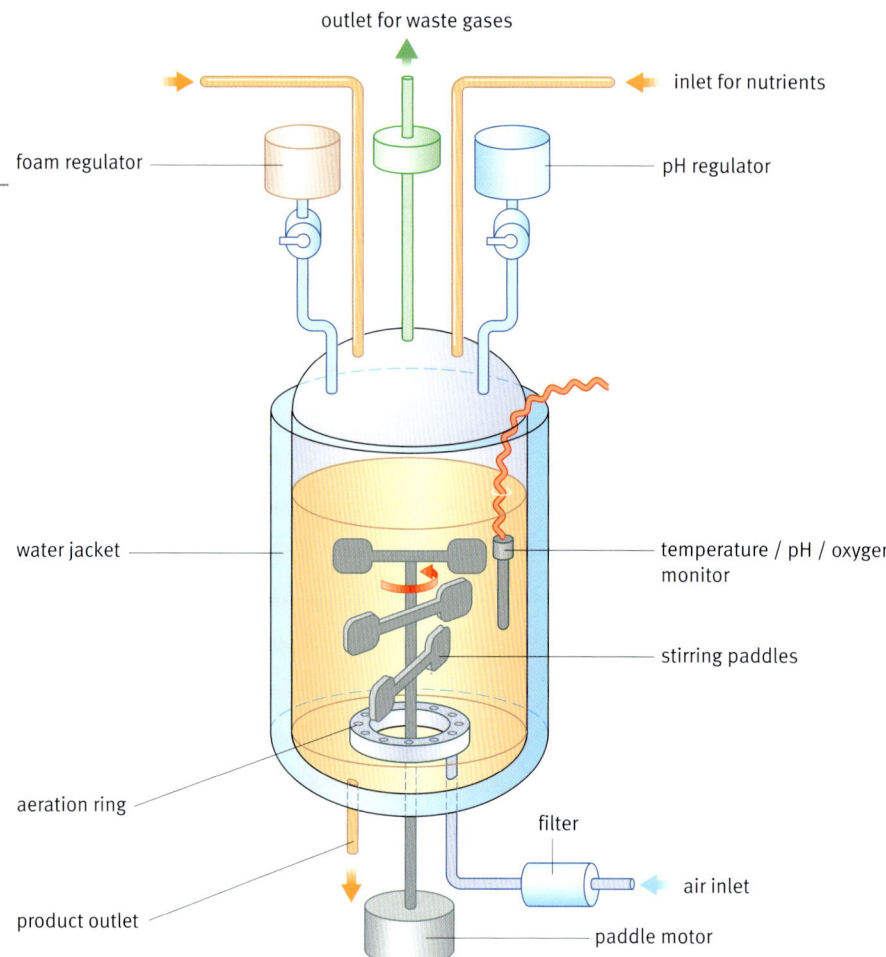

outlet for waste gases

inlet for nutrients

foam regulator

pH regulator

water jacket

temperature / pH / oxygen monitor

stirring paddles

aeration ring

filter

air inlet

product outlet

paddle motor

Requirements for growth

Microorganisms need a source of energy and raw materials for growth. The organism being grown may have special requirements. In wine fermentation, the grape juice provides all the nutrients needed by the yeast, but extra sugar can be added as a source of energy.

Environmental conditions

Conditions are chosen to give the best growth and best yield of product. Most organisms have a preferred temperature and pH. Oxygen concentration is also important; some fermentations need the presence of oxygen while others only produce the right end-product if air is excluded.

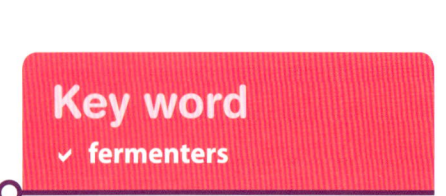

Key word
✓ **fermenters**

THE SCIENCE

Aerobic respiration

Many microorganisms can only live if oxygen is present. They obtain their energy by **aerobic respiration**.

glucose + oxygen $\longrightarrow$ carbon dioxide + water + *energy*

$C_6H_{12}O_6$ + $6O_2$ $\longrightarrow$ $6CO_2$ + $6H_2O$ + *energy*

Anaerobic respiration

Some microorganisms can live with or without oxygen. Others can *only* survive in conditions without oxygen. In these cases energy is obtained directly from glucose, but other end-products vary. The **anaerobic respiration** of yeast produces alcohol. This is also known as alcoholic fermentation. The alcohol produced in this way can be used for fuel (bioethanol) or drinks.

glucose $\longrightarrow$ alcohol + carbon dioxide + (less) *energy*

$C_6H_{12}O_6$ $\longrightarrow$ $2C_2H_5OH$ + $2CO_2$ + (less) *energy*

In yoghurt production, the anaerobic respiration of *lactobacilli* bacteria produces lactic acid.

glucose $\longrightarrow$ lactic acid + *(less) energy*

$C_6H_{12}O_6$ $\longrightarrow$ $2C_3H_6O_3$ + *(less) energy*

Measuring growth

Bacteria and yeast 'grow' by increasing in number. Growth measurements are made by measuring the number or the activity of the organisms. Moulds grow by producing hyphae and mycelium.

Aseptic technique

When a particular organism is being grown, whatever the scale of the process, it is often important to exclude other organisms. Because spores and microorganisms are so widespread, equipment has to be sterilised and special procedures used to prevent contamination. Working in this way is known as using the **aseptic technique**.

Measuring living organisms

A **colony count** measures organisms by growing diluted samples on a medium of nutrient agar. Each living organism grows into a small colony that can be seen and counted.

It is also possible to estimate the number of living organisms by using a chemical test to measure some activity of the organisms.

Measuring total organisms (living and dead)

Biomass is a measure of all the organisms present when they have been separated from the growing medium. The cloudiness of a liquid can also indicate how many organisms are present. A cloudy liquid is **turbid**.

A colony count. The diluted sample contained living organisms. They grew into a colony on a nutrient agar plate and can be counted.

Mould growing on bread.

THE SCIENCE

Questions

1 What do microorganisms use as a source of energy for growth?

2 List the ways that aerobic and anaerobic respiration in yeast are:
   a similar   b different

3 Explain why it is often important to use aseptic technique when growing microorganisms.

Key words

✓ aerobic respiration
✓ anaerobic respiration
✓ colony count
✓ biomass
✓ turbid
✓ aseptic technique

'Brewers only make wort. Yeast makes the beer.'

Microorganisms break down complex molecules in food to provide the energy for their growth. The by-products of fermentation can give foods a good taste, texture, and smell. They can also cause changes that inhibit the growth of undesirable food microbes, improving food's storage life and safety.

Nowadays fermentations are used to make a wide range of food and drink. Microbes are involved in the production of foods in other ways too, and without their help our diet would be very dull.

Fermented foods

For many years fermentation was a hit-or-miss process that relied on naturally occurring microbes getting into our food. For example, yeasts in the air would land in fruit juice and ferment the natural sugars in it, producing wine.

Microbes and fermentation are now better understood and each process is carefully controlled from start to finish. Special cultures of microbes are used in the fermentations to give the product the required properties.

The process of fermentation can be either anaerobic (without oxygen), aerobic (with oxygen), or in many cases where a combination of microbes are involved in the process there can be both aerobic and anaerobic stages.

Mycoprotein fermentation

Mycoprotein is a protein-rich meat substitute made from a fungus that can be grown by fermentation. A very large fermenter is set up containing nutrients and a pure culture of the fungus. The process is aerobic, so air is pumped steadily through the mixture in the fermenter. This aeration system also stirs and mixes the liquid.

The whole system is designed to avoid contamination with other microorganisms. Automatic controls maintain the right environmental conditions. After a time, some of the contents are removed and more nutrients are added.

This type of **continuous** culture is often used to grow and harvest a microorganism itself. It requires a complicated control system and the fermenters are expensive to set up. However, the fermenters do not have to be emptied, cleaned, and refilled very often.

This is different from a **batch** culture where microorganisms grow in an enclosed vessel, until they have used up the available nutrients. Then the vessel is emptied to harvest the products, and cleaned thoroughly before using again.

Growth curve

Bacteria

Bacteria can be grown in an enclosed vessel (a batch culture). The culture must be in a medium containing a supply of necessary nutrients, at the optimum temperature and aerated to supply oxygen, if required. As they grow the organisms will use the nutrients in the medium, secrete metabolites, and excrete wastes. This will change the composition of the culture medium, which in turn will affect bacterial growth. During fermentation, the bacterial population changes, as described below.

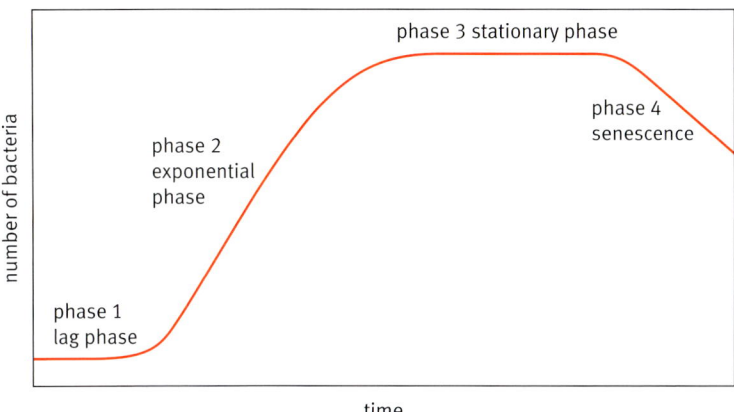

Phases of microbial growth.

Phase 1 – Lag phase

Immediately after inoculation, the bacteria are metabolically active, but do not reproduce: time is required to synthesise the enzymes necessary for growth. The length of lag time can depend upon factors such as the type of culture the inoculums has come from and the medium in which it was grown.

Phase 2 – Exponential or log phase

During this phase, cells reproduce at their fastest. Their generation time is short (eg, cells of *E. coli* divide once every 21 minutes in optimum conditions) and the number of cells increases exponentially.

Phase 3 – Stationary phase

The culture medium changes due to the growth and accumulation of cells: pH falls due to the excretion of acids, and carbon dioxide and nutrients are used up and can start to become limiting. This causes some cells to die, but the number of cells remains constant because the growth rate and the death rate are the same.

Phase 4 – Senescence (death phase)

During this phase, more bacteria die than are produced. The number of living bacterial cells decreases exponentially.

Questions

1   Make a chart or table to compare the difference between batch and continuous culture systems.

2   How could Phase 2 of bacterial growth be extended?

Key words

✔ **mycoprotein**
✔ **continuous culture**
✔ **batch culture**
✔ **lag phase**
✔ **exponential phase**
✔ **stationary phase**
✔ **senescence**

THE SCIENCE

Controlling bioreactors

Instrumentation systems

When running large industrial processes it is important to know what is going on. Scientists and engineers use instrumentation systems to monitor what is happening.

First the operator must set up the system to achieve the desired results. For example, they might set the temperature and the pH. These target settings are the conditions needed for the system to work best.

Sensors are used to provide data. Sensors in a bioreactor provide data on the temperature and pH within the system. This data is displayed so that action can be taken by the people operating the system, if conditions change from the targets.

Automatic control systems are a development of instrumentation systems, which do not require people to make adjustments. Automatic control systems monitor conditions and make changes to control the conditions.

System diagrams

Instrumentation systems can be complex so diagrams are used to show how they operate. Both instrumentation and automatic control systems can be represented by system flow diagrams.

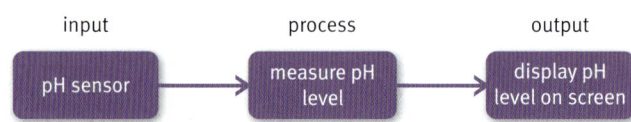

System flow diagram.

In system flow diagrams it is the function, not how it is achieved, that is important. System flow diagrams show:
- **inputs**, such as sensors or switches
- the **process** of the system
- **outputs**, such as a graphical display, numerical display, or alarm.

Instrumentation system.

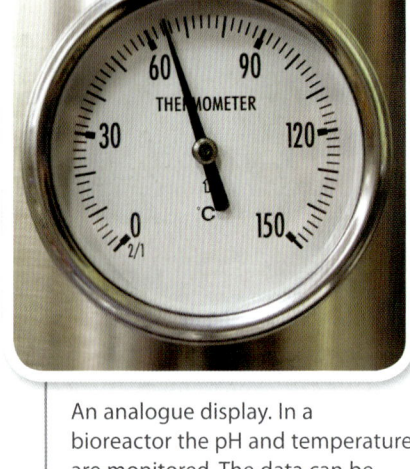

An analogue display. In a bioreactor the pH and temperature are monitored. The data can be displayed in analogue, digital, or graphical forms.

Data loggers record changes over a period of time. They often have graphical displays. This shows what is happening over a period of time.

Feedback for automatic control

Many instrumentation systems use **comparators**. A comparator compares two inputs, for example, the reading on a temperature sensor and the target temperature, to see if they are the same or if one is higher. The output of the comparator is a signal indicating the state of its inputs.

If the difference between the sensor reading and the target setting is too large the comparator changes its output signal. This is used to set alarms for high and low values.

Feedback loops

Feedback loops are used to control bioreactors. They use data from sensors to make adjustments to the operation of the bioreactor. The feedback system is used to ensure that the temperature and pH remain at the required levels. The output from the sensor goes to a comparator, which in turn controls a heater or a valve.

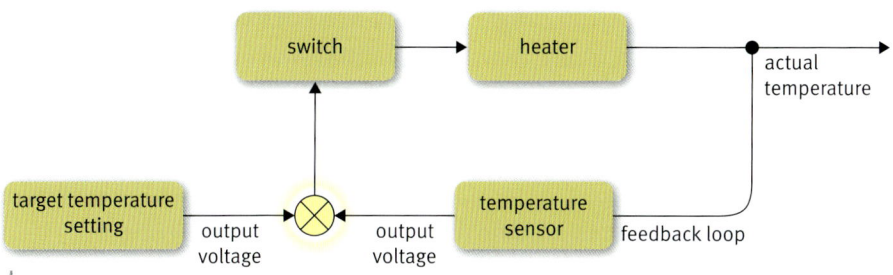

Process control diagram for temperature.

If the temperature in the bioreactor falls too low the heater is switched on. The temperature goes up. When the temperature reaches the required value, the heater is switched off. This type of control is called **negative feedback**; any change in the system results in an action that reverses the change.

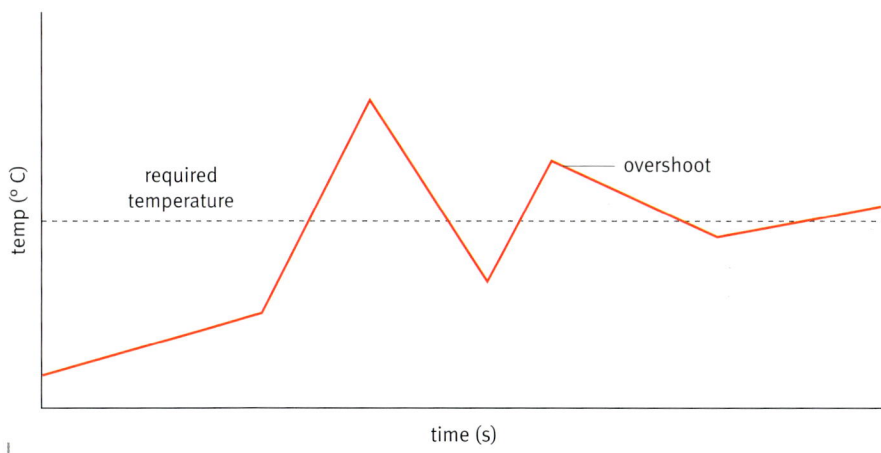

Temperature control graph.

Genetic modification

The food industry uses products from **genetically modified** (GM) organisms. GM microorganisms can produce food ingredients and enzymes used in food processing.

Microorganisms used in industrial processes are carefully selected from the wide range of naturally occurring organisms. However, it is now possible to modify the **genes** of a microorganism to produce a new organism for a particular purpose.

Useful products

The rennet traditionally used to set milk in cheese-making is an extract from the stomach lining of young calves. Its activity is due to an enzyme, **chymosin**. There are not enough calves being killed to supply all the rennet needed by cheese-makers.

Enzymes are **protein** molecules with very particular biological activities. Other natural sources of enzymes that will clot milk have been found (in bacteria and fungi) but they are not exactly the same as calf chymosin. They give slightly different results when used to make cheese.

A yeast that can provide a supply of chymosin has been produced by genetic modification. The GM yeast produces chymosin that is identical to calf chymosin. Chymosin made in this way has been thoroughly tested and approved for use in cheese-making. It has been found to give more consistent results than rennet extracts. The final product does not contain any GM yeast. Yoghurt can also be produced using GM microorganisms.

The Health and Safety Executive regulates these 'indoor' or contained organisms. Stringent laboratory regulations mean that they are unlikely to threaten the environment.

Key words

- chymosin
- genes
- enzyme
- protein
- genetic modification
- DNA

Questions

1 Suggest why some people prefer cheese made with chymosin from the GM yeast, while others are against the use of GM yeast.

2 Food containing GM organisms has to be labelled. Why does cheese made with chymosin from the GM yeast not have to be specially labelled?

3 Explain why the chymosin produced by the GM yeast is identical to calf chymosin.

1 The chymosin gene

The genetic material in all cells is **DNA**. The sequence of chemical units along the DNA carries coded information. The DNA is organised into thousands of sections called genes. One gene carries the coded information to make one protein. In cattle, one of these genes is the chymosin gene. When the gene is activated in calf stomach cells, the protein chymosin is produced.

2 Copying the gene

Special enzymes were used to find and make exact copies of the DNA sequence of the chymosin gene in cattle.

3 Inserting the gene

A suitable yeast was chosen to receive the gene. Different enzymes were used to insert the chymosin gene into the DNA of the yeast. This gives the yeast the coded information to make the chymosin protein.

4 Screening the GM organism

Many safety and suitability tests were carried out on the GM yeast.

5 The fermentation process

The modified yeast can now be grown in an industrial fermentation process. Chymosin is produced by the growing yeast.

6 Purification

Chymosin is separated from the fermentation mixture and purified. It is now ready to be supplied to cheese-makers.

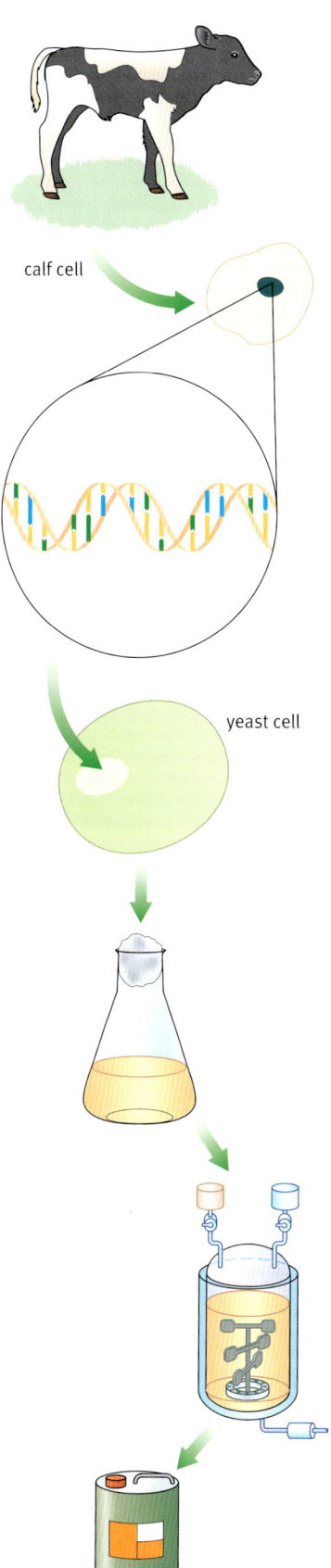

calf cell

yeast cell

How a yeast is made to produce chymosin.

Some microorganisms cause illness when people eat food in which the microorganisms are growing. The microorganisms feed on the foodstuff and produce waste products that make people ill, or simply make the food unpleasant to eat. Most foodborne illnesses are caused by bacteria. Food that is incorrectly produced, stored, or cooked may contain a large number of the sort of bacteria that cause food poisoning.

The food-safety laws are intended to safeguard against poor handling and storage of food at every stage of production, transport, and sale.

Problems with microorganisms

Bacteria are single-celled microbes that can reproduce by splitting in two – often very rapidly. In the right conditions of warmth, acidity, and moisture they can divide every 20 minutes, producing millions of cells in a few hours.

Some bacteria form spores that are resistant to heat and drying. They can survive cooking and will start to grow again.

Food-poisoning bacteria cause illness in different ways:
- They grow in the food and produce a toxin. Toxins can survive cooking even though the bacteria that produced them are killed.
- They are eaten with food where they multiply and they may also produce toxins.

Fungi can also produce mycotoxins that do not cause food-poisoning symptoms but are still dangerous in our food. Aflatoxins are produced by moulds that grow on nuts and cereals that have been grown in damp conditions. Many aflatoxins are carcinogenic so foods are tested to make sure aflatoxin levels are safe.

Recognising food spoilage

Food becomes slimy when there are so many bacteria that they touch one another.

Foods may go sour when the microorganisms produce acids, for example, sour milk. Some meat products become sour if packaged in certain types of plastic.

Food may become discoloured from microbiological growth. Some moulds have coloured spores, for example, black mould on bread, or blue and green mould on citrus fruit. Sometimes meat becomes green due to the growth of microorganisms.

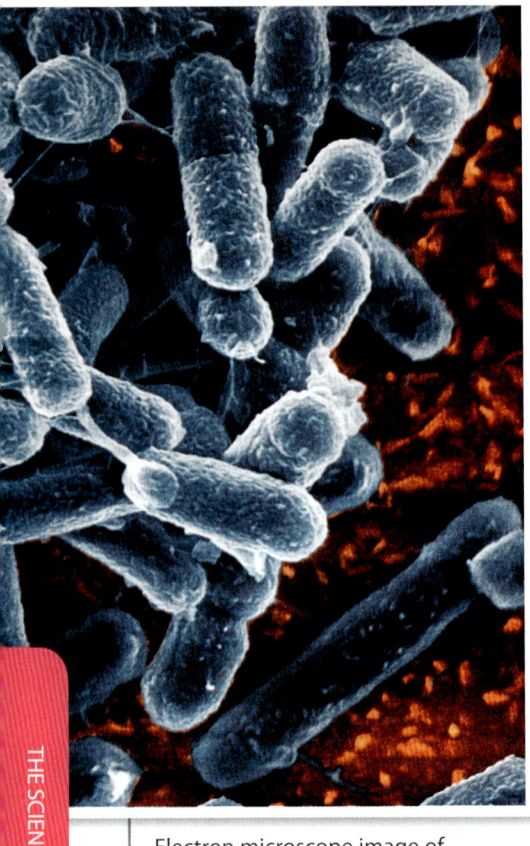

Electron microscope image of bacteria that can cause food poisoning, magnified 35 000 times.

THE SCIENCE

Questions

1 Give two examples of ways in which microorganisms can be:
   a useful
   b harmful.

2 Give three examples of food made with the help of microorganisms.

3 Describe one method of stopping food 'going off'. Explain why the method works.

Bacteria often produces gas as a by-product, which affects food. This means that:

- meat becomes spongy in texture
- packages, cans, or vacuum packs swell
- rotten smells develop from the breakdown of food by bacteria.

Preserving food

There is a range of methods of preventing microorganisms spoiling food by providing conditions that slow down their growth.

The various methods of food preservation aim to prevent or delay microbial and other forms of food spoilage, and to guard against food poisoning.

These methods therefore retain nutritional value of food products, extend their shelf-life, and keep them safe for human consumption. Preservation techniques include refrigeration, packaging, acidity, chemicals such as nitrite or metabisulfite, and heat treatment.

Sterilisation prevents spoilage by killing microorganisms, but packaging is needed to prevent recontamination.

Fungi growing on crops or on stored food can produce aflatoxins. Health authorities in ports and local authorities test food samples regularly to check that levels of aflatoxins are not above the legal limits. Independent agencies also carry out surveys to monitor levels of aflatoxins in foods.

Testing the freezing point of milk

Principle

Milk and water have different freezing points. Milk freezes below 0 °C. If milk contains added water, it raises its freezing point nearer to 0 °C. The freezing point of milk is compared with that of pure water and expressed as the freezing-point depression (FDP). A low FDP indicates that added water is present.

Procedure

1   Freeze a sample of milk containing a thermometer or temperature probe.

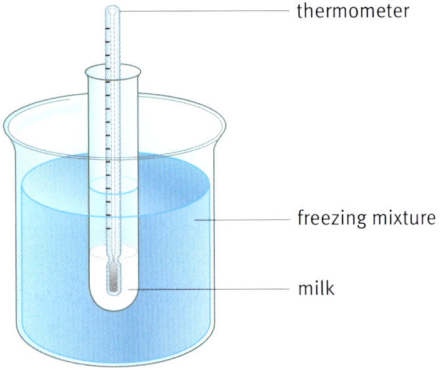

2   Remove the sample from the freezing mixture and record the temperature at regular intervals as the milk thaws.

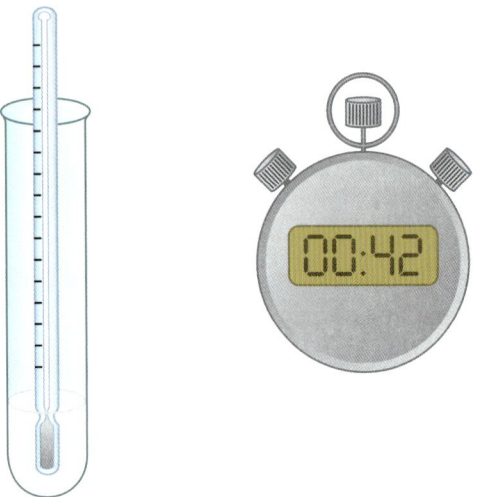

3   Plot a graph of temperature against time.

4   Read the freezing point from the graph. The graph flattens out at the freezing point.

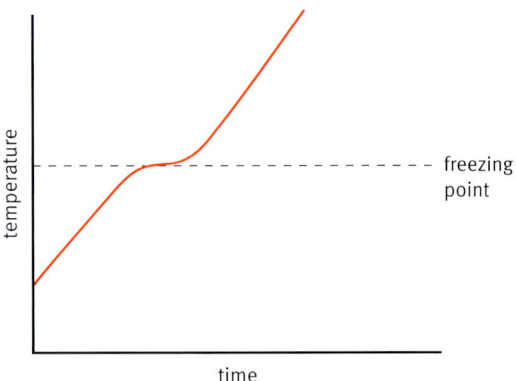

5   Repeat the procedure with pure water instead of milk.

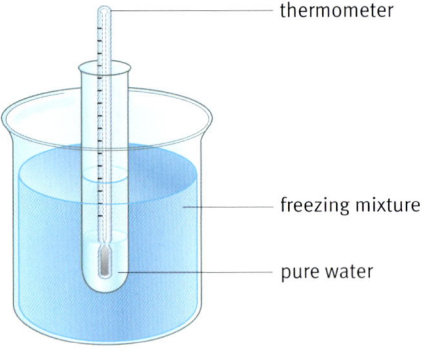

6   Calculate the FDP of the milk sample. An example is shown below:

The same thermometer was used to take the following measurements:
Freezing point of a milk sample: = −0.53 °C
Freezing point of a pure water sample: = 0.01 °C
The freezing-point depression (FDP) is the difference between the two readings:
FDP = freezing point of water sample − freezing point of milk sample
FDP = 0.01 − (−0.53) = 0.54 °C

Testing milk quality

Principle

Milk is a good source of food for microbes because it contains water, protein, sugar, fat, and vitamins. Pasteurisation kills most but not all of the pathogenic and spoilage microbes as well as much of the natural milk microbes. Some microbes survive this process and can cause the milk to 'go off'. The number of microbes in the milk is a measure of its quality. This can be estimated by finding the rate of change of colour of **resazurin** dye.

A solution of resazurin dye changes from blue to pink to clear depending on how little oxygen is present. In fresh clean milk, with few bacteria, it will change colour slowly. In contaminated milk, bacteria use up the oxygen and the colour change is fast.

Use sterilised equipment.

CHECK SAFETY
Never work unsupervised

CAUTION: Precautions should be taken when incubating samples. Work must be supervised by someone with suitable training.

Procedure

1 Prepare fresh resazurin solution.

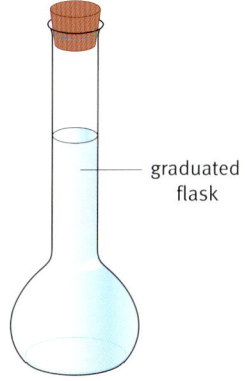

graduated flask

2 Pipette 1 cm³ of resazurin solution into a labelled test tube.

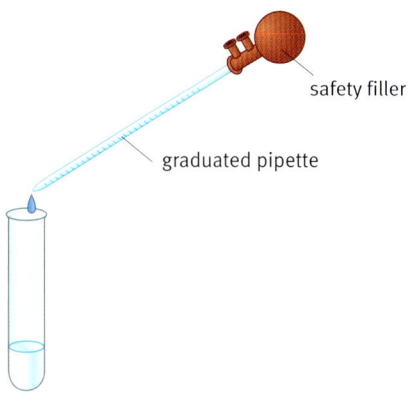

safety filler

graduated pipette

3 Add 10 cm³ of milk. Insert a bung and turn the tube upside down three times to mix.

Milk

x3

4 Place in a water bath at 36 °C for 1 hour.

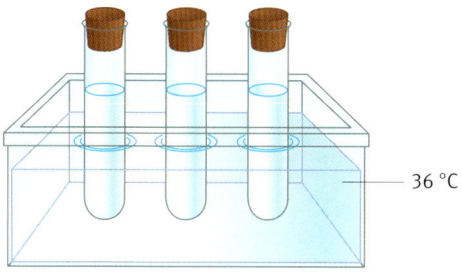

36 °C

5 Classify the sample according to its colour. Do not open the tube.

Colour of sample	Quality
blue (no change)	excellent
blue to mauve	good
mauve to pink	fair
pink to very pale pink	poor
white	bad

Growing yeast

Principle

Yeasts will grow if they are provided with nutrients and the right conditions. Yeast cells bud and form new cells, so the number of individual cells increases.

Baker's yeast can easily be grown for a short time in a flask of liquid containing suitable nutrients that is kept in a warm place.

To get the best growth rate the yeast must be provided with its optimum conditions. For baker's yeast, this is aerobic conditions at a temperature of 25–30 °C. These conditions can be provided in the laboratory using a special bioreactor.

A bioreactor containing a liquid glucose nutrient medium is sterilised to destroy any microorganisms. A pure culture of baker's yeast is added. Filtered air is pumped through the system, a stirrer keeps the system mixed, and a heater maintains the right temperature.

Procedure to show yeast growth

1 Put 200 cm³ of glucose nutrient broth into a sterilised flask. Seal the flask with sterile cotton wool.

2 Add 2 g of dried yeast to the flask. Replace the cotton wool. Gently swirl the mixture.

3 Fit a rubber bung with an airlock to the flask. Half fill the airlock with sterile water.

4 Leave the flask in a warm place. Observe the liquid in the flask over a few days. Do not open the flask.

Procedure to set up a laboratory bioreactor

1   Put glucose nutrient broth and a magnetic flea into a bioreactor. Sterilise the whole apparatus (by heating to 121 °C for 20 minutes). NOTE: This step will be done for students.

2   Add a pure yeast culture through the inlet tube.

3   Start the magnetic stirrer and the heating block. Attach the air inlet filter to a pump.

4   If the bioreactor has probes fitted (pH, temperature), attach them to a data recorder.

5   Observe the fermentation over a period of a few days. If suitable precautions are taken, samples can be removed for testing from the sampling port.

CHECK SAFETY
Never work unsupervised

CAUTION: Precautions should be taken when culturing microorganisms. Work must be supervised by someone with suitable training.

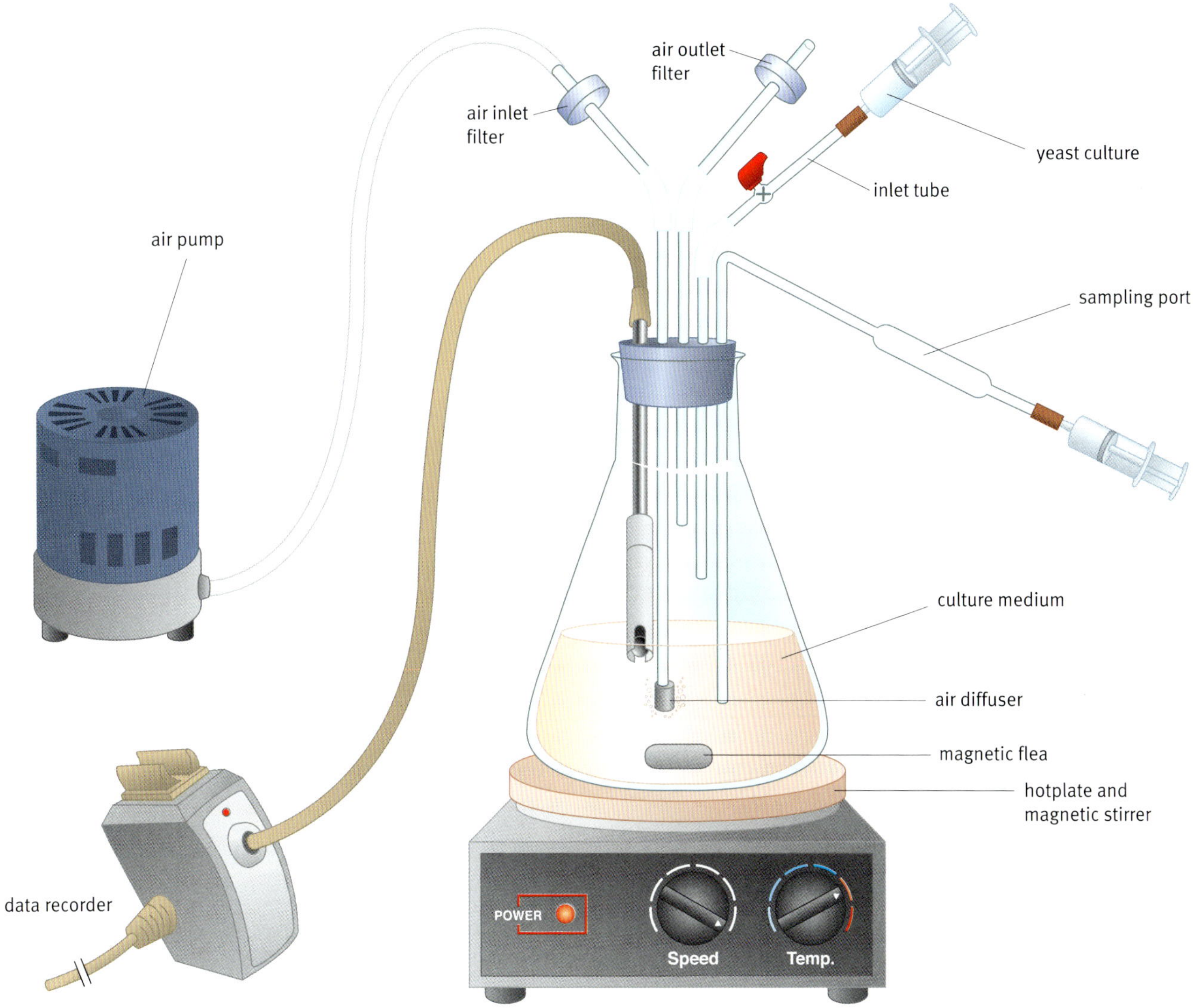

air outlet filter

air inlet filter

yeast culture

inlet tube

air pump

sampling port

culture medium

air diffuser

magnetic flea

hotplate and magnetic stirrer

data recorder

POWER

Speed   Temp.

A laboratory-scale fermenter for growing yeast.

Preserving food

Principle

Most of the food we buy contains small numbers of microbes. These are usually harmless and do not spoil the food. If they multiply, they may cause the food to 'go off'. Food preservation methods aim to stop or slow down this process.

Procedure

1 Label eight test tubes A to H. Write your name and the date on each one. Using the forceps, put three peas in each tube.

2 Add nothing to tubes A and B. Partly fill tubes C–H as follows:

Tube	Content
A	none
B	none
C	distilled water
D	dilute salt solution
E	concentrated salt solution
F	sugar solution
G	vinegar
H	sodium nitrite solution

3 Plug each tube with cotton wool. Tube A will be kept in a refrigerator until the next lesson. Tubes B–H will be incubated at room temperature.

Next lesson…

4 Examine your tubes and record the appearance of the peas and the solutions.

5 Do your results confirm your predictions? If not, suggest reasons for any differences.

From your results, describe the effect of temperature on the growth of microbes.

Reproduced from Practical Microbiology for Secondary Schools, published by the Society for Microbiology, 2008.

CHECK SAFETY
Never work unsupervised

DO NOT OPEN THE TUBES

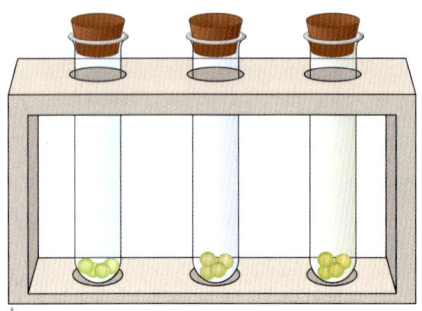

Test tubes containing peas under different preservation conditions.

PROCEDURES & TECHNIQUES

Flaming a wire loop

Principle

Wire loops are sterilised using red-hot heat in a Bunsen flame before and after use. They must be heated to red hot to make sure that any contaminating bacterial spores are destroyed. The handle of the wire loop is held close to the top, as you would a pen, at an angle that is almost vertical. This leaves the little finger free to take hold of the cotton wool plug/screw cap of a test tube/bottle.

The flaming procedure is designed to heat the end of the loop gradually because after use it will contain culture, which may 'splutter' on rapid heating with the possibility of releasing small particles of culture and aerosol formation.

Procedure for flaming a loop

1 Position the handle end of the wire in the light-blue cone of the flame. This is the cool area of the flame.

2 Draw the rest of the wire slowly upwards into the hottest region of the flame, (immediately above the light-blue cone).

3 Hold there until it is red hot.

4 Ensure the full length of the wire receives adequate heating.

5 Allow to cool then use immediately.

6 Do not put the loop down or wave it around.

7 Re-sterilise the loop immediately after use.

 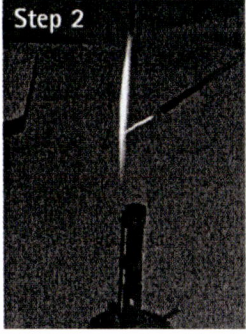  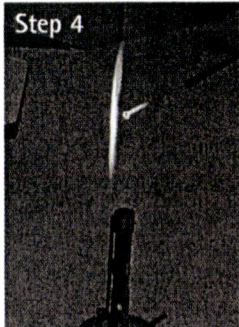 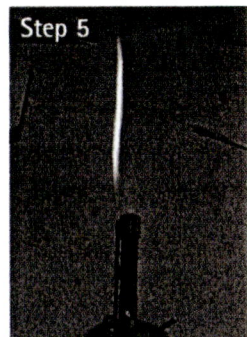

Stages in the procedure for flaming a wire loop.

Reproduced from *Practical Microbiology for Secondary Schools*, published by the Society for Microbiology, 2008.

Culturing microorganisms on agar plates

Principle

Nutrient agar plates can be used to culture microorganisms. The plates are sterilised before samples are put on the surface of the agar. Plates are kept for one to three days at a steady temperature. Microorganisms present multiply and grow into colonies that can be seen.

For a streak plate, a loop is dipped in the sample and then spread across the surface of the plate in streaks. At some point along the streaks individual microorganisms will grow into separate colonies. Streak plates are used to check if a sample is able to grow to form colonies or if it is pure and forms only one kind of colony.

On a colony count plate, single drops from dilute samples are cultured. If one shows colonies that can be counted, the concentration of microorganisms in the original sample can be calculated.

CHECK SAFETY
Never work
unsupervised

CAUTION: Precautions should be taken when culturing microorganisms. Work must be supervised by someone with suitable training.

Procedure for a streak plate

1   Gather together in a clean workspace: a sterile nutrient agar plate, the sample container, a lit Bunsen burner, and a metal spreading loop.

2   Flame the loop (see previous page) and allow it to cool.

3   Dip the loop in the sample.

4   Partially lift the lid of the agar plate and smear the sample backwards and forwards across the first section.

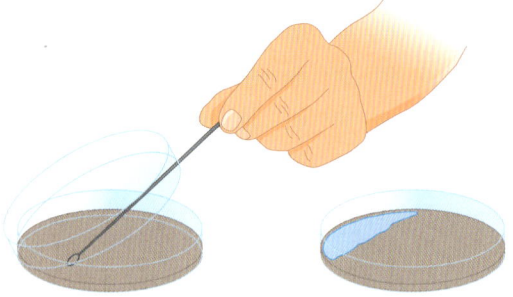

5   Lower the lid. Reflame the loop and allow it to cool. Turn the plate a quarter turn, partially open it and streak three lines across from the last smear.

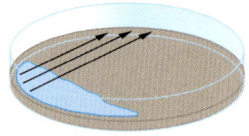

6   Repeat step 5 twice more. Zigzag the last streak to the centre.

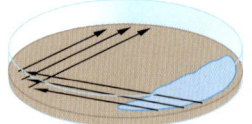 

7   Seal the plate with four pieces of tape. Label the plate clearly. Incubate the plate upside down at 25–30 °C.

8   Examine the plate for growing colonies.

Procedure for a colony count plate

1   Gather together in a clean workspace: a set of sterile containers for the dilute sample, a sterile nutrient agar plate, the sample container, a lit Bunsen burner, a 10 cm^3 syringe, and sterile pipettes.

2   Pipette 5 cm^3 of sample into tube A. Add 5 cm^3 of distilled water to it.

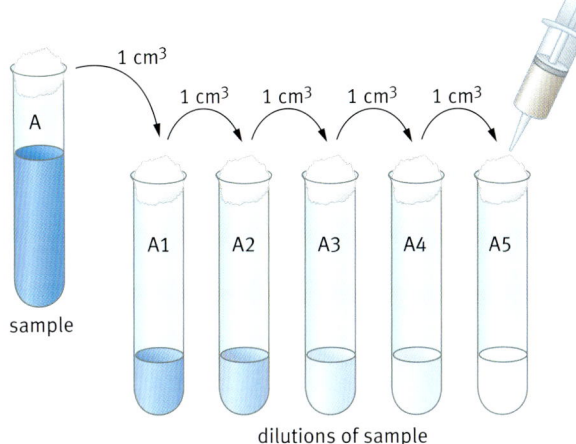

3   Label 5 tubes A1, A2, A3, A4, and A5. Use the 10 cm^3 syringe or pipette to place 9 cm^3 distilled water in each one.

4   Using a 1 cm^3 syringe or pipette, mix the contents of tube A thoroughly by filling and emptying the syringe/pipette several times. Then transfer 1 cm^3 sample from tube A to A1. Mix the solution. Now transfer 1 cm^3 from A1 to A2. Mix thoroughly after each transfer. Repeat to complete the series of dilutions.

5   Label the underneath of an agar plate as shown.

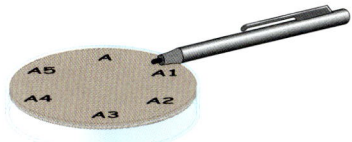

6   Use a pipette of known drop size to place a single drop of each dilute sample in its position. Start with the most dilute sample, A5.

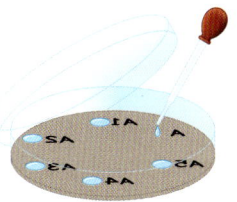

7   Allow the drops to soak into the agar. Seal the plate with four pieces of tape. Label the edge of the plate clearly.

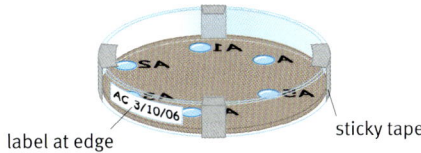

label at edge   sticky tape

8   Incubate the plate upside down at 25–30 °C until next lesson.

9   Examine your agar plate and count the number of colonies visible at A, A1, A2, A3, A4, and A5. Record your results in a table.

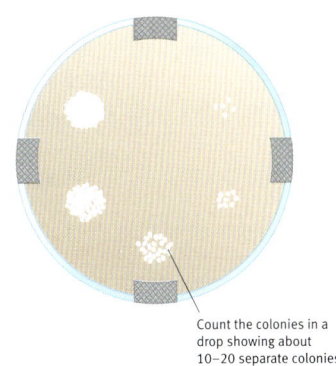

Count the colonies in a drop showing about 10–20 separate colonies.

10   Use the drop size and dilution factor to calculate the concentration of microorganisms in the sample:

If the drop size is 0.02 cm^3
number of drops in 1 cm^3 = $\dfrac{1}{0.02}$ = 50
Assume each colony grew from a single bacterium that was originally present.
If 25 colonies were counted in 1 drop, number of bacteria per cm^3 in the diluted sample = 25 × 50 = 1250
If the dilution for that drop was 1:1000, no. of bacteria cm^3 in original undiluted sample = 1250 × 1000 = 1 250 000

Principle

Good-quality milk is incubated with a starter culture of lactic acid bacteria. This quickly reduces the pH, so that harmful bacteria do not grow. An enzyme is added to coagulate the milk protein. The resulting curds are cut, heat treated, and strained to remove the liquid whey. Salt is added to the curds. Then they are pressed into moulds to remove liquid and form the cheese. The cheese is matured by storage in controlled conditions for several months.

Procedure

1 Warm some milk to 30 °C and add starter culture (about 1% by volume). Keep at this temperature for about 45 minutes until the correct acidity level is reached (0.16–0.18% lactic acid).

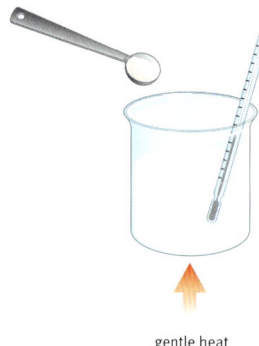

gentle heat

2 Add rennet or vegetarian rennet (about 0.2% by volume). Keep warm for another 30 minutes.

gentle heat

3 When the curd is ready cut it into small cubes.

4 Stir the curds and whey and warm to about 45 °C.

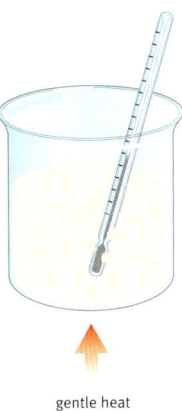

gentle heat

5 Separate the curds from the whey by straining.

6 Break up the curds and add salt.

SALT

7 Pack the curds into a mould. Press under a weight for 24 hours.

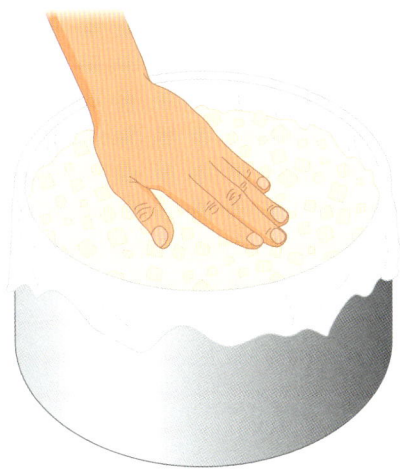

8 Store in cool conditions and allow to mature.

Principle

The ability of seeds to germinate depends on both genetic differences and environmental factors. Seed batches may vary in quality due to their age or the weather conditions when they were harvested. One measure of seed quality is the **germination rate**.

A known number of seeds are given standard conditions for germination. The seeds that germinate are counted. Germination rates are usually worked out as a percentage:

$$\text{germination rate} = \frac{\text{number of seeds that germinated}}{\text{total number of seeds}} \times 100\%$$

Procedure

1   Place a circle of filter paper in a Petri dish. Scatter a known number of seeds evenly over the surface of the paper.

2   Add 15 cm^3 of water. Carefully add water until all the filter paper is moist.

3   Label the Petri dish lid and cover the seeds. Wrap the covered Petri dish in cling film to prevent it drying out. Leave it in a warm room.

4   Examine the seeds regularly. Continue for at least a week after the first signs of germination. Record the total number of seeds that have grown a healthy looking shoot. Calculate the germination rate.

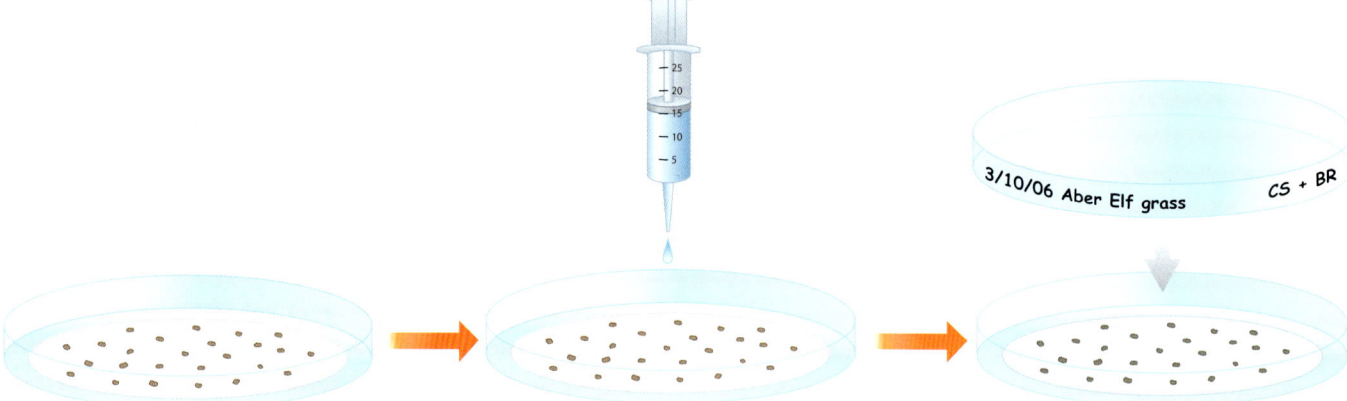

Principle

There are two ways of measuring crop yields:

- Wet mass measures the fresh plant material. This includes a variable amount of water in the plant tissue.
- **Dry mass** measures the plant material after the water has been removed by drying.

Crop yields can be described as harvestable matter. The way this is measured varies with the crop.

- Hay is measured as the dry mass of the whole crop, as it is all used in animal feed.
- Potatoes are measured as the wet mass of the clean tubers.
- Cereal yields are usually measured as the mass of the grain after it has been dried enough to slow the growth of moulds.

These measure quantities, but the quality is also an important factor in the value of a crop.

Procedure for measuring total wet mass

1 Decide on a chosen number of plants, or a chosen area of growing plants. Remove the plants from the ground with as much root as possible.

2 Wash the soil off the roots. Dry off excess water with a tissue.

3 Weigh the plant material. Record the wet mass of plant material.

The wet mass of a plant includes water, so this is not a very reliable measure of growth.

Procedure for measuring total dry mass

1 Carry out steps 1–3 as above.

2 Dry the plant material in an oven at 60 °C for several hours.

3 Weigh the plant material. Replace in the oven and heat again. Weigh again.

4 Record the dry mass as the final mass when heating the plant material again no longer reduces the mass.

The dry mass gives a better indication of growth than wet mass, but the plant is destroyed.

Module Summary

People and organisations

- about the work people do in the food industry, including Environmental Health Officers, food technologists, and inspectors
- that producing food involves people working between the farm or factory and our homes, at every stage of growing, transporting, processing, storing, and delivering food
- that regulations control work in the food industry to make sure animals are cared for, the environment is protected, and there are no threats to public health
- that regulations ensure that industry workers follow health-and-safety guidelines, and that consumers are safe

The science

- the best conditions for producing wheat (soil, water supply, nutrients, and pH) and the features of different varieties of wheat (bread wheat, durum wheat, winter wheat, and spring wheat)
- the stages in wheat production (soil preparation, sowing, use of chemicals, harvesting, drying, and storage) and what is important about each stage
- the risks and benefits of using chemicals such as insecticides, fungicides, herbicides, and fertilisers and the difference between organic and inorganic methods of producing wheat crops
- about the costs of producing wheat and   about the stages in milk processing
- the features of good dairy cattle and how artificial insemination and selective breeding help farmers to improve a herd's productivity
- what affects animal growth – temperature, shelter, food, water, and disease – and how hormones can control the timing of reproduction in animals
- that yeasts, bacteria, and viruses are microorganisms; some cause disease (pathogens) and make food go 'off'
- how a population of microorganism grows
- word and symbol equations for aerobic and anaerobic respiration in yeast and lactobacilli
- how microorganisms are used to help make useful products such as alcohol, enzymes, fermented foods, bread, yeast extract, and mycoprotein
- how fermentation is involved in producing coffee and soy sauce
- that DNA is the genetic material in organisms, that each gene codes for a particular protein, and that genetically modified organisms produce the protein of an introduced gene
- two examples of the use of GMOs to make a useful food product
- how to label flow diagrams describing how bioreactors are monitored and controlled
- how feedback is used to control a bioreactor automatically
- what a test of milk quality tells you
- how to calculate germination rate and crop yields (including dry mass)
- how to interpret information about outbreaks of food poisoning
- how to make calculations of population growth for microorganisms, including bacteria, from colony counts, biomass measurements, and turbidity

Standard procedures

- how to test milk for freshness using Resazurin
- why aseptic techniques are important when cultivating microorganisms
- how fermenters are used to produce large quantities of culture, and the advantages and disadvantages of batch and continuous processes.

Review Questions

1 Otto buys a pint of milk from the shop. Describe the chain of food production between the farm and the shop.

2 Explain the role of factory inspectors in the chain of food production.

3 Pete grows wheat on his farm. He wants to increase the yield. He uses the table below to choose the best fertiliser for his wheat crop.

Fertiliser	Application (kg/ha)	Cost (£/kg)	Yield (tonnes/ha)
growmuch	100	0.60	5.2
wheatmore	200	0.40	4.7
maxyield	400	0.25	6.3

a Explain the meanings of each of the headings in the table.

b Which is the cheapest fertiliser? Justify your answer.

c Suggest why Pete should consider using one of the more expensive fertilisers.

4 Yoghurt is a food made from cows' milk. Describe the production of **another** food made from milk.

5 Making food by fermentation requires microorganisms to respire anaerobically.

a Write down the word equation for the anaerobic respiration of yeast when it converts sugar into alcohol.

b Describe the production of **another** food by fermentation.

c Explain why standards of hygiene are particularly important where foods are produced by fermentation.

6 Genetic modification is an important technology in the food industry.

a Describe two examples of the use of genetic modification in the food industry.

b Explain the role of a gene in the cell of an organism.

c Explain how genetic modification makes an organism with different characteristics.

7 The pH in a bioreactor is monitored with these electronic components.

   alarm   display   processor   sensor

a Describe the function of these components in monitoring the pH.

b Here is part of a chart from a bioreactor monitor.

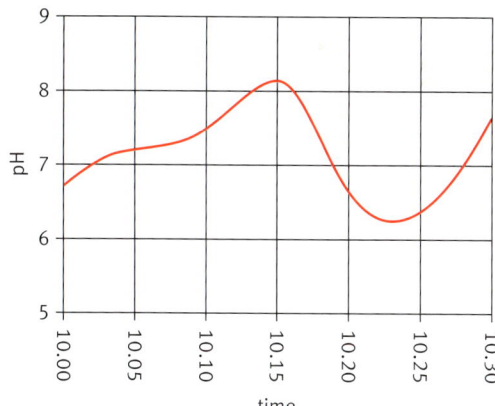

At what time was extra acid added to the bioreactor? Justify your answer.

pasteurisation

selection

breed of cattle

artificial insemination

choices

hormones to control reproduction

milking

processing

UHT

stages

science of milk production

skimming

water

conditions

disease

food

shelter

temperature

health and safety

public health

regulation

protects environment

animal welfare

food technologists

people

Environmental Health Officers

inspectors

farmers

interpreting data

food-poisoning outbreaks

crop yield

dry mass

milk freshness

bioreactor monitors

germination rate

effects of fertilisers

microorganism population

WORKING IN AGRICULTURE, BIOTECHNOLOGY, AND FOOD

science of biotechnology
- fermentation
  - soy sauce
  - coffee
  - bioreactor
    - conditions
    - control
    - feedback
    - flow diagram
  - respiration equations
    - anaerobic
    - aerobic
    - word
    - symbol
- varieties
  - bacteria
  - yeasts
  - moulds
  - viruses
  - pathogens
  - food spoiling
- growth stages
  - lag
  - experimental
  - senescence
- GMOs
  - DNA
  - gene
- products
  - alcohol
  - enzymes, eg, chymosin
  - food
    - bread
    - mycoprotein
    - cheese

standard procedures
- fermenters
  - batch
  - continuous
  - aseptic techniques
- food tests
  - quality
  - safety
  - milk
  - Resazurin

science of growing wheat
- costs
  - labour
  - chemicals
    - fertilisers
    - pesticides
- varieties
  - bread
  - durum
  - winter
  - spring
- conditions
  - water supply
  - soil
  - pH
  - nutrients
- stages
  - soil preparation
  - sowing
  - using chemicals
  - dry
  - store
  - harvest

B4 Making chemical products

Why study how to make chemical products?

Almost all aspects of everyday life are supported by chemical products in one way or another. Some of the chemicals come from natural sources. Some are synthetic. Food manufacturers use chemicals to improve the texture and flavour of what we eat while also giving foodstuffs a longer shelf life. People decorate and protect themselves and their surroundings with the chemicals in cosmetics, paints, and printing inks. Chemicals are also the essential ingredients of drugs and medicines. The UK chemical industry is one of the largest manufacturing industries in the UK.

What you already know

- most chemicals are compounds of two or more elements
- chemists use symbols and equations to describe reactions
- acids neutralise alkalis to make salts
- some compounds are soluble in water, others are insoluble.

Find out about

- how to make pure samples of chemicals from acids
- how to precipitate insoluble chemicals for use as pigments
- how to mix chemicals to make useful products.

The Science

Scientists extract chemicals from natural sources. They make new chemicals by reactions such as the neutralisation of an acid with an alkali. They mix chemicals to create new products, taking care to ensure that what they make is fit for purpose and safe to use.

What's in beauty products?

The search for beauty and health

Men and women have used scented oils and ointments to clean and soften their skin and mask body odour for tens of thousands of years. They have also used dyes and paints to colour the skin and hair.

Some of the old cosmetics contained poisonous compounds of elements such as lead and arsenic. People risked their health, and even their lives, by using them. Today men and women all over the world spend large amounts of money on cosmetics while hoping that good science and government regulations will protect them from products that might be hazardous.

Cosmetic creams

Cosmetic creams are emulsions of oil and waxes with water. Some consist of small droplets of the oils dispersed in water. Others have small droplets of water dispersed in the mixture of oils. Oils and water do not mix. This means that emulsifiers must be added to prevent the small droplets joining together and the oil and water separating. Creams may also include other ingredients such as coloured pigments, perfumes, and preservatives.

Perfume

Perfumes are solutions. The solvent is alcohol mixed with some water. Dissolved in the solvent are scented oils. Some of the oils come from plants and some from animals. Today many of the scented ingredients are made by the chemical industry but are often identical to the chemicals from natural sources. The scents are blended so that some evaporate more quickly for immediate effect, whilst others stay on the skin longer to give a lingering scent.

Nail polish

Nail polish is based on a solution of colourless polymers and plasticisers in a solvent. Coloured pigments are mixed with the solution. The solvent

A coloured X-ray of a handbag showing a range of cosmetics.

is chosen so that it evaporates quickly when spread over a fingernail. As the solvent disappears, the polymers form a flexible film on the nail, which traps the colour.

Deodorants

The active ingredients in deodorants, or antiperspirants, are often dissolved in alcohol because it evaporates quickly and feels cool on the skin. The main active ingredients are aluminium salts. These dissolve in sweat and then form a thin coating of gel over the sweat glands, cutting down the flow of sweat. Deodorants are complex mixtures with many other ingredients including perfume, oils to soften the skin, moisturisers, silica to mop up oiliness and an antioxidant to act as a preservative. Some deodorants also contain a detergent to make them easier to wash off at the end of the day.

Nanocosmetics

In recent years the big cosmetic companies have been exploring the use of tiny nanoparticles in cosmetics. Nanoparticles are tens of thousands times smaller than the width of a human hair. It is easier to spread sunscreens and wrinkle creams if the specks or droplets in them are very small. Once applied, the nanocosmetics are invisible on the skin.

Consumer organisations are worried that there may be unknown dangers in the use of nanoparticles in cosmetics. The tiny particles may cause harm if they are able to get into the body through the skin. More research is needed to ensure that these cosmetics are safe.

Taking your medicine

People often take medicines when they don't feel well. Medicines contain active ingredients called drugs.

Drug design

New drugs go through a long and expensive process of testing. Scientists test new drugs to make sure they are effective and that they are safe. They carry out chemical tests and clinical trials. New drugs must be approved by the Medicines and Healthcare Products Regulatory Agency (MHRA).

What formulation?

Medicines are usually formulations – mixtures of more than one ingredient made by following a 'recipe'. Usually only one or two of the ingredients are drugs, which treat the symptoms. Other ingredients are there to make the medicine look and taste better, or to hold it together as a tablet.

How pure?

Drugs in medicines must meet the specifications published in a large reference book called the *British Pharmacopoeia*. Standard procedures to measure the purity of ingredients are described in the *Pharmacopoeia*. It also gives the level of purity (or grade) that is required for each ingredient.

The letters BP after the name of the active ingredient show that it meets the standard of purity specified in the British Pharmacopoeia.

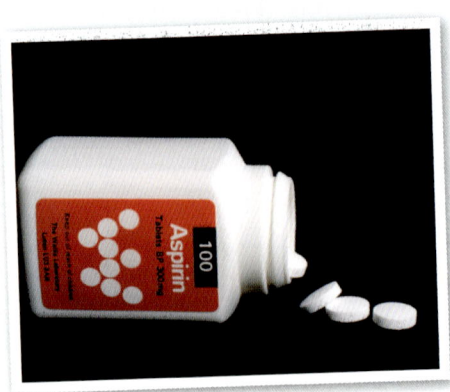

Designer colours

In the 1920s, Ford painted their cars 'any colour you like, so long as it's black'. Nowadays our world is much more colourful, thanks to chemistry – and physics, because colour depends on chemicals that absorb and reflect different colours of light. We really can make things any colour we like, and any shade of that colour.

The pigments in paints are insoluble chemicals. Paint-makers mix them with liquids to make thick suspensions that spread smoothly over surfaces.

It takes more than just a good pigment to make a paint. The other ingredients:
- allow the colour to be applied successfully
- make sure that it dries to an even film
- stick the paint to the surface
- protect against wear and tear.

A paint contains four main ingredients:
- pigment – the colouring material
- binder – holds the pigment particles together and forms a tough layer when paint 'dries'
- solvent – dissolves the binder so the paint can be spread; may be water or an organic solvent
- additives – such as fungicides to stop mould growing.

Producing the right mixture is an example of the art of formulation.

It is not only the colour that matters. The paint must spread evenly, dry quickly, and form a durable surface. The technician in protective clothing is putting the finishing touches to the paintwork on a new truck.

The chemical industry

The chemical industry converts raw materials, such as crude oil, natural gas, minerals, air, and water, into useful products. The products include chemicals for use as drugs, fertilisers, detergents, paints, and dyes.

Chemicals made on a large scale are called **bulk chemicals**. Thousands or even millions of tonnes of these chemicals are produced every year. Examples are ammonia, sulfuric acid, and sodium hydroxide.

Large chemical plants are usually located near to where the raw materials they need are found. For example, chemical plants that turn salt (sodium chloride) solution into sodium hydroxide and chlorine are often located near a salt mine. This is to keep down the cost of transporting the raw materials.

Chemicals made on a much smaller scale are called **fine chemicals**. Examples are pharmaceutical drugs, food additives, and fragrances.

Maintenance workers help to keep chemical plants running. This technician is inside a chemical reactor.

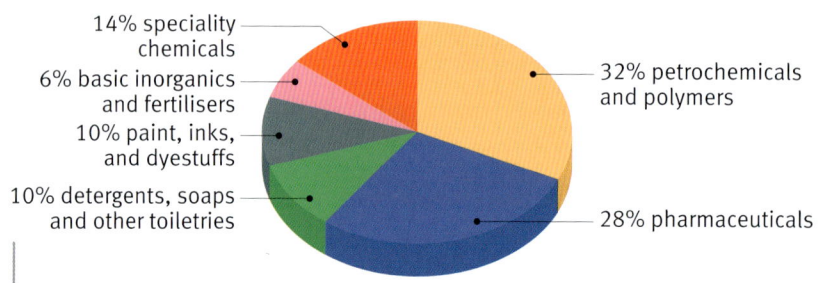

14% speciality chemicals

6% basic inorganics and fertilisers

10% paint, inks, and dyestuffs

10% detergents, soaps and other toiletries

32% petrochemicals and polymers

28% pharmaceuticals

The range of products made by the chemical industry in the UK by value of sales.

Products from the chemical industry

Basic inorganics including fertilisers

Bulk inorganic chemicals include:

- ammonia – made from natural gas, water, and air
- sulfuric acid – made from sulfur, water, and air
- sodium hydroxide – made from sodium chloride solution.

Ammonia reacts with sulfuric acid to make the fertiliser ammonium sulfate. Fertilisers provide nutrients to help plants grow. This increases food production and helps to support the world's growing population. Sodium hydroxide is used to make soaps and in chemical processes to neutralise acids.

Petrochemicals and polymers

Petrochemical plants use hydrocarbons from crude oil to make a wide variety of products, including polymers.

Paint, inks, and dyestuffs

Modern dyes are also made from petrochemicals. Chemists have designed colourful dyes that stick fast to natural and synthetic fibres.

A scientist carrying out research into new drugs designed to treat diseases of the nervous system.

Pharmaceuticals

The **pharmaceutical** industry produces drugs and medicines to treat, cure, or prevent disease. These fine chemicals are very valuable.

Metals

Steel is used in vehicles, buildings, pipelines, and tools. Iron ore is imported to the UK where it is processed to extract iron metal. The iron metal is mixed with small amounts of carbon to make steel.

Copper is used to make wires and cables that carry electricity, electric motors, and water pipes. Copper is very expensive to extract from its ore and so much of it is recycled after use.

Speciality chemicals

Speciality chemicals are examples of fine chemicals used by manufacturers to make other products. They include food additives, fragrances, and the liquid-crystal chemicals in LCD televisions.

People in the chemical industry

Research and development chemists work in laboratories to find new processes and develop new products. The industry needs new processes so that it can be more competitive and more sustainable.

Marketing and sales departments work out whether a new product is wanted. If a new product is promising it may first be tried out by making a small amount of it in a pilot plant.

Financial experts estimate the value of the new product in the market and then compare this with the cost of making it.

Chemical engineers scale up the process and design a full-scale plant.

Transport workers carry the chemicals to the industry's customers.

Chemical-plant managers and administrators control the whole operation.

Service departments look after the needs of the people working there. This includes medical and catering staff, and training and safety officers.

To become a research and development chemist, formulation chemist, chemical engineer, or plant manager you need a degree in the chemical sciences or chemical engineering. You will also need to have project-management skills, be able to manage resources, and understand health-and-safety requirements.

You can become a chemical engineering technician by taking an Advanced Apprenticeship. You must be good at chemistry and maths, have good communication skills, and enjoy working in a team. You will need a minimum of four GCSEs (A*–C), or the equivalent, including English, maths, and science or technology.

Key words
- ✔ bulk chemicals
- ✔ fine chemicals
- ✔ pharmaceuticals
- ✔ formulation

Questions

1 Give the name and chemical formula of a bulk chemical.

2 What percentage value of products of the chemical industry in the UK are:
   a used as cleaning products and toiletries?
   b used for medical diagnosis and treatment?
   c made from the hydrocarbons from crude oil?

3 Draw up a table to summarise who works in the chemical industry. Use these headings:
   Type of worker
   What they do

4 When the chemical reactor in the picture on page 268 is in use it is filled with a reaction mixture. Suggest the purpose of:
   a the rotating paddle in the centre of the tank
   b the network of pipes around the edge of the tank.

Developing a new medicine

PEOPLE & ORGANISATIONS

Laboratory research and testing

Research and development is particularly important in the pharmaceutical industry. Most medicines are very expensive because developing new drugs is a long and costly process. It can take over 15 years and £500 million to develop a new drug from an intial idea to being on sale. Thousands of new compounds are invented, made, and tested. Only a few are chosen for trials with patients. Eventually one may be manufactured and used in a medicine.

Clinical research

The compounds chosen for trials are given to patients to find out how effectively each drug treats a particular disease. This is clinical testing. Researchers also check for harmful side-effects. On the basis of the trials, the most effective drug with few or no side-effects is chosen.

Process research

The company has to find a way to manufacture a new product economically and safely. The methods used in the research laboratory are not usually suitable for large-scale production. Industrial chemists find an economic, safe, and sustainable synthetic route for scaling up the production.

Formulation

Each dose of a medicine may only contain a few milligrams of the active ingredient. The job of the pharmacologists is to discover the best way to get the drug to the right part of the body. This might be as a liquid for injection, a cream to be absorbed through the skin, or a coated tablet to be swallowed. The challenge is to find a **formulation** that is effective, acceptable to patients, and also keeps well.

Manufacture and quality assurance

Eventually the medicine goes into production. Plant managers and operators ensure the smooth running of the process. Analytical scientists check that both the chemicals used as reactants, and the final products, meet the necessary standards.

discovery phase

Research into cause of disease. Thousands of compounds synthesised and tested as possible drugs.

thousands of compounds

The responses of enzymes, cells, or organs to the compounds are tested.

fewer than 100 active compounds

The responses of the animals to the compounds are tested.

2 or 3 effective compounds

Safety testing on animals. **development phase**

application to the Medicines and Healthcare Products Regulatory Agency

Human testing – testing on healthy human volunteers.

Clinical trials – tests on willing patients for whom the drug was designed.

application to the Medicines and Healthcare Products Regulatory Agency

Marketing authorisation.

Researching, developing, and testing a new drug is a long and expensive process.

Questions

1 Explain why new medicines are expensive.

2 Suggest a reason why companies give their medicines brand names.

The work of laboratory technicians

Laboratory technicians carry out a wide range of skilled tasks. They work in laboratories in industry, in hospitals, in government-funded research institutions, and in environmental agencies.

Sampling, testing, and analysis

Technicians prepare formulations or samples for analysis and testing. They follow **standard procedures** as they carry out laboratory tests. In this way they help to produce the precise and accurate data that is needed for scientific investigations and research.

Operating and maintaining equipment

Technicians maintain and operate laboratory equipment. It is important that equipment such as electrical meters, colourimeters, blood-pressure monitors, pH meters, and so on are checked and well looked after so that they are always in working order and give accurate results.

Caring for plants and animals

Many technicians are needed to help with research to improve food supplies, discover new medical treatments, and check on the safety of new chemicals. These technicians have to be skilled in caring for plants and animals.

Maintaining stocks

Laboratories can only run smoothly if stocks of chemicals and other resources are well looked after and checked. Technicians are responsible for keeping records of stocks and ordering replacements when needed. In this work, as in many other areas, they have to be familiar with computer systems including spreadsheets and databases.

Adhering to health-and-safety regulations

Technicians have to work safely and be aware of health-and-safety regulations. They must know how to deal with waste and what to do in case of fire or other accidents.

Questions

1 Why is it important to have technicians in hospitals?

2 Why do industries that manufacture products need technicians?

3 Make a list of some of the tasks carried out by technicians in school laboratories.

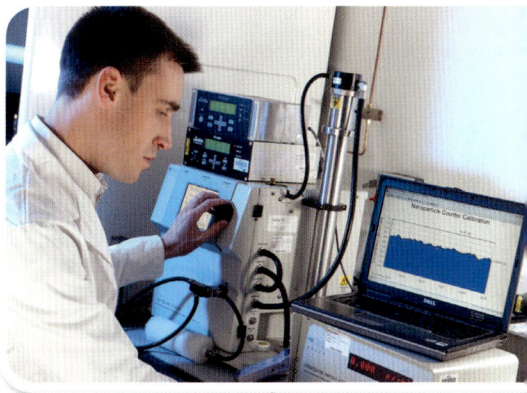

A technician measuring the concentration of nanoparticles in the air at the National Physical Laboratory.

A technician checking on potato seedlings in an agricultural research laboratory. The laboratory uses genetic research to find new varieties of the crop that give good yields but resist diseases.

Technician checking stocks of highly scented plant oils in a laboratory that supplies the oils to manufacturers of consumer products.

PEOPLE & ORGANISATIONS

Key word

✔ **standard procedure**

Barry Taylor in his laboratory.

Preparing, separating, and purifying chemicals is only the first step in making a usable product. Most everyday products are complex mixtures, made to precise **formulations**.

Trade secrets

Danisco makes ingredients for the food and drink industry. Among its products are concentrates that can be diluted to produce carbonated drinks. Food companies buy concentrates, dilute them with carbonated water, then pasteurise and bottle the drinks. After stamping with a sell-by date the bottles are sent to the shops.

Barry Taylor has been in the food industry all his working life. He specialises in flavourings and soft drinks for Danisco. Most of Barry's formulations are secret, but the diagram on the next page shows how a typical pineapple-and-grapefruit-flavour fizzy drink is made.

Testing formulations

Danisco perform tests on their formulations to make sure that they meet their own quality-control standards and all the required national and international safety standards.

Food standards

Manufactured products must pass strict quality-assurance tests to make sure they are 'fit for purpose' (do what they are designed to do) and are safe if used correctly. For most products there are national and international standards. For example, the type and amount of each chemical additive used is regulated together with the way it is listed on the food label.

Local authorities sometimes carry out tests to check that food on sale in the shops is safe. These tests are performed by public analysts. Tests follow standard procedures, so that they are the same wherever the tests are done.

Questions

1 Using the information on the page opposite, make a table to show the ingredients of 100 litres of fruit concentrate. Use these headings:
Ingredient   Quantity   Reason for adding the ingredient

2 Look at the ingredients of the fruit concentrate.
   a What is the concentration in g/litre of lutein in the final product?
   b What mass of sodium benzoate is present in 206.0 ml of a 20% solution?

Formulating 100 litres of concentrate for a fruit drink

Sweeteners

Aspartame (137.0 g) Aspartame is 200 times sweeter than sugar. It is most stable in slightly acidic solutions (pH 4.3).

Sodium saccharin (90.7 g) Saccharin is 300 times sweeter than sugar, but has a distinctive aftertaste that not everyone likes. Saccharin itself is not all that soluble, so the more soluble compound sodium saccharin is used.

Artificial sweeteners are used instead of sugar as this is a low-calorie drink.

Colouring

Lutein (50 g) Colouring is added to 'maximise the enjoyment of the beverage', or in other words to make the drink look good. Only compounds that have been proved to be safe and non-toxic can be used. Barry recommends lutein, a natural extract of the Aztec marigold flower, to get the right sort of yellow colour.

Preservative

Sodium benzoate, 20% solution (206.0 ml)
This is a preservative to stop the growth of moulds and bacteria. It would not be necessary if the drink was for immediate consumption. Benzoic acid is found naturally in some fruit (such as cranberries), but it is not very soluble in water. The salt sodium benzoate is more soluble and also acts as a preservative.

Fruit and flavouring

Pineapple and grapefruit compound (12.5 litres)
This is a complicated mixture of pineapple and grapefruit juice, fruit flavouring, citric acid (found naturally in grapefruit), and ascorbic acid (vitamin C). The acids add flavour and also make the drink more refreshing. The fruit flavouring is added to intensify the taste. The peel of citrus fruit contains highly flavoured oils that do not mix with water, so citrus flavouring is made by mixing the oils with an **emulsifying agent** that disperses the oil in the mixture.

Water

Purified water (to make the volume up to 100 litres) Barry recommends soft water, so tap water is treated to remove any dissolved minerals. It is then passed through a series of filters and treated with UV radiation to remove suspended organic material and kill any remaining microorganisms.

Key word
✓ **emulsifying agent**

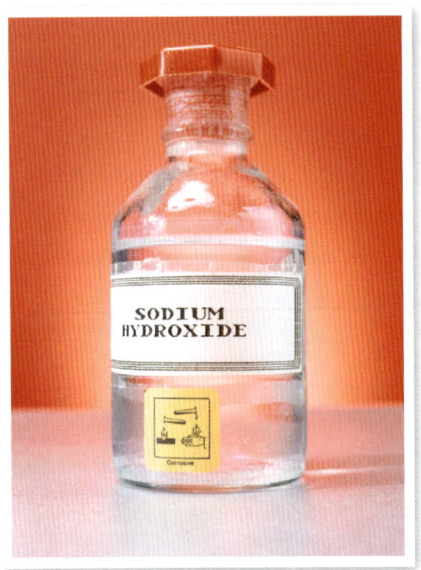

Pure sodium hydroxide is a white solid. It is soluble in water and used in solution as a laboratory alkali. It is corrosive.

Hazards of chemicals

Some chemicals are extremely hazardous. For example, concentrated acids are **corrosive**. Dilute solutions of acids in water are much less hazardous. Dilute hydrochloric acid is often used in solutions that are so dilute that they are not even classified as an **irritant**.

Dilute sulfuric acid is an irritant. Nitric acid is more hazardous. Even quite dilute solutions can be corrosive.

Some alkalis attack skin and flesh. Examples are sodium hydroxide and potassium hydroxide. They used to be called caustic alkalis because of the way they attack living tissue. Even dilute solutions of these compounds can be hazardous, especially in the eyes. They are corrosive.

If you rub a small drop of dilute sodium hydroxide solution between the tips of two fingers they will soon feel soapy as the alkali attacks the grease on the skin.

Some salts are very poisonous. They are **toxic**. The soluble salts of barium and lead are examples of toxic chemicals.

HARMFUL
a substance that if inhaled, ingested, or taken in through the skin, may involve limited health risks

IRRITANT
a non-corrosive substance that through immediate, prolonged, or repeated contact with the skin or eyes, may cause inflammation or lesions

TOXIC
a substance that if inhaled, ingested, or taken in through the skin, may involve serious acute or chronic health risks and even death

CORROSIVE
a substance that on contact with living tissues may destroy the tissues

OXIDISING
a substance that produces a reaction giving off great heat when in contact with other substances, particularly flammable substances

HIGHLY FLAMMABLE
a liquid that may easily catch fire or a solid that burns after brief contact with a flame or that gives off flammable gases in contact with water

EXPLOSIVE
a substance that may explode under the effect of flame or heat, or that is sensitive to friction and shocks

Hazchem symbols. Substances harmful to health should display recognised warning symbols. These are some of the hazard symbols used.

PEOPLE & ORGANISATIONS

Worker safety

The Health and Safety Executive (HSE) aims to protect people's health and safety by making sure risks in the workplace are properly controlled. Its staff come from a range of different backgrounds, including administrators, lawyers, statisticians, inspectors, scientists, technicians, engineers, doctors, and nurses.

The HSE works with large companies and small businesses in three main areas:

- accident investigation – they collect information about workplace accidents, to help draw up new advice on how to reduce risk
- guidance – they advise companies on health-and-safety issues
- inspection – they check safety standards and make sure companies obey the law.

The regulations

The HSE enforce many regulations designed to protect people at work. Below are two examples of regulations.

Control of Substances Hazardous to Health (COSHH)

Employers must do a risk assessment for chemicals or mixtures that could be hazardous to health, and make sure that proper controls and precautions are followed. This could mean changing procedures to use less hazardous substances.

Personal Protective Equipment at Work

If hazards cannot be removed completely the employer must provide suitable personal protective equipment (PPE) and give training in its use. PPE includes protective clothes, boots, gloves, safety helmets, and goggles.

These workers in a factory are wearing overalls and rubber gloves to protect their skin from chemicals. Their masks filter the air they breathe, and also protect their eyes.

Questions

1 Which organisation ensures that companies follow safety regulations?

2 Suggest a reason why the workers in the photograph have not tucked their trousers into their boots.

3 What PPE is available in your school science laboratory?

Key words

- ✓ explosive
- ✓ harmful
- ✓ irritant
- ✓ toxic
- ✓ corrosive
- ✓ oxidising
- ✓ flammable
- ✓ hazchem

Acids

pH

14 — dilute sodium hydroxide
13 —
12 — limewater
11 —
alkaline
10 — some brands of toothpaste
9 —
8 —
neutral 7 — blood — pure water
— fresh cows' milk
6 — distilled water
5 —
4 —
acidic 3 — vinegar
2 — lemon juice
1 — digestive fluids in the stomach
0 — dilute hydrochloric acid

The pH scale.

Geologists use hydrochloric acid to test rocks. If the rock fizzes it shows that the rock is a carbonate such as limestone or dolomite.

Three common laboratory **acids** are sulfuric acid (H_2SO_4), hydrochloric acid (HCl), and nitric acid (HNO_3). These are inorganic acids. Acids are also vital to life and there are many different acids in living things. These are called organic acids. Two examples are citric acid and acetic acid.

Acids and indicators

The easiest way to recognise an acid is to test a solution with an **indicator**. Litmus turns red in any acid. Full-range or universal indicator turns to shades of orange and red in acid depending on the type of acid and the **concentration** of the solution.

The **pH scale** is a number scale from 0 to 14 that shows the acidity or alkalinity of a solution in water. Solutions of acids have a pH lower than 7.

Acids and metals

Metals such as magnesium, zinc, and iron react with dilute solutions of acids. The mixture of metal and acid fizzes as the reaction produces hydrogen gas. This is shown by the **word equation**:

metal + acid $\longrightarrow$ **salt** + hydrogen

Some metals are unreactive with dilute acids. Examples are copper and lead.

Acids and carbonates

Another reaction that produces a gas is the reaction of a **carbonate** with an acid. Limestone is an example of a carbonate. Limestone and marble are both forms of calcium carbonate. These minerals fizz when mixed with dilute hydrochloric acid.

metal carbonate + acid $\longrightarrow$ salt + carbon dioxide + water

Acids and metal oxides and hydroxides

Most metal **oxides** and metal hydroxides dissolve when they react with an acid to make a salt. No gas forms when metal oxides or metal **hydroxides** neutralise an acid.

metal oxide (or hydroxide) + acid $\longrightarrow$ salt + water

Questions

1 Write a word equation for the reaction of dilute sulfuric acid with:
   a zinc b magnesium carbonate c sodium hydroxide

2 Write balanced symbol equations for the reactions in question 1.

Alkalis

Pharmacists sell **antacids** to control heartburn and indigestion. The chemicals in these medicines are the chemical opposites of acids. They are designed to **neutralise** excess hydrochloric acid produced in the stomach – hence the name 'antacids'.

Milk of Magnesia is an antacid used to treat too much acidity in the stomach. It is supplied either as tablets or as a milky suspension of magnesium hydroxide in water.

Some chemical antacids are soluble in water to give solutions with a pH above 7. Chemists call them **alkalis**. Common alkalis are sodium hydroxide and potassium hydroxide. These alkalis are not safe to swallow but they can be used in powerful household cleaners to break down grease in ovens.

Toothpaste is mildly alkaline. It neutralises the acids that attack teeth. These acids form when bacteria in the mouth act on sugars in food.

When an alkali neutralises an acid, a salt is formed.

Alkalis are used in a range of domestic products. Alkalis help to remove greasy dirt.

THE SCIENCE

Key words
- acids
- indicator
- neutralise
- alkalis
- concentration
- pH scale
- word equation
- salt
- carbonates
- oxides
- hydroxides
- corrosive
- irritant

Ammonia

Another example of an alkali is ammonia, a pungent-smelling gas. Ammonia is an important chemical because it is used to make fertilisers. The fertiliser ammonium sulfate is made by a neutralisation reaction. Ammonium sulfate is a salt. When ammonia reacts with an acid, a salt is the only product.

ammonia   +   sulfuric acid $\longrightarrow$ ammonium sulfate

$2NH_3(g)$ + $H_2SO_4(aq)$ $\longrightarrow$ $(NH_4)_2SO_4(aq)$

Question

1 Milk of Magnesia contains magnesium hydroxide. Write a word equation to show what salt will form in your stomach if you take this antacid medicine.

Salts form when acids are neutralised by a metal oxide or hydroxide, so every salt can be thought of as having two parents. Salts are related to a parent acid and to a parent metal oxide or hydroxide.

⌀ Sodium chloride, NaCl, is a salt with two parents.

Soluble or insoluble?

Chemists need to know whether a salt is **soluble** or **insoluble** in water before making it or using it. You can look up the solubility of a chemical in data tables, but it is useful to be familiar with the patterns of solubility of common salts.

Salt	Soluble in water	Insoluble in water
nitrates	all soluble	none
chlorides	mostly soluble	silver and lead chlorides
sulfates	mostly soluble	barium and lead sulfates; calcium sulfate is very slightly soluble
carbonates	sodium and potassium carbonates	mostly insoluble

Making soluble salts

Soluble salts can be made by the reactions of acids. **Filtration** is used to remove any solid from the reaction mixture, and the solution **(filtrate)** is heated gently to evaporate some of the water. When the concentrated solution is left to cool, salt crystals form by **crystallisation**. The crystals are larger if the solution cools slowly.

As crystals form, water molecules may become part of the structure. Chemists call this water of crystallisation. They include the water in the formula of the salt, for example, copper sulfate, $CuSO_4.5H_2O$.

Examples of soluble salts

Iron is an important mineral in the diet. If someone has a very low level of iron their doctor may advise them to take a mineral supplement such as iron sulfate. Iron sulfate is made by the reaction

$$iron \quad + \quad sulfuric\ acid \quad \longrightarrow \quad iron\ sulfate \quad + \quad hydrogen$$
$$Fe(s) \quad + \quad H_2SO_4(aq) \quad \longrightarrow \quad FeSO_4(aq) \quad + \quad H_2(g)$$

Zinc is vital for growth and healing in the human body. People who do not get enough zinc in their diet can take zinc sulfate as a mineral supplement. Zinc sulfate is made by the reaction

$$zinc\ oxide \quad + \quad sulfuric\ acid \quad \longrightarrow \quad zinc\ sulfate \quad + \quad water$$
$$ZnO(s) \quad + \quad H_2SO_4(aq) \quad \longrightarrow \quad ZnSO_4(aq) \quad + \quad H_2O(l)$$

THE SCIENCE

Key words
- ✓ **soluble**
- ✓ **insoluble**
- ✓ **filtration**
- ✓ **filtrate**
- ✓ **crystallisation**

Questions

Look at page 290 for help with the chemical formulae.

1 Give the chemical formula for potassium nitrate.

2 Give the name and chemical formula of two salts that are insoluble in water.

3 Give the name and chemical formula of two salts that are soluble in water.

Epsom salts are used as a laxative, a treatment for boils, and a muscle soak. Epsom salts are mainly magnesium sulfate, made by the reaction

magnesium + sulfuric acid ⟶ magnesium + water + carbon
carbonate   sulfate   dioxide

$$MgCO_3(s) + H_2SO_4(aq) \longrightarrow MgSO_4(aq) + H_2O(l) + CO_2(g)$$

Copper sulfate is used in agriculture to control the growth of fungus on plants. It is a fungicide. It is made by the reaction

copper oxide + sulfuric acid ⟶ copper sulfate + water

$$CuO(s) + H_2SO_4(aq) \longrightarrow CuSO_4(aq) + H_2O(l)$$

Ammonium sulfate is used in agriculture as a fertiliser. When ammonia reacts with an acid the only product is an ammonium salt.

ammonia + sulfuric acid ⟶ ammonium sulfate

$$NH_3(g) + H_2SO_4(aq) \longrightarrow NH_4SO_4(aq)$$

Making insoluble salts

Insoluble salts are made by mixing two solutions of soluble salts. The 'metal' and the 'acid' parts of each salt rearrange and two new salts form.

Magnesium carbonate is used as an ingredient in cosmetics such as face powders. It can be made by mixing two solutions of soluble salts.

magnesium + sodium ⟶ magnesium + sodium
sulfate   carbonate   carbonate   sulfate

$$MgSO_4(aq) + Na_2CO_3(aq) \longrightarrow MgCO_3(s) + Na_2SO_4(aq)$$

The table of solubilities can be used to predict whether the salts that form are soluble or insoluble. Magnesium carbonate is insoluble, so it is a solid (s). Sodium sulfate is soluble, so it stays in **solution** (aq).

The reaction of magnesium sulfate with sodium carbonate is an example of a **precipitation** reaction. The insoluble magnesium carbonate appears as a solid in the reaction mixture. A precipitate can be separated from a reacting mixture by filtration.

Examples of insoluble salts

Calcium carbonate is used as an ingredient in cosmetic face powders and toothpaste.

calcium + sodium ⟶ calcium + sodium
chloride   carbonate   carbonate   chloride

$$CaCl_2(aq) + Na_2CO_3(aq) \longrightarrow CaCO_3(s) + 2NaCl(aq)$$

Copper carbonate is used as a pigment in artists' paints.

copper sulfate + sodium carbonate ⟶ copper carbonate + sodium sulfate

$$CuSO_4(aq) + Na_2CO_3(aq) \longrightarrow CuCO_3(s) + Na_2SO_4(aq)$$

Cobalt violet is another pigment. Its chemical name is cobalt phosphate.

cobalt + sodium ⟶ cobalt + sodium
chloride   phosphate   phosphate   nitrate

$$3CoCl_2(aq) + 2Na_3PO_4(aq) \longrightarrow Co_3(PO_4)_2(s) + 6NaCl(aq)$$

A precipitation reaction. Two colourless solutions are mixed and an insoluble precipitate forms.

Key words

✓ **solution**
✓ **precipitation**
✓ **precipitate**

Questions

4 Name two soluble salts that can be made by the reaction of a metal oxide with an acid.

5 What is a precipitate?

6 Name two soluble salts that would react to form the insoluble salt zinc carbonate (use the tables on pages 278 and 290 to help you).

7 What equipment would you need to separate a precipitate from a reaction mixture?

Chemical quantities

Chemists need to be able to work out how much of each reactant to mix together to make the quantity of the product they need. To do this they need to be able to turn the symbols in the balanced chemical equation into masses in grams. This is possible if we know the masses of the atoms relative to each other.

Relative atomic masses

A hydrogen atom is the lightest atom. On the scale used by chemists, hydrogen has a relative mass of 1. All other atoms have more mass. The **relative atomic mass** of carbon is 12. This means that a carbon atom is 12 times heavier than a hydrogen atom.

The relative atomic mass of magnesium is 24, so magnesium atoms are twice as heavy as carbon atoms and 24 times as heavy as hydrogen atoms.

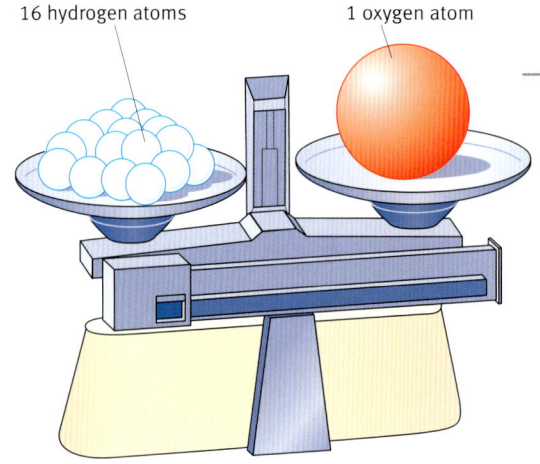

16 hydrogen atoms   1 oxygen atom

One oxygen atom has the same mass as 16 hydrogen atoms. The relative atomic mass of oxygen is 16.

Relative formula masses

Adding up the relative atomic masses for all the atoms in the formula of a compound gives the **relative formula mass.**

Question

1 Use these relative atomic masses to answer the questions:
   H = 1, C = 12,
   O = 16, Na = 23,
   S = 32, Ca = 40

   a How many times heavier is a sodium atom than a hydrogen atom?
   b How many times heavier is a sulfur atom than an oxygen atom?
   c What is the relative formula mass of sodium hydroxide?
   d What is the relative formula mass of calcium hydroxide?

Worked example

The relative formula mass of calcium carbonate

The formula of calcium carbonate is $CaCO_3$

The relative atomic masses are Ca = 40, C = 12, and O = 16

The relative formula mass of the compound = $40 + 12 + (3 \times 16) = 100$

(Note that these are relative masses so you do not include units here.)

Reacting masses

Given the relative atomic and formula masses, it is possible to work out the **reacting masses** from a balanced equation.

Rules for working out reacting masses

STEP 1   Write down the balanced symbol equation.

STEP 2   Work out the relative formula mass of each reactant and product.

STEP 3   Write the relative reacting masses under the balanced equation, taking into account the numbers used to balance the equation.

STEP 4   Convert to reacting masses by adding the units (g, kg, or tonnes).

STEP 5   Scale the quantities to amounts actually used in the synthesis or experiment.

Follow these rules for calculating reacting masses.

Worked example

What are the reacting masses when sodium hydroxide reacts with sulfuric acid?

Step 1

$2NaOH + H_2SO_4 \longrightarrow Na_2SO_4 + 2H_2O$

Step 2

The relative atomic masses are Na = 23, H = 1, S = 32, and O = 16

Relative formula mass of $NaOH = 23 + 16 + 1 = 40$

Relative formula mass of $H_2SO_4 = (2 \times 1) + 32 + (4 \times 16) = 98$

Relative formula mass of $Na_2SO_4 = (2 \times 23) + 32 + (4 \times 16) = 142$

Relative formula mass of $H_2O = (2 \times 1) + 16 = 18$

Step 3 and 4

$2NaOH$	$+$	H_2SO_4	$\longrightarrow$	Na_2SO_4	$+$	$2H_2O$
$2 \times 40 = 80$		98		142		$2 \times 18 = 36$
80 g		98 g		142 g		36 g

The reacting masses have the same number value in grams as the relative masses. Note that the proportions stay the same even working on a different scale. So 80 tonnes of sodium hydroxide react with 98 tonnes of sulfuric acid to make 142 tonnes of sodium sulfate and 36 tonnes of water.

This is another way of showing that the equation is balanced. No atoms are created or destroyed and so 80 g + 98 g = 178 g of reactants gives 142 g + 36 g = 178 g of products. The masses of reactants and products are equal.

Key words

✔ relative atomic mass
✔ relative formula mass

Question

2 Use these relative atomic masses to work out the reacting masses for the reactions:
H = 1, C = 12,
N = 14, O = 16,
Na = 23, S = 32,
Cl = 35.5, Ca = 40,
Cu = 64

  a $NaOH + HCl \longrightarrow NaCl + H_2O$

  b $CuO + H_2SO_4 \longrightarrow CuSO_4 + H_2O$

  c $CaCO_3 + 2HNO_3 \longrightarrow Ca(NO_3)_2 + CO_2 + H_2O$

The **yield** is the quantity of product obtained from known amounts of starting materials. The quantities used in a chemical synthesis are often different from those in the balanced equation, so we scale the figures up or down.

Actual yield and theoretical yield

The actual yield is the mass of product after separating it from the mixture, purifying, and drying it.

The **theoretical yield** is the mass of product expected if the reaction goes exactly as shown in the balanced equation, with no by-products and no losses while transferring chemicals from one container to another. The actual yield is always less than the theoretical yield.

Key words
- ✓ yield
- ✓ theoretical yield
- ✓ technical grade
- ✓ laboratory grade
- ✓ analytical grade

Questions

1 What is the theoretical yield of magnesium carbonate when 60 g of magnesium sulfate ($MgSO_4$) reacts with excess sodium carbonate? (Relative atomic masses: C = 12, O = 16, S = 32, Mg = 24)

2 What is the theoretical yield of zinc sulfate when 12.5 g of zinc carbonate ($ZnCO_3$) reacts with excess sulfuric acid? (Relative atomic masses: C = 12, O = 16, S = 32, Zn = 65)

3 Which grade of salt (sodium chloride) would you use to melt ice on roads in winter, technical, laboratory, or analytical? Explain your answer.

Worked example 1

The theoretical yield of silver chloride made from 6.8 g of silver nitrate

Silver chloride precipitates as a solid on mixing solutions of silver nitrate and sodium chloride. In this example it is only necessary to work out the reacting masses of the silver nitrate and the silver chloride. Sodium chloride is cheap and added in excess.

Relative atomic masses:
Ag = 108, N = 14, O = 16, Cl = 35.5

silver nitrate + sodium chloride $\longrightarrow$ silver chloride + sodium nitrate

$AgNO_3$ + NaCl $\longrightarrow$ AgCl + $NaNO_3$
170   143.5

This shows that, theoretically, 170 g of silver nitrate should give 143.5 g of silver chloride. In the lab we would use a much smaller quantity of silver nitrate, only a fraction of the reacting mass in the equation. The same fraction of the reacting mass of silver chloride forms.

So, if we use 6.8 g of silver nitrate this should give $\frac{6.8 \text{ g}}{170 \text{ g}} \times 143.5 \text{ g} \times 5.74 \text{ g}$.

Purity of Chemicals

Suppliers of chemicals offer a range of grades of chemicals, including **technical**, general **laboratory,** and **analytical grades**. Technical grade is the least pure, and analytical grade is the most pure. Laboratory grade is suitable for most general laboratory uses.

Purifying a chemical costs money, so the higher the purity the more expensive the chemical. Manufacturers therefore buy the grade most suitable for their purpose.

Controlling reaction rates

Some chemical reactions seem to happen in an instant. An explosion is an example of a very fast reaction.

Other reactions take time – seconds, minutes, hours, or even years. Rusting is a slow reaction and so is the rotting of food.

Rate of reaction

Your pulse rate is the number of times your heart beats every minute. The production rate in a factory is a measure of how many articles are made in a particular time. Similar ideas apply to chemical reactions. Chemists measure the **rate of a reaction** by finding the quantity of product produced or the quantity of reactant used up in a fixed time.

For the reaction

magnesium + hydrochloric acid $\longrightarrow$ magnesium chloride + hydrogen

$Mg(s)$ + $2HCl(aq)$ $\longrightarrow$ $MgCl_2(aq)$ + $H_2(g)$

the rate can be found quite easily by measuring either the disappearance of the magnesium or the formation of hydrogen gas.

In most chemical reactions the rate changes with time. The graph shows the volume of hydrogen formed against time for the reaction above. The graph is steepest at the start, showing that the rate of reaction was greatest at that point. As the reaction continues the rate decreases until the reaction finally stops. The steepness of the line is a measure of the rate of reaction.

What affects reaction rates?

The rates of chemical change can be affected by:
- changing the surface area of solid reactants
- changing the **concentration** of reactants in solution
- changing the temperature
- adding a catalyst.

What is a catalyst?

A catalyst is a chemical that speeds up a chemical reaction. It takes part in the reaction but is not used up. At the end of the reaction there is the same quantity of catalyst as there was at the beginning. Catalysts are essential in many industrial processes. They make many processes more cost effective.

An explosion is an example of a very fast chemical reaction, as this fireball from a detonation of gunpowder shows.

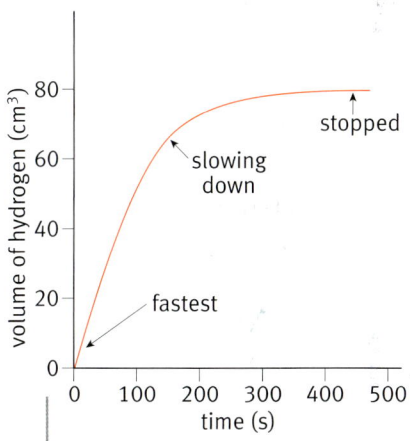

A plot of the volume of hydrogen formed against time for a reaction of magnesium with hydrochloric acid.

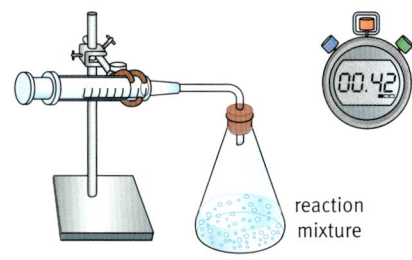

The rate of the reaction between magnesium and hydrochloric acid can be found by measuring the volume of hydrogen with a gas syringe.

Explaining reaction rates

Atoms and molecules can only react if they bump into each other. The more they collide and the harder they bump into each other, the faster the reaction is likely to go.

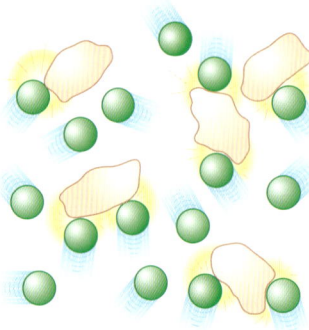
Questions

1 How would you account for the fact that:
   a sliced bread goes stale more quickly than unsliced bread?
   b there is a danger of explosions in flour mills?

2 How is it possible to control conditions to slow down or stop these changes?
   a the rusting of iron
   b milk going sour
   c the reaction of calcium carbonate with sulfuric acid.

3 How is it possible to control conditions to speed up these changes?
   a the cooking of an egg
   b the setting of an epoxy glue
   c the reaction of magnesium with hydrochloric acid.

4 Give two advantages of using a catalyst in an industrial process.

several small lumps   one big lump

Grinding up a solid into smaller pieces increases its surface area. If the surface area is bigger, more molecules can collide with the solid and react. So the smaller the solid particles, the bigger the surface area and the faster the reaction.

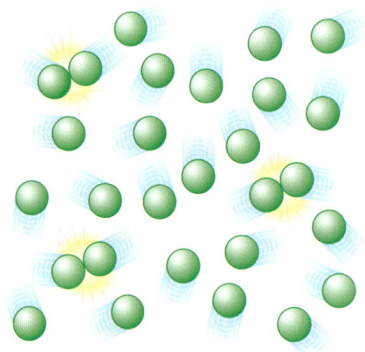 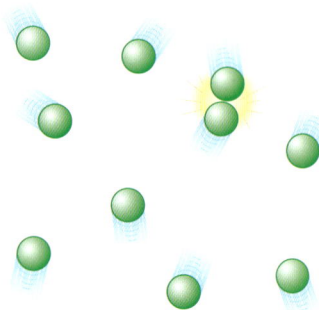

higher concentration   lower concentration

If the reactants are more concentrated, the atoms or molecules are closer together. They collide more often and the reaction goes faster. The higher the concentration, the faster the reaction.

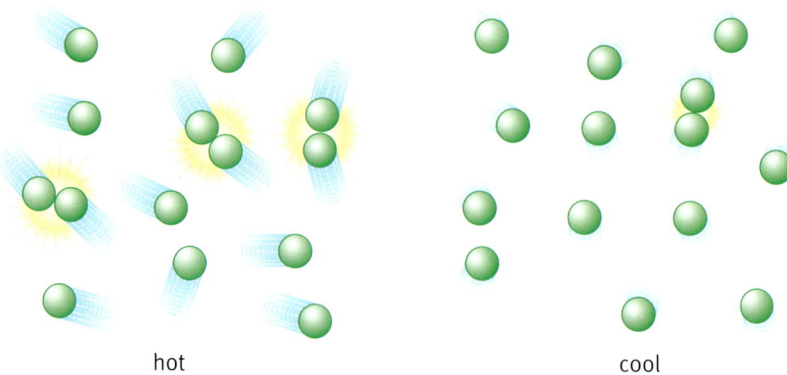

hot   cool

The hotter the reactants, the faster their atoms or molecules collide. When hotter they collide more often and with more energy. The higher the temperature, the faster the reaction.

THE SCIENCE

Choosing a synthetic route

Often there is more than one way to make a chemical. Scientists have a choice of **synthetic routes** and must decide which to choose.

Sodium hydroxide, for example, can be made by heating sodium carbonate strongly, then adding water. Alternatively it can be made by electrolysis of a sodium chloride solution (brine).

The Ineos Chlor chemical plant in Runcorn. Sodium hydroxide and chlorine are made by the electrolysis of brine.

What is the optimum route? How does a synthetic chemist decide which route to choose? There are a number of factors to consider, including:

* the availability and cost of starting materials – the materials should be easy to obtain, cheap, and renewable if possible
* the energy requirements and their costs – the route should need as little energy input as possible
* the efficiency of the route – the route should give a high yield of the required product and the minimum of by-products
* the hazards associated with the chemicals and the process – where possible the reacting chemicals should not be toxic, highly flammable, or corrosive, and the process should not be hazardous to the operators
* the disposal or recycling of by-products – ideally by-products should be recycled or, if this is not possible, be disposed of easily, safely, and cheaply.

Often there is conflict. For example, one route might use the cheapest starting materials but give poor yield. Another route might give high yield but require a high energy input.

Scaling up

What is zinc chelate?

Traces of zinc are essential for healthy plant growth. Farmers can improve yields of some crops by spraying the leaves with the soluble zinc compound called zinc chelate.

Laboratory-scale preparation of zinc chelate

Chemists make zinc chelate from two solids (zinc sulfate and EDTA) and a solution of an alkali in water (sodium hydroxide). The starting chemicals come from bottles on the laboratory shelves; the quantities are small and easily transferred. Technical-grade zinc sulfate can be used as long as the impurities are removed later by filtering.

The reaction

- Heat the EDTA and water on a hotplate while stirring with a glass rod.
- Add the zinc sulfate and stir.
- Add the sodium hydroxide solution and continue stirring.

Filtration

- Pour the hot solution into the filter funnel lined with filter paper.
- Reduce the pressure in the flask to speed up the filtration. The **residue** of unreacted zinc sulfate is trapped. Collect the filtrate.

Separation of the solid product

- Evaporate the filtrate to concentrate the solution.
- Set aside to cool and crystallise. Slow cooling allows bigger crystals to grow.
- Weigh the dry product.

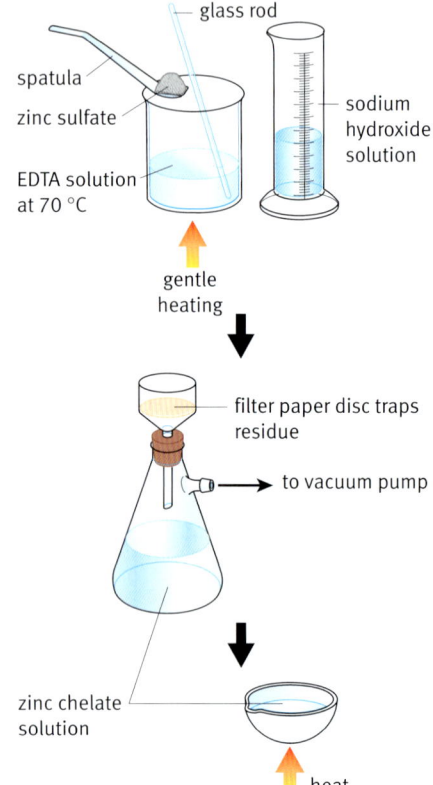

glass rod

spatula

zinc sulfate

sodium hydroxide solution

EDTA solution at 70 °C

gentle heating

filter paper disc traps residue

to vacuum pump

zinc chelate solution

heat

THE SCIENCE

Industrial-scale preparation of zinc chelate

Scaling up from a laboratory to an industrial process does not change the chemistry but it does change the way the operations are carried out. The EDTA and zinc sulfate are bought in 25 kg bags; the water comes from the mains, while sodium hydroxide solution flows in a pipe from a large storage tank.

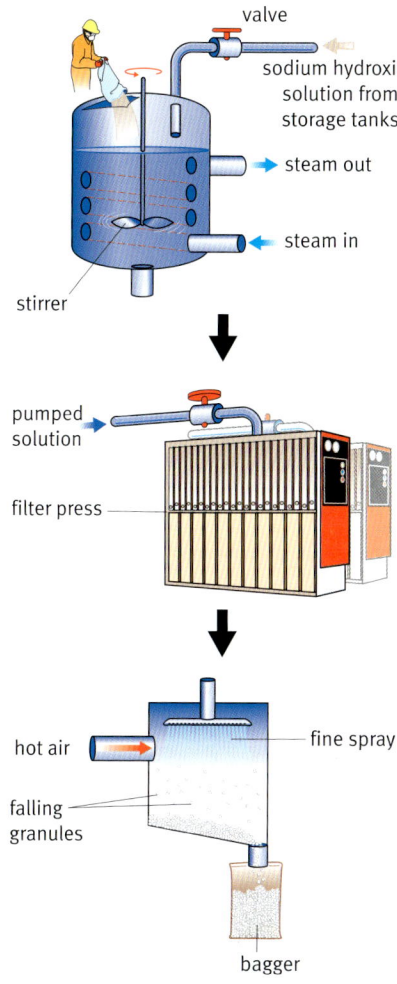

The reaction

- Open the valve to let water be pumped into the reaction vessel. Switch on the large motor-driven stirrer, open the valve to allow steam to flow through the pipes, and heat the reaction vessel. Tip in 40 bags of EDTA.
- Add 29 bags of zinc sulfate.
- Open the valve to run in sodium hydroxide solution. The reaction is exothermic. It gives out energy and so the reaction mixture gets hot. Stop the flow of steam so that the mixture does not get too hot.

Filtration

- Pump the hot solution from the reaction vessel into a filter press with cloth filters.
- Use pressure to force the solution through the filters trapping the residue. Collect the filtrate in a heated storage tank.

Separation of the solid product

- Spray the hot solution into the drying chamber. As the droplets fall through the blast of hot air from the gas burner, the water evaporates.
- Collect the dry granules from the bottom of the drying chamber and pack in 25 kg bags.

Batch or continuous?

The demand for zinc chelate is seasonal. It is only needed when crops are growing, so the manufacturer makes it in a **batch process**. The reaction vessel, filter, and drying chamber are used to make other salts when demand for zinc chelate is low.

A **continuous process** could be used if the product was sold all over the world, so that the demand was more constant. A lot of money is needed to set up an automated continuous process, but once installed it can be kept running continuously with lower labour costs and higher productivity.

Key words
- ✔ scaling up
- ✔ batch process
- ✔ continuous process
- ✔ residue

A chemical language

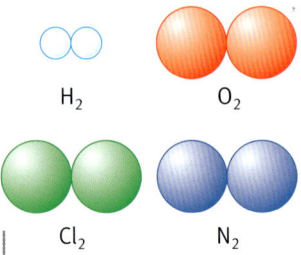

H_2 O_2

Cl_2 N_2

Some non-metals are made up of small molecules. These elements are usually represented by the symbols for these molecules. In any element all the atoms are the same. In these four elements the atoms are joined up in pairs.

Key words

- ✔ **elements**
- ✔ **chemical formula**
- ✔ **compound**
- ✔ **word equations**
- ✔ **balanced symbol equations**
- ✔ **reactants**
- ✔ **products**
- ✔ **state symbols**

Questions

1 Give the chemical names of these molecules:

   a H_2 c Cl_2

   b O_2 d N_2

2 Choose the correct chemical formula from this list for each answer.

   Na_2CO_3 Cl_2
   C_2H_4O $C_2H_4O_2$

   Which chemical

   a is an element?
   b contains sodium?
   c contains carbon, oxygen, and hydrogen atoms in the ratio 2:1:4?

Symbols

Chemists use symbols for the chemical **elements**. One advantage of symbols is that they are the same all over the world.

Metal element	Symbol	Non-metal element	Symbol
calcium	Ca	carbon	C
magnesium	Mg	chlorine	Cl
potassium	K	hydrogen	H
sodium	Na	nitrogen	N
zinc	Zn	oxygen	O
iron	Fe	sulfur	S

Formulae

Every compound has a **chemical formula**. The formula shows the elements that make up the **compound**. The formula also shows the number of atoms of each type in the formula. For example:

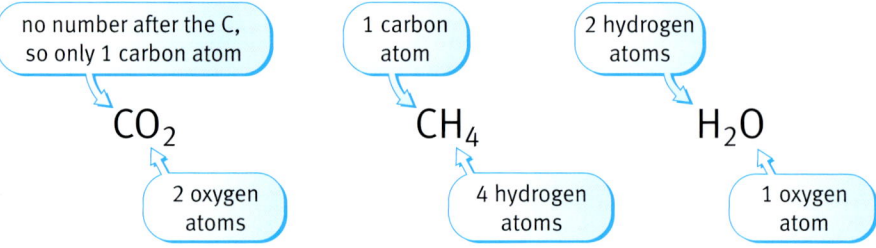

no number after the C, so only 1 carbon atom

CO_2

2 oxygen atoms

1 carbon atom

CH_4

4 hydrogen atoms

2 hydrogen atoms

H_2O

1 oxygen atom

Chemists use symbols and formulae to describe chemical changes. They write two kinds of equation:

- **word equations**
- **balanced symbol equations.**

Word equations

In a word equation the rule is to write the names of the **reactants** on the left and the **products** on the right. An arrow (⟶) between them is short for 'goes to', meaning 'changes into' or 'becomes'.

Worked example

The word equation for methane burning

Methane burns by reacting with oxygen in air. The equation may be written:

methane + oxygen ⟶ carbon dioxide + water

THE SCIENCE

Balanced symbol equations

Writing an equation in symbols makes it much easier to see what is happening chemically during a reaction. Atoms are not created or destroyed during reactions. They are simply rearranged. This means that the number of atoms of each type must be the same on both sides of the equation. In this sense the two sides of the equation are 'equal' and that is why a chemical equation must always 'balance'.

State symbols

State symbols are used to show whether the chemicals in an equation are solid (s), liquid (l), gaseous(g), or dissolved in water (aq).

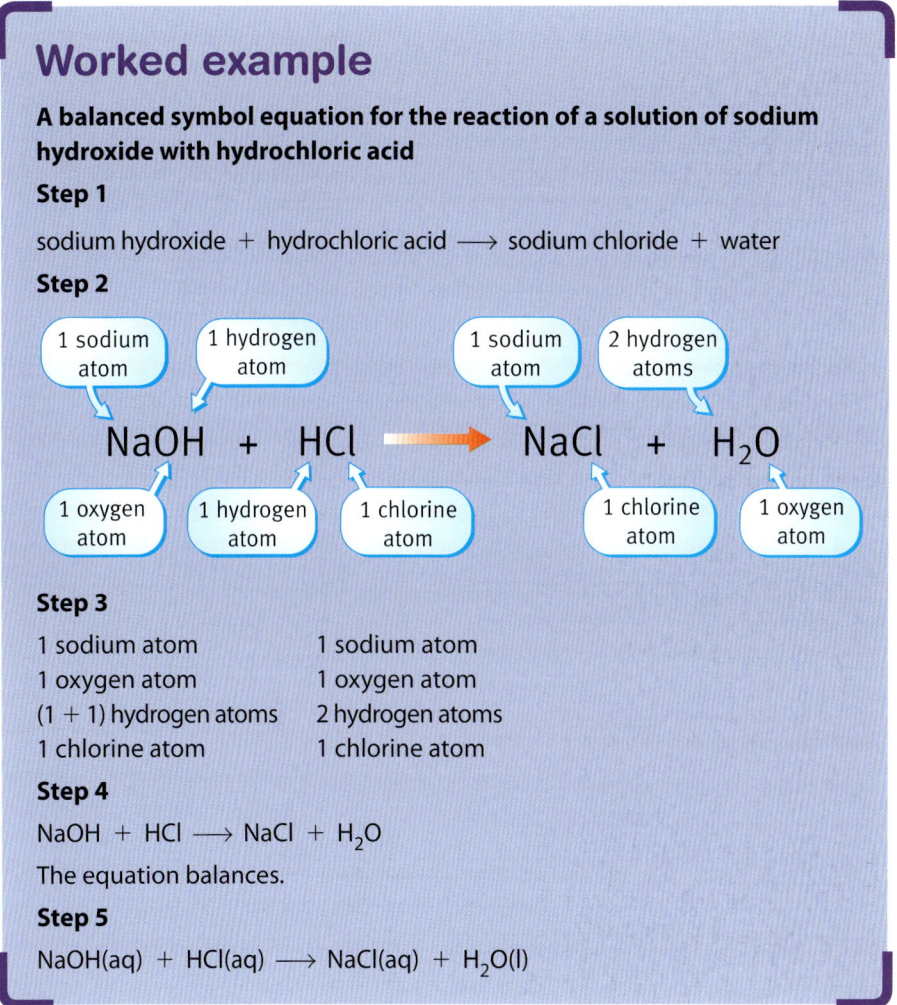

Worked example

A balanced symbol equation for the reaction of a solution of sodium hydroxide with hydrochloric acid

Step 1

sodium hydroxide + hydrochloric acid ⟶ sodium chloride + water

Step 2

| 1 sodium atom | 1 hydrogen atom | | 1 sodium atom | 2 hydrogen atoms |

$NaOH + HCl \longrightarrow NaCl + H_2O$

| 1 oxygen atom | 1 hydrogen atom | 1 chlorine atom | 1 chlorine atom | 1 oxygen atom |

Step 3

1 sodium atom	1 sodium atom
1 oxygen atom	1 oxygen atom
(1 + 1) hydrogen atoms	2 hydrogen atoms
1 chlorine atom	1 chlorine atom

Step 4

$NaOH + HCl \longrightarrow NaCl + H_2O$

The equation balances.

Step 5

$NaOH(aq) + HCl(aq) \longrightarrow NaCl(aq) + H_2O(l)$

Rules for writing balanced equations

STEP 1 Write down a word equation.

STEP 2 Underneath write down the correct formula for each reactant and product.

STEP 3 Check to see if the equation needs balancing.

STEP 4 Balance the equation if necessary, by putting numbers in front of the formulae.

STEP 5 Add **state symbols**.

NEVER change the formula of a compound or element to balance the equation.

Question

3 Use the five steps shown in the worked example to write balanced symbol equations for the following reactions: (see page 290 for chemical formulae)

a potassium hydroxide solution with dilute nitric acid (the salt formed is KNO_3)

b magnesium metal with dilute hydrochloric acid (the salt formed is $MgCl_2$)

c solid zinc carbonate with dilute sulfuric acid (the salt formed is $ZnSO_4$)

d calcium hydroxide with dilute hydrochloric acid (the salt formed is $CaCl_2$).

Chemical formulae

Salts are related to a parent acid, and to a parent metal oxide, hydroxide, or carbonate.

Parent acid	Related salts	Examples
hydrochloric acid, HCl	chlorides	sodium chloride, NaCl magnesium chloride, $MgCl_2$
sulfuric acid, H_2SO_4	sulfates	potassium sulfate, K_2SO_4 copper sulfate, $CuSO_4$
nitric acid, HNO_3	nitrates	potassium nitrate, KNO_3 calcium nitrate, $Ca(NO_3)_2$

Parent metal oxide, hydroxide, or carbonate	Related salts	Examples
sodium hydroxide, NaOH	sodium salts	sodium chloride, NaCl sodium sulfate, Na_2SO_4 sodium nitrate, $NaNO_3$
potassium hydroxide, KOH	potassium salts	potassium chloride, KCl potassium sulfate, K_2SO_4 potassium nitrate, KNO_3
magnesium oxide, MgO magnesium hydroxide, $Mg(OH)_2$ magnesium carbonate, $MgCO_3$	magnesium salts	magnesium chloride, $MgCl_2$ magnesium sulfate, $MgSO_4$ magnesium nitrate, $Mg(NO_3)_2$
calcium carbonate, $CaCO_3$	calcium salts	calcium chloride, $CaCl_2$ calcium sulfate, $CaSO_4$ calcium nitrate, $Ca(NO_3)_2$
copper oxide, CuO	copper salts	copper chloride, $CuCl_2$ copper sulfate, $CuSO_4$ copper nitrate, $Cu(NO_3)_2$
zinc oxide, ZnO zinc carbonate, $ZnCO_3$	zinc salts	zinc chloride, $Zn\,Cl_2$ zinc sulfate, $ZnSO_4$ zinc nitrate, $Zn(NO_3)_2$

The 'OH' part of a hydroxide is a pair of atoms that go together. In the hydroxides of calcium and magnesium there are two OHs for every metal atom. So in the formula the OH appears in brackets with a 2 after it.

Ammonium salts form when ammonia (NH_3) reacts with an acid. Ammonium salts include:
- ammonium chloride, NH_4Cl
- ammonium sulfate, $(NH_4)_2SO_4$
- ammonium nitrate, NH_4NO_3

1 magnesium atom

the '2' applies to everything inside the brackets, so there are 2 oxygen and 2 hydrogen atoms

$Mg(OH)_2$

Concentrations

Scientists often work with chemicals in solution. The dissolved chemical is the **solute**. It is often important to produce solutions with exactly the right **concentration**.

Volumes of solutions

A volume of one **litre** is the same as 1000 ml (millilitres). Millilitres (ml) are the units usually shown on volumetric glassware, medicine spoons, and everyday liquids bought in supermarkets.

One millilitre is the same as one cubic centimetre, and so 1 ml = 1 cm^3. In laboratories, scientists often prefer to write 1 cm^3 instead of 1 ml. So 1 litre = 1000 ml = 1000 cm^3.

Concentration of solutions

Chemists find the concentration by measuring the mass of chemical in one litre of solution. They weigh out the solid and then dissolve it in water. Next they transfer the solution to a volumetric flask and add water until the volume of solution, after mixing, is exactly the volume stated on the flask.

A technician using a pipette to prepare dilutions for analysis in a colourimeter. Accurate working makes sure that the concentration of each solution is exactly known.

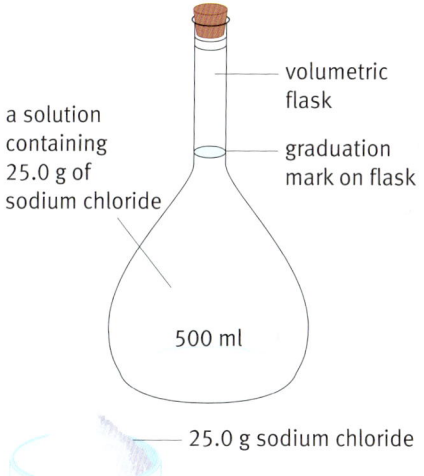

A solution made by dissolving exactly 25.0 g of sodium chloride in water and making the volume of solution up to 500 ml in a volumetric flask.

Questions

1. What is the concentration of the solution in the volumetric flask in the diagram:
   a in g/litre?
   b in g/cm^3?

2. What is the mass of copper sulfate in 10 cm^3 of a solution of the salt if the concentration is 30 g/litre?

3. What is the mass of potassium nitrate in 40 cm^3 of a solution of the salt if the concentration is 0.2 g/cm^3?

Worked example

What is the mass of magnesium sulfate in 50 cm^3 of a solution of the salt if the concentration is 40 g/litre?

The volume of the sample = 50 cm^3 = $\dfrac{50}{1000}$ litres

The concentration shows that there are 40 g of the salt in 1 litre of the solution

The mass of magnesium sulfate in the 50 cm^3 of the solution

= 40 g/litre × $\dfrac{50}{1000}$ litres

= 2 g

THE SCIENCE

Key words
- solute
- concentration
- dilution
- litre

Try to imagine what the world would be like with no clouds, no brilliant sunsets, and no fog on city streets. Think how different your diet would be with no butter or milk, no ice cream or jelly, no ketchup, or salad cream. Could you live without cosmetics, medical creams, or paint?

What all these things have in common is that they consist of one chemical very finely dispersed (spread) in another.

Emulsions

Milk is an example of an **emulsion**. It consists of tiny droplets of an oily liquid dispersed in a watery liquid.

Many emulsions tend to separate into two layers unless an **emulsifying agent** is present in the mixture. You can see this in a salad dressing made with oil and vinegar. Shaking the two liquids together breaks up the oil into droplets. But if you leave the dressing to stand, the oil droplets soon join up to form bigger drops and this continues until the two liquids separate into two layers again.

The oil and vinegar in this bottle have not been shaken.

The bottle has been shaken, creating a temporary emulsion of oil and vinegar.

The shaken bottle has been allowed to stand; the two layers of oil and vinegar have begun to re-form.

Mayonnaise is an emulsion. It is made with egg yolk. Egg yolk contains lecithin, which is a natural emulsifying agent. It stops the mayonnaise from separating.

THE SCIENCE

Unlike water and oil, which are transparent, emulsions of water and oil tend to have a cloudy appearance. This is because light bends as it passes from oil to water, or from water to oil. The tiny droplets in an emulsion mean that this happens many times and the light is scattered. A substance is only transparent if light can pass through it in a straight line.

Many cosmetics are emulsions. These are of two types: oil-in-water and water-in-oil emulsions.

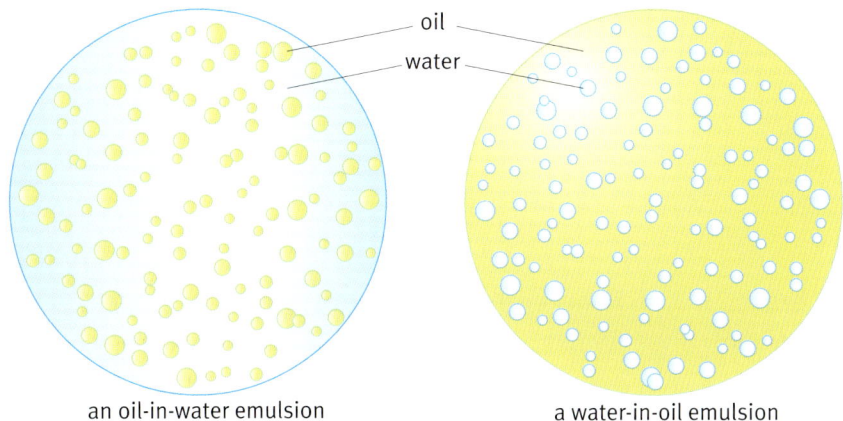

an oil-in-water emulsion   a water-in-oil emulsion

The properties of an emulsion are different to the properties of their ingredients. Skin creams are emulsions that have been carefully formulated to spread easily by hand. On its own the oil in a skin cream would be too greasy and would not be easily absorbed. The water part would run off skin and be difficult to use.

Suspensions

A soluble solid dissolves in a liquid to give a solution. Colourless or coloured, all solutions are transparent.

An insoluble salt does not dissolve. Shake or stir an insoluble solid with a liquid and it forms a **suspension**. A suspension consists of small specks of a solid dispersed in a liquid. Suspensions are cloudy. They may be quite runny, like Milk of Magnesia, or thick like toothpaste.

The insoluble solid in a suspension may settle out quickly. Other suspensions last longer and may take a very long time to separate.

Solutions are transparent. Suspensions are not. This person can see through the solution, but not through the suspension.

Questions

1 Give two examples of emulsions.

2 Give two examples of suspensions.

3 What is the difference between an emulsion and a suspension?

4 Explain why an emulsion looks cloudy when its ingredients are transparent.

Key words
- ✓ emulsion
- ✓ emulsifying agent
- ✓ suspension

Making a soluble salt

Principle

These three general reactions of acids can be used to make **soluble** salts.

acid + metal $\longrightarrow$ salt + hydrogen

acid + metal oxide or hydroxide $\longrightarrow$ salt + water

acid + carbonate $\longrightarrow$ salt + carbon dioxide + water

The procedure below is used to make salts where the metal does not react with water, or the metal oxide, metal hydroxide, or carbonate does not dissolve in water.

CHECK SAFETY
Never work
unsupervised

Procedure

Method 1 – **Reacting an acid with an insoluble solid**

1 Measure the required volume of acid into a beaker. Add the insoluble metal oxide, metal hydroxide, or carbonate bit by bit until no more dissolves in the acid. Warm if necessary. Add a slight excess of the solid to ensure that all the acid is used up.

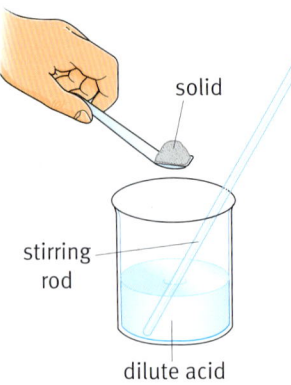

solid

stirring
rod

dilute acid

2 **Filter** off the excess solid, collecting the solution of the salt (filtrate) in an evaporating basin. The **residue** on the filter paper is the excess solid.

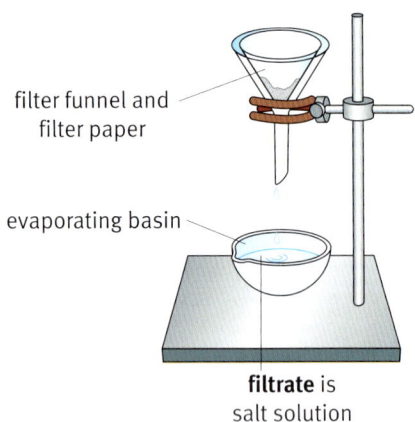

filter funnel and
filter paper

evaporating basin

filtrate is
salt solution

3 Heat gently to **evaporate** some of the water. Evaporate until crystals form when a droplet of solution picked up on a glass rod cools.

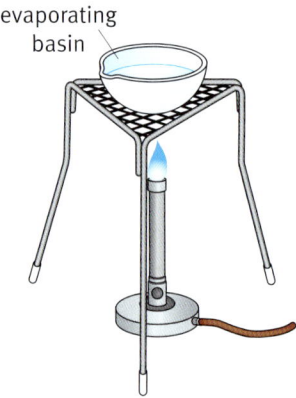

evaporating
basin

4 Pour the concentrated solution into a labelled Petri dish and set it aside to cool slowly and **crystallise**. The crystals are larger if they form slowly.

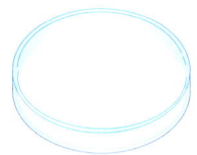

Principle

This procedure is suitable for an alkali or a carbonate that is soluble in water.

Procedure

Method 2 – **Reacting an acid with an alkali or soluble carbonate**

1 Measure the required volume of a solution of the alkali or carbonate into a beaker. Add the acid gradually to the solution in the beaker. Mix well with a stirring rod.

DILUTE ACID

alkali or carbonate

2 Use the stirring rod to take a drop of the mixture and test it with a small piece of indicator paper on a white tile. Continue adding acid until the solution in the beaker is just neutralised.

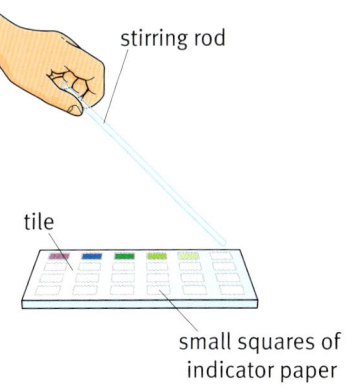

stirring rod

tile

small squares of indicator paper

3 Pour the neutral solution of the salt into an evaporating basin. Heat gently to evaporate. Continue heating until crystals form when a droplet of solution picked up on a glass rod cools.

evaporating basin

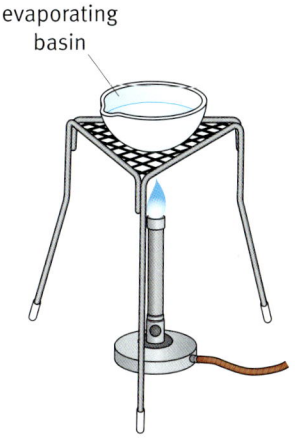

4 Pour the concentrated solution into a labelled Petri dish and set it aside to cool slowly and crystallise. The crystals are larger if they form slowly.

Making an insoluble salt

Principle

Two solutions of soluble salts may react to produce one **insoluble** salt and a new soluble salt. The insoluble salt separates as a solid **precipitate**. The new soluble salt stays in solution.

soluble	+	soluble	→	insoluble	+	soluble
salt A		salt B		salt		salt C
(aq)		(aq)		(s)		(aq)

Procedure

1 Mix the solutions of two soluble salts. A precipitate of the insoluble salt forms immediately.

salt solution **A**

salt solution **B**

2 Filter off the precipitate.

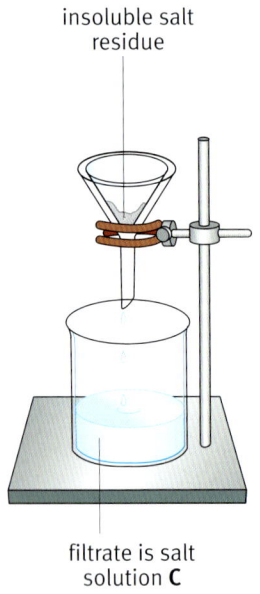

insoluble salt residue

filtrate is salt solution **C**

3 Wash the precipitate on the filter paper with pure water. This removes the soluble salts.

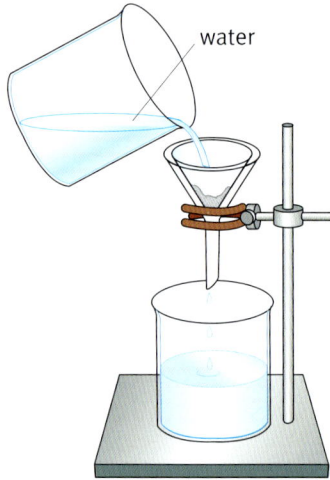

water

4 Open the filter paper and leave the precipitate to dry at room temperature (or in an oven).

filter paper

5 Scrape the dry solid into a weighed sample tube. Re-weigh. Label the container to show the name of the solid and the date.

name of salt

Measuring reaction rates

CHECK SAFETY
Never work
unsupervised

Principle

Chemists measure the **rate of a reaction** by finding the quantity of product produced or the quantity of reactant used up in a fixed time. There are several methods of measuring the rate of reaction, including those shown here and using a gas syringe (see page 283).

Procedure

Collecting and measuring a gas product:

Collecting a gas in a measuring cylinder

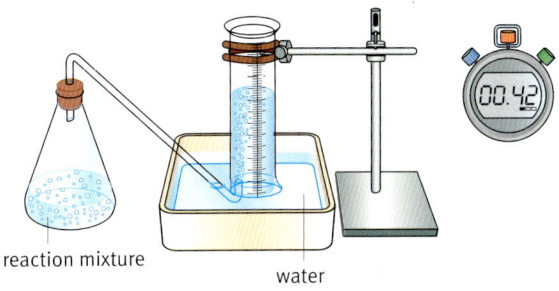

reaction mixture

water

Measuring the loss of mass as a gas is formed

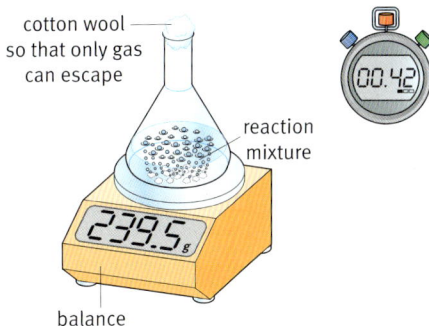

cotton wool
so that only gas
can escape

reaction
mixture

239.5 g

balance

Record the mass at regular intervals, such as every 30 or 60 seconds.

Timing how long it takes for a solution to turn cloudy: reactions that form a precipitate.

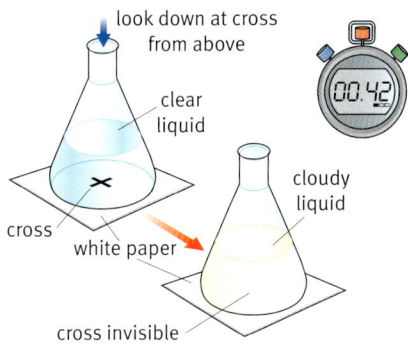

look down at cross
from above

clear
liquid

cloudy
liquid

cross

white paper

cross invisible

Mix the liquids in the flask and start the stopwatch. Stop it when you can no longer see the cross.

Timing how long it takes for a solid reactant to dissolve

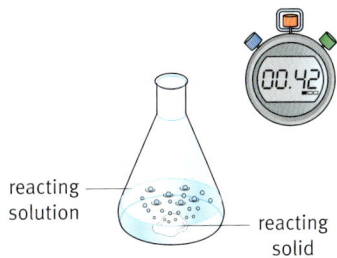

reacting
solution

reacting
solid

Mix the solid and liquid in the flask and start the stopwatch.
Stop it when you can no longer see any solid.

Preparing a solution with a known concentration

Principle

A solution consists of a **solute** (usually a solid) dissolved in a **solvent** (often water).

It is important to be able to make up a solution with a concentration that is accurately known.

Units of concentration are often given in g/litre or g/cm^3.

CHECK SAFETY
Never work
unsupervised

Procedure

1 Accurately weigh the solid.

2 Dissolve the solute in a small amount of solvent, warming if necessary.

stirring rod

3 Transfer the solution to a volumetric flask, letting the liquid run down a glass rod to make sure it does not spill.

stirring rod

paper wedge

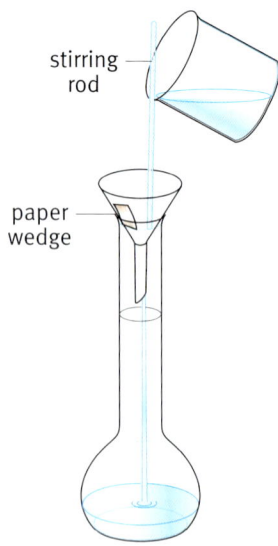

4 Rinse all the solution from the beaaker into the flask with more solvent.

wash bottle

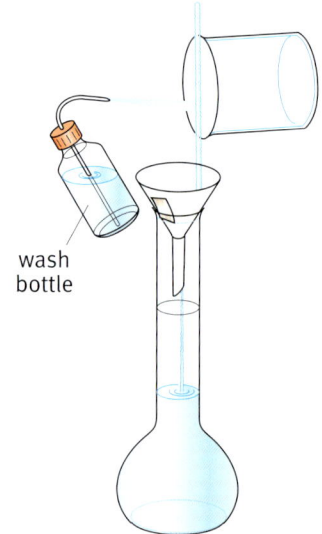

5 Carefully add solvent up to the mark on the flask, drop by drop.

6 Stopper and mix well by turning upside down at least six times.

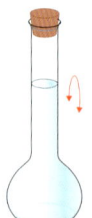

7 Pour the solution into a clean, dry container labelled to show the type of solution, the date the solution was made, and its concentration (in grams per litre).

NaCl solution 20 g/l

Making an emulsion

Principle

An **emulsion** is a mixture of two liquids that do not dissolve in each other. In an emulsion one of the liquids is very finely dispersed in the other. Without an **emulsifying agent**, emulsions tend to separate into two layers. Sometimes the emulsifying agent forms by a chemical reaction during mixing.

Procedure

Procedures for making emulsions vary according to the ingredients. This is an example of a procedure to make a cosmetic emulsion.

1 Measure out the oily part of the emulsion. In this example this is stearic acid.

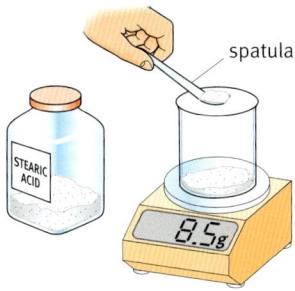
spatula

STEARIC ACID

8.5g

2 Warm and stir the stearic acid until it melts and reaches the required temperature.

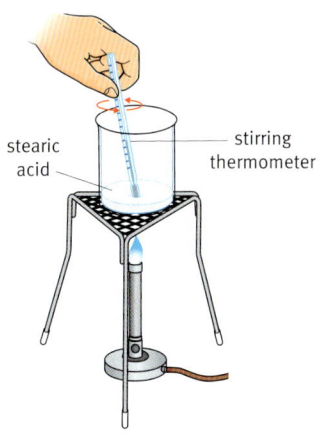
stearic acid
stirring thermometer

3 In a second beaker warm and stir a mixture of glycerol with a very dilute solution of alkali. This makes the watery part of the emulsion.

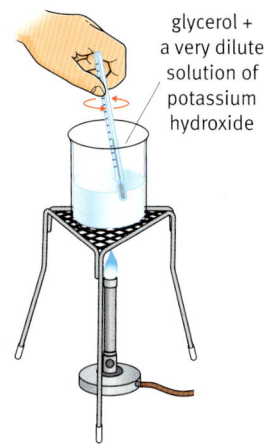
glycerol + a very dilute solution of potassium hydroxide

4 Pour the solution of glycerol in dilute alkali into the molten stearic acid. Keep stirring. Some of the organic acid reacts with the alkali to make the emulsifying agent. Keep stirring as the mixture cools and forms a thick emulsion.

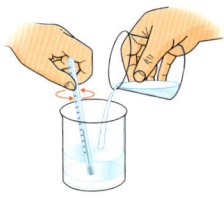

5 Add a small amount of preservative and a few drops of perfume.

6 Transfer the emulsion to a labelled container.

Module Summary

People and organisations

- that a technician's role can include making product formulations, sampling, testing and analysis, using and maintaining equipment, and storing and handling chemicals safely
- how the chemical industry employs skilled chemists, chemical engineers, plant managers, business support staff, and safety officers
- why the chemical industry has important local and national economic impacts
- that the Health and Safety Executive regulates the risks to health and safety arising from the chemical industry
- why harmful chemicals need to be clearly labelled with the relevant hazchem symbols
- why pharmaceuticals need an extensive programme of development and testing over many years
- that product formulations are routinely tested to ensure they conform to national and international standards.

The science

- examples of bulk chemicals and metals that are produced in large quantities
- examples of fine or speciality chemicals that are produced in smaller amounts
- how soluble salts can be made by neutralising an acid
- how insoluble salts can be made by precipitation reactions
- how the rate of a chemical reaction will depend on particle size, concentration of reactants, temperature, and the presence of a catalyst
- that emulsions have different properties to their ingredients and appear cloudy because the dispersed droplets scatter light
- that producing chemicals on a commercial basis involves scaling up laboratory processes and examining yields, rates, and benefits
- how solubility data can be used to predict chemicals that can be made by precipitation
- how the theoretical yield of a reaction can be calculated if the equation and relative formula masses are known
- how data on product formulations will reveal if they conform to the required standards.

Standard procedures

- that chemicals are available in technical, laboratory, and analytical grades, depending on their purity; technicians make solutions of known concentrations
- how soluble salts are made by reacting an acid with an alkali or metal; fertilisers, food ingredients, and medicines are made in this way
- how suspensions are made by dispersing a solid in a liquid, whereas emulsions are made by dispersing one liquid in another liquid.

Review Questions

1. State one chemical made on a large scale. Explain its importance to society.

2. Asif is a laboratory technician.
   a Describe what Asif might be expected to do at work.
   b Asif finds a bottle with this label. What precautions should he take?

3. Explain what happens in each of the following processes:
   a crystallisation
   b evaporation
   c precipitation
   d filtration

4. Use examples to explain the difference between the production of 'bulk' chemicals and 'fine' chemicals.

5. Ralph makes a sample of the soluble salt potassium chloride by dissolving potassium carbonate (K_2CO_3) in hydrochloric acid.
   a Write word and symbol equations for this preparation of potassium chloride (KCl).
   b Describe a procedure to make large pure crystals of potassium chloride this way.

6. Sue has been asked to make samples of these different calcium salts.

Salt	Melting point (°C)	Solubility (g/100 ml)
calcium sulfate	1460	0.21
calcium nitrate	561	121
calcium chloride	772	75

   Which will she have to make by precipitation? Justify your answer.

7. Explain the difference between a suspension and an emulsion. Give an example of each.

8. Trevor needs to prepare a 250 ml solution of sodium nitrate that has a concentration of 170 g/l.
   a Calculate the mass of sodium nitrate that Trevor will dissolve in water to make 250 ml of solution.
   b Describe the procedure that Trevor should follow to prepare the solution.

9. Jim has been working in the laboratory to test the production of a new chemical.
   a Describe the issues that must be considered when the company decides to scale up to make the chemical in much larger quantities.
   b What are the issues they should consider when choosing between a continuous process and a batch process?

10. Magnesium sulfate is made when 16.8 g of magnesium carbonate reacts with excess sulfuric acid.

    $$MgCO_3 + H_2SO_4 \longrightarrow MgSO_4 + CO_2 + H_2O$$

    a Calculate the theoretical yield of magnesium sulfate.
    b Explain why, in practice, the actual yield from the experiment will be less than that calculated in part a.

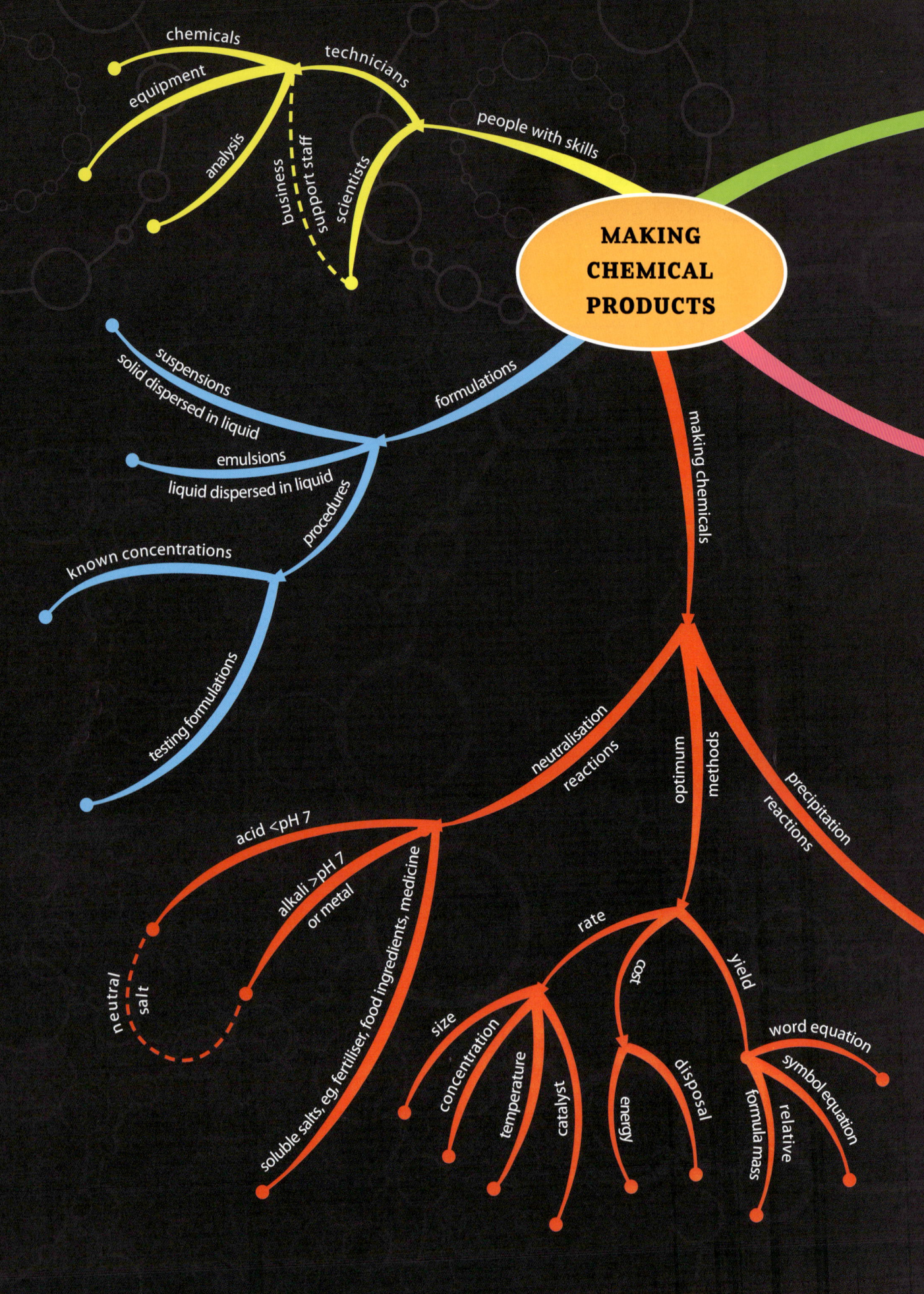

MAKING CHEMICAL PRODUCTS

chemicals
equipment
technicians
analysis
business
support staff
scientists
people with skills

formulations
suspensions
solid dispersed in liquid
emulsions
liquid dispersed in liquid
procedures
known concentrations
testing formulations

making chemicals

neutralisation
reactions
optimum
methods
precipitation
reactions

acid <pH 7
alkali >pH 7
or metal
salts, eg, fertiliser, food ingredients, medicine
neutral
salt
soluble

rate
cost
yield

size
concentration
temperature
catalyst
energy
disposal
formula mass
relative
word equation
symbol equation

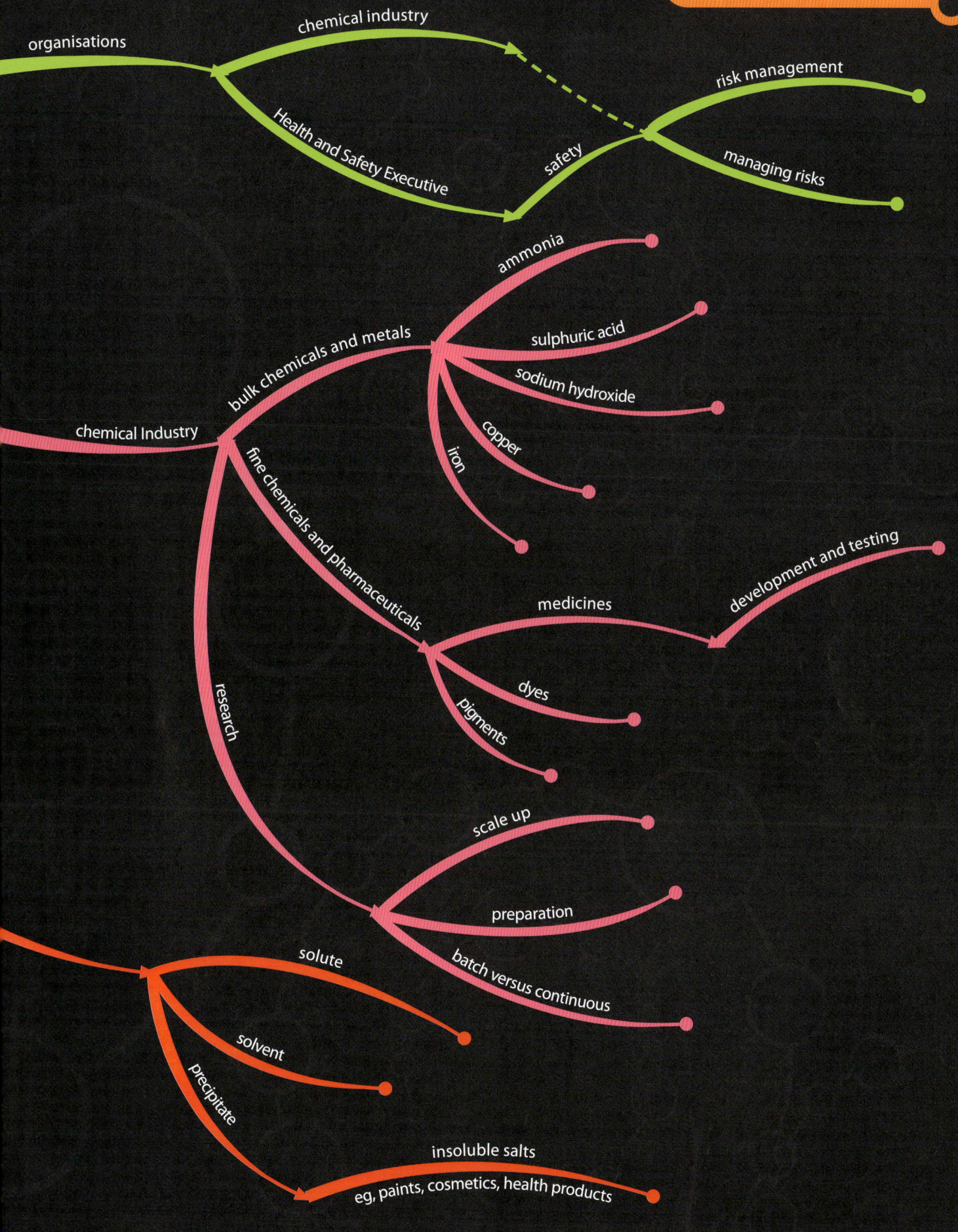

organisations

chemical industry

Health and Safety Executive

safety

risk management

managing risks

chemical Industry

bulk chemicals and metals

ammonia

sulphuric acid

sodium hydroxide

copper

iron

fine chemicals and pharmaceuticals

medicines

development and testing

dyes

pigments

research

scale up

preparation

batch versus continuous

solute

solvent

precipitate

insoluble salts

eg, paints, cosmetics, health products

Glossary

absorb   Sound, light, or vibrations are absorbed when they are converted to heat within a material, and are not reflected or transmitted.

accreditation   A system to recognise the standards and reliability of testing laboratories.

accuracy   A measure of how close a result is to the true value.

acid   A compound that dissolves in water to give a solution with a pH lower than 7. An acid can be neutralised by a base.

acoustic   Relating to sound and hearing.

aerobic exercise   Any exercise of moderate to low intensity that involves large groups of muscles.

aerobic fitness   The ability to keep exercising at low to moderate intensity.

aerobic respiration   The process of releasing energy by using oxygen from the air and breaking down glucose to carbon dioxide and water.

aerodynamic   An object with an aerodynamic shape has less drag force as the object moves through the air.

aesthetic   Assessing the aesthetic quality of an environment means assessing how good it looks to a human observer.

agriculture   The science and practice of growing crops and rearing animals.

airways   Parts of the body through which air enters and leaves the lungs.

alcohol   A family of organic chemicals that contain the OH functional group.

alkali   A compound that dissolves in water to give a solution with a pH higher than 7. An alkali can be neutralised by an acid.

alveolus (plural alveoli)   One of many tiny air-filled sacs at the end of each bronchiole through which gases are exchanged in the lungs.

amniocentesis   A test that is done to check a baby's health. It involves studying some cells from the amniotic fluid in the amnion.

amnion   A membrane that encloses the fetus as it develops in the womb.

amniotic fluid   A fluid that fills the space in the womb where the fetus develops. Amniotic fluid protects the fetus from drying out, and cushions it against injury.

amplifier   A circuit component that increases the amplitude of an electrical signal.

amplitude   Amplitude is the maximum voltage, pressure, or other quantity in a signal.

anaemia   A condition in which either the amount of haemoglobin in blood, or the number of red blood cells, is lower than normal. The person is pale and feels tired.

anaerobic respiration   The process of releasing energy by partially breaking down glucose without using any oxygen from the air.

analytical grade   The most pure version of a chemical available. Pure enough for precise quantitative experiments.

antacid   A chemical that neutralises an acid, such as a metal oxide or a metal carbonate. A soluble antacid is an alkali.

antenatal   Any action taken before birth is antenatal.

antioxidant   Chemicals that 'mop up' the products of metabolism that can damage the body.

aperture   The size of the opening that lets light into a camera when the shutter is open.

APGAR score   Newborn babies are tested for Activity, Pulse, Grimace, Appearance, and Respiration, and given an APGAR score.

aqueous solution   Chemicals dissolved in water.

area   A measure to compare the size of flat rectangular spaces. Calculated by multiplying the lengths of the sides.

artery   Blood vessels that carry blood from the heart to the rest of the body.

artificial insemination (AI)   Inserting semen collected from a stud animal into the uterus of a female animal to bring about fertilisation without sexual intercourse.

aseptic technique   Methods of handling sterile equipment and microorganisms so that cultures of chosen microorganisms are made with no unwanted microorganisms.

atrium (plural atria)   One of the upper chambers of the heart. The two atria pump blood to the ventricles.

automatic control   A control system that does not need frequent action by people is automatic.

bacteria   Microorganisms that consist of simple cells without a nucleus. Some bacteria are harmful; many are useful.

balanced symbol equation   An equation representing a chemical reaction using symbols and formulae, showing equal numbers of each type of atom each side of the equation.

batch culture   A process of growing microorganisms in a batch, then the fermenter is emptied and cleaned out before starting another batch.

batch process   A process that produces a certain quantity of a chemical in one operation. An industrial batch process is a large-scale version of a synthesis carried out in laboratory glassware. After producing a batch of a chemical, the apparatus is cleaned and used again.

beats per minute (bpm)   Units for measuring a pulse rate – the average number of times your heart beats in a minute.

bioinformation   Detailed personal information including physical appearance and distinguishing features, fingerprints, and DNA profile.

biomass   A measure of the mass of living material in a sample.

biotechnology   The use of biological materials in technological processes.

birth   The point at which a developing fetus leaves the mother's womb and begins to breathe and feed outside the mother's body.

blood pressure   Pressure (force per unit area) exerted by blood on the walls of a blood vessel. It is usually expressed as two figures: systolic blood pressure over diastolic blood pressure.

body mass index (BMI)   An estimate of body composition based on a weight-to-height ratio.

bone   The hard, living tissue that makes up our skeleton.

bread wheat   A cultivated wheat species used for making bread.

breathing rate   The number of breaths taken in a given time, usually per minute.

breathing system   The airways and lungs aided by the diaphragm and ribs.

brittle   A brittle material snaps cleanly when it breaks as the body of the material cracks. A small deformation is enough to break it.

bronchiole   Small air tubules leading off from the bronchi.

bronchus (plural bronchi)   Tube leading from the windpipe (trachea) to the right or left lung.

bulk chemicals   Chemicals made by the chemical industry on a scale of thousands or millions of tonnes per year, such as ammonia, sulfuric acid, and sodium hydroxide.

caesarean section   A surgical operation used to remove a baby from the womb when birth through the vagina is not possible.

calibrated   Checked for accuracy with a suitable standard.

calibration   Calibration is the comparsion of a measuring instrument with an accurate standard.

calibration graph   A graph plotted for known concentrations of a coloured solution using a colourimeter allowing the concentration of unknown solutions to be calculated.

cancer   A general term used to describe a malignant growth in body tissue. Cancer is caused when cells divide in an abnormal or uncontrolled way.

capillary   Narrow blood vessels between arteries and veins. Capillaries have very thin walls, only one cell thick. This allows gases, nutrients, and waste products to be exchanged between blood and tissues.

carbonate   A chemical with a CO_3 group.

cardiovascular system   The organ system that circulates blood around the body. It consists of the heart and blood vessels.

cartilage   A connective tissue that is tough and flexible. Cartilage is found at the ends of bones, in joints, in the nose, and in the external ear.

catalyst   A chemical that speeds up a chemical reaction without itself being used up in the reaction.

ceramic   A material such as brick or pottery that is hard, strong in compression, and weak in tension.

cervix   The opening of the uterus that projects into the vagina. The cervix dilates (opens wider) during childbirth.

chain of food production   Every stage of growing, transporting, storing, and delivering food from farm to home. Not to be confused with 'food chain' in ecology, which is a list of living things showing the flow of energy by feeding relationships in an ecosystem.

charge   An electrical properly that can be positive or negative. A positive charge attracts a negative charge. A negative charge attracts a positive charge.

cheese   A protein-rich food made by fermenting milk.

chemical formula   A way of describing a chemical using symbols for atoms. It gives information about the number of different types of atom in the chemical.

chromatogram   The result of chromatography, showing the positions of the separated components.

chromatography   An analytical technique in which the components of a mixture are separated by the movement of a mobile phase through a stationary phase.

chymosin   An enzyme that makes milk protein form clots.

climate change   The steady change in conditions of temperature, humidity, cloud cover, and wind movements across the planet.

coating   A very thin layer of one material covering another material.

colony counts   A method of measuring the number of bacteria in a sample by spreading the sample on a plate of culture medium and counting the number of colonies.

colourimeter   An analytical instrument that measures colour intensity.

comparator   An electronic device that compares two or more electronic input levels and gives different outputs depending on the result.

complementary   Materials with complementary properties have different properties that combine to do a particular job.

composite   A material made up of two or more materials combined, which combines the properties of the constituent materials.

compound   A chemical made of two or more elements bonded together.

compression   A material is in compression when it is being squashed by forces pushing together on it.

compressive strength   The maximum load that a material of standard shape and size can withstand before crumbling.

concentration   The quantity of a chemical dissolved in a stated volume of solution.

consistency (of a product)   The proportion of a manufacturer's products that fully meet the specification (one aspect of product standards).

contamination   Contamination of a sample means something extra has got into the sample, so it is no longer representative.

continuous culture   A process of growing microorganisms continuously, with nutrients added and products removed at intervals so the process does not have to stop.

continuous process   A process for manufacturing chemicals on a large scale in industry that operates 24 hours a day. Raw materials are constantly fed into the plant and products continuously removed.

contrast   A measure of how easy it is to distinguish dark areas from lighter areas.

converging lens   A lens that bends light rays so that they are closer together. A converging lens is thicker in the middle than at the edges.

core body temperature   The temperature of our vital organs. Core body temperature usually stays constant even if our extremities change.

coronary arteries   The blood vessels that supply the heart.

coronary heart disease (CHD)   A malfunction of the heart caused by a blockage in one or more of the arteries supplying blood to the heart.

corrosive   A corrosive substance may destroy living tissues on contact.

counselling   An opportunity to talk to a healthcare professional and get advice. Offered when difficult decisions have to be made.

crystallisation   Producing crystals from a solution by evaporation or cooling.

crystallise   The process of forming crystals. Larger crystals will be formed where the solution is allowed to cool slowly.

culture medium   A preparation of nutrients suitable for growing microorganisms.

dairy   Dairy cattle produce milk. A dairy is a place where milk is processed. Dairy products, such as butter, yoghurt, and cheese, are made from milk.

damping   The use of an absorbent material to stop vibrations.

data logging   The process of keeping a record of data, such as temperature level in a bioreactor, usually electronically.

decibel   Units of the intensity of sound.

deform   To change the shape of something.

degrees Celsius (°C)   Units used to measure temperature. 0 °C is the temperature at which water freezes and 100 °C is the temperature at which water boils.

density   The mass of a material per unit volume.

deoxyribonucleic acid (DNA)   Deoxyribonucleic acid is a polymer: the sequence of units in the polymer forms a code that carries genetic information.

depth of field   The distance between the nearest and the furthest objects that are in focus in an image. When depth of field is narrow, only objects at a particular distance are in focus.

deterioration   A sample has changed over time and is no longer representative.

develop   Treat a chromatogram to reveal previously invisible spots.

diabetes   A disease in which there is excessive excretion of urine. The body cannot control the blood glucose level.

diaphragm   A muscular sheet of tissue that separates the chest cavity from the abdomen.

diastolic blood pressure   The blood pressure when all parts of the heart muscle are relaxed and the heart is filling with blood.

dilution   The addition of solvent to make a weaker solution.

dimmer   A dimmer allows the electric current in a circuit to be controlled to control the brightness of a lamp.

dissolve   When a solid dissolves in water, the molecules separate from one another and mix completely with the water molecules. The dissolved solid will not settle out.

diverging lens   A lens that bends light rays so that they spread further apart. Also called a concave lens, because of its shape.

DNA profile   Analysis of DNA in order to identify individuals.

Down's syndrome   The result of a fertilised egg developing with one extra chromosome. People with Down's syndrome have some degree of learning difficulties and may have other problems such as heart defects.

drugs   Chemicals used in medicine for the treatment, relief, diagnosis, and prevention of disease. People also take drugs for stimulation and relaxation.

dry mass   Mass after water has been removed by drying.

durability   Resistance to abrasive wear or weathering.

durum wheat   A variety of wheat with high levels of gluten.

elastic   An elastic material returns to its original shape after being deformed.

electrocardiogram (ECG)   A graph showing electrical events during the heartbeat. It can be used to reveal irregular heartbeats or damage to heart muscle.

electrode   Electric conductor used to pass an electric current through a liquid or gel.

electrophoresis   An analytical technique that separates charged particles according to their size and the size of their electric charge.

element   A chemical that cannot be broken down chemically into a simpler one. Each element contains only one type of atom.

embryo   A fertilised egg in its very early stages before human characteristics appear.

emphysema   A disease in which the tissues of the airways lose elasticity. The alveoli walls break down, reducing the surface area for gas exchange and causing breathlessness.

emulsifying agent   A chemical that is added to an emulsion to prevent the two liquids separating into layers.

emulsion   A mixture in which one liquid is dispersed through the other in tiny droplets.

environmental health officer   A job role with responsibility for monitoring food at all stages of production to ensure public health and safety.

enzyme   A biological catalyst. Enzymes are very specific for the reaction catalysed. Enzymes are protein molecules.

epidural   An injection for pain relief during birth.

evaporate   A liquid turning into its vapour, often without boiling. Cooling takes place. (Wet things dry by evaporation at room temperature.)

evaporation   The process of liquid turning to vapour below its boiling point. The evaporation of sweat is the body's main cooling system during exercise.

explosive   A material containing a large amount of stored chemical energy that can suddenly detonate.

exponential growth   A stage in the growth of a population of bacteria where the size of the population is doubling rapidly.

extension   The change in length of an object when loaded in tension (how far it stretches).

eyepiece lens   The microscope lens closest to the eye.

Fallopian tubes   The tubes that carry an egg from the ovary to the uterus.

feedback loops   Connections in control systems that make sure the systems react quickly to changes.

fermentation   The process of growing microorganisms in fermenters.

fermenter   A large vessel in which microorganisms are grown to produce a useful product.

fertilisation   The point at which a male sex cell (sperm) fuses with a female sex cell (ovum).

fertilised   A fertilised ovum has fused with a sperm cell.

fertilisers   Chemicals or mixtures of chemicals that are put on the soil to help plants grow better.

fetus   A human embryo once human features become recognisable. This occurs about eight weeks into pregnancy.

fibres   Fibres are threads of material often used in composite materials to stop cracks spreading through the matrix.

filter   A light filter allows some parts of the spectrum to pass through and absorbs all other parts of the spectrum.

filter   A material with fine perforations that will trap coarse particles whilst letting dissolved substances and liquids through.

filtrate   The liquid and dissolved substances that pass through a filter.

filtration   The process of passing material through a filter.

fine chemicals   Chemicals made by the chemical industry in smaller quantities than bulk chemicals, such as drugs and pesticides.

flammable   A material that catches fire easily often by producing flammable vapour at around room temperature.

flexible   A flexible material is easily deformed when a force is applied.

flow diagram   A simplified diagram to show how a complex machine is controlled using boxes to represent the components and arrows to show how they are connected.

fluid-filled dampers   Structures used on bridges and buildings to absorb vibrations.

fluorescent lamp   A fluorescent lamp produces light without generating a lot of heat.

focal length   The distance from the centre of a lens to its focus.

focal plane   The plane in which a camera focuses all parts of an image.

focus   An image is in focus if the edges of objects in the image are sharp and clear.

food poisoning   Illness caused by eating food containing harmful microorganisms or toxins produced by microorganisms.

force   A push or a pull.

forensic   Using scientific methods in the investigation of crime.

formulation   A particular mixture designed to make a product suitable for its use.

frequency   The number of vibrations a second.

fungi   Microorganisms that live by absorbing nutrients from their surroundings. Some fungi are harmful; many are useful.

fungicide   Any chemical that could be used in agriculture to kill fungi that might make a crop mouldy.

gene   A length of DNA that codes for a particular protein.

general practitioner (GP)   A qualified doctor who is also the first point of contact with the National Health Service for most people. A GP provides a range of care within the local community.

genetic modification (GM)   The production of a new combination of genetic material by direct alteration of a section of DNA.

germination   The start of growth of a seed after a period of not growing.

germination rate   The number of seeds that successfully start to grow as a percentage of the total number of seeds tested.

gestational diabetes   A condition that can arise during pregnancy when the mother's body fails to control her glucose level.

graduation   A line on a container or ruler that marks a measurement.

growth chart   A chart comparing a child's length (or height) and weight with other children of the same age. It also shows how quickly a child is growing.

haemoglobin   The protein molecule in red blood cells. Haemoglobin gives blood its red colour and transports oxygen round the body.

hardness   The resistance of a material to scratches or dents.

harmful   A chemical that can cause damage to the body through contact with the eyes, skin, or inhalation.

harvest   A gathered crop or other product.

hazchem   Warning labels found on chemical containers that indicate the potential hazard that they present.

health and safety   The main concerns of those responsible for staff in a workplace. All workplaces must be safe places for people to work and not cause ill health.

health centre   The local place where a community can access healthcare services.

heartburn   A condition in which acid from the stomach irritates the oesophagus in the upper chest, close to the heart. Common in pregnancy.

herbicide   Any chemical that could be used in agriculture to kill other plants (weeds) that compete with the crop and reduce the yield.

Hertz   The unit used for frequency.

hormone   A chemical produced in an animal's body that controls the activity of another body part, for example, reproductive hormones control fertility.

hormone treatment   In IVF, hormone treatment encourages the ovaries to produce many eggs at once, instead of the normal one per month.

howl   Howl is the unpleasant loud sound (also called feedback) that is produced when sound from the loudspeaker leaks back to the microphone.

hydroxide   A chemical with an OH group.

hypothermia   An abnormally low body core temperature, below 36 °C.

image   An image is formed when light from an object is reflected from a mirror or passes through a lens.

implantation   A fertilised egg becomes embedded into the lining of the uterus.

in vitro fertilisation (IVF)   A process in which egg cells are fertilised by sperm in a glass container. The five stages of IVF are stimulation, egg collection, fertilisation, implantation, and pregnancy. Children conceived by IVF are sometimes called test-tube babies.

incandescent lamp   The light of an incandescent lamp comes from a white-hot strip of metal.

indicator   A solution that gives a distinct colour at a specific pH or in the presence of a particular chemical.

indicator organisms   Organisms that are sensitive to conditions in an environment such as levels of oxygen or sulfur dioxide.

indicator solution   A solution that gives a distinct colour at a specific pH or in the presence of a particular chemical.

infrared (IR)   Electromagnetic radiation beyond the red end of the visible spectrum. It can be detected by its heating effect.

input   An electronic device that sends a signal giving information about conditions at some point of a controlled system.

insecticide   A chemical that kills insects.

insoluble   A chemical that does not dissolve in a certain liquid.

instrument   A measuring instrument is any device used to make measurements, such as time, length, mass, electric current.

intensity   In colour matching and colourimetry, intensity is the brightness or depth of a colour.

intensity   The loudness of a sound, or how much energy reaches your ear per second. Intensity depends on amplitude.

intercostal muscle   A set of muscles between the ribs. The intercostal muscles contract during breathing to draw air into the lungs as you breathe in, and force air out as you breathe out.

irritant   An irritant substance is not corrosive but may cause inflammation or lesions to the skin or eyes on contact.

jaundice   When the body does not get rid of coloured waste products; for example, when the kidneys are not working well, the skin can appear yellow. This is jaundice.

joint   A point of contact between two or more bones.

justifiable   Backed by sound, reliable evidence.

kidney   Organs that remove wastes from the body, and regulate the levels of water and salts in the body.

labelling   Marking a sample with information to allow you to identify it fully.

laboratory grade   A chemical that is pure enough for general laboratory work.

lactic acid bacteria   A group of bacteria that can use the sugar lactose as a source of energy. They produce lactic acid as a product of fermentation.

lag phase   A stage in the growth of a population of bacteria where the size of the population is not increasing.

laser   A light source that emits a narrow intense beam of light of a single colour.

lens   Lenses are usually made of glass. They are shaped to change the direction of the rays of light that pass through.

ligament   Connective tissue that joins two bones together.

light sources   Light sources emit light and can be seen in the dark, for example, the Sun, light bulbs, and LEDs.

light-sensitive surface   A light-sensitive surface at the back of a camera detects the light from an object and produces an image on film or a screen.

linear scale   A measurement scale with equally spaced lines where the spacing represents equal quantities.

litigation   The conduct of a case in court against someone who has done something wrong.

litmus   An indicator chemical that is red in acidic solutions and blue in alkaline solutions.

litre (l)   1000 cm³ or 10 cl or 1 cubic decimetre. One litre of pure water weighs one kilogram.

loudness   How you perceive a sound's intensity. As well as the energy of the sound, it also depends on the frequency and how long the sound lasts.

loudspeaker   A loudspeaker produces a louder copy of a sound.

lungs   A pair of organs in the chest cavity. They have a large surface area because of their millions of alveoli and so allow gaseous exchange to take place. The lungs are linked to the air by a system of tubes, the airways.

magnification   A measure of how much an image has increased in size compared to the object it shows.

magnifying power   The factor by which the size of the image in a microscope is greater that the size of the object. Also known as magnification.

maintenance   The process of monitoring, cleaning, repairing, and regularly servicing equipment to ensure it works well.

manure   The droppings of farm animals used as fertiliser.

matrix   The matrix in a composite material is a polymer solid that is strengthened by the inclusion of fibres.

medical history   A medical history is a record of any illnesses, conditions, and medical procedures a person experiences.

meniscus   The water surface in a narrow tube curves to form a meniscus.

menopause   The time in a woman's life when she stops ovulating and menstruating.

menstrual cycle   The cycle of changes in a woman's body associated with ovulation. The cycle takes approximately 28 days.

menstruation   The shedding of the uterus lining each month as part of the menstrual cycle.

metabolic rate   The rate at which all the chemical reactions in the body take place.

metal   A material that is shiny and a good conductor of heat and electricity. Metals are elements on the left-hand side of the Periodic Table.

micrograph   A microscope image recorded as a photograph.

microorganism   Very small living organisms; microorganisms may be fungi (including yeasts and moulds), bacteria, or viruses.

microphone   A microphone converts sound into an electrical signal.

midwife   A healthcare professional who deals with pregnancy and postnatal care in our communities.

millimetres of mercury (mm Hg)   Units used to measure blood pressure.

mobile phase   The solvent that carries chemicals from a sample through the stationary phase during chromatography.

moulds   Types of fungi that grow on damp organic matter.

multiple birth   Any birth of more than two babies at one time.

muscle   A tissue that contracts to allow movement in the body.

mycoprotein   A protein made by fungi.

National Health Service (NHS)   The organisation in the UK that provides healthcare to our citizens.

negative feedback   A control system where a condition is monitored and the system reacts to reverse any change in that condition, for example, to cool the system if the temperature rises and to heat the system if the temperature drops.

neutralise   To cancel out the acidity or alkalinity of a solution until pH 7 (neutral) is reached.

non-aqueous   A liquid other than water.

nutrients   Chemicals that are needed for organisms to grow well.

obese   Term used to describe somebody with an unhealthy excess of fat stored in the body.

object   An object is seen when light from a source is reflected from the object towards our eyes.

objective lens   The main magnifying lens of a microscope, nearest the object being examined.

oesophagus   The muscular tube connecting the mouth to the stomach.

opaque   An opaque material absorbs light and does not allow it to pass through.

organic farming   Farming using crop rotation, natural fertilisers, and natural pest control systems while avoiding artificial pesticides and herbicides.

outliers   Measurements that appear inconsistent.

output   An electronic device that produces an electrical signal that can be detected and interpreted.

ovaries   A pair of organs in a woman that produce eggs.

ovulation   The release of an egg from an ovary, ready for fertilisation.

ovum (plural ova)   An egg: the female reproductive cell.

oxide   A compound of an element with oxygen.

oxidising   A chemical that can supply oxygen, allowing another chemical to burn more vigorously, even in an air-free environment.

palpated   Examined using the sense of touch (felt).

paramedic   Someone who can provide medical care for people who are ill or who have had an accident in an emergency situation before they reach hospital.

pasteurisation   The heat-treatment of milk and milk products, which kills bacteria without affecting the flavour or mineral and vitamin content.

pathogen   Organism that causes disease.

pesticides   Any chemical that could be used in agriculture to kill pests that might damage or destroy a crop.

pH   A measure of acidity or alkalinity.

pH scale   A 14-point scale that shows the acidity or alkalinity of a solution in water. Acids have a pH less than 7, alkalis have a pH greater than 7.

pharmaceuticals   A medicine used in the treatment or prevention of a disease.

pitch   How high or low a sound is. Pitch depends on frequency.

placenta   A spongy structure that connects an embryo or fetus to the mother during pregnancy. Nutrients, gases and waste products are exchanged between the fetus and mother through the placenta.

plastic   A plastic material does not return to its original shape after being deformed.

platelets   Cell fragments found in blood. Platelets play a role in clotting.

polymer   A material made up of very long molecules. The molecules are long chains of smaller molecules.

postnatal   Any action taken shortly after birth, in caring for babies or their mothers, is postnatal.

precipitate   A solid product that comes out of solution in a chemical reaction.

precipitation   The process of forming a precipitate.

precision   A measure of the certainty that a reading is close to the true value.

pre-eclampsia   A condition that can arise during pregnancy in which the mother's blood pressure rises to dangerous levels and protein is excreted in her urine.

pregnancy   The period of time from the fertilisation of an egg to the birth of a baby.

primary colours   The primary colours are red, blue, and green; other colours are made by mixing these three colours.

primary healthcare   Care provided within the community at the first point of contact with a health team.

primary standard   A primary standard instrument is an instrument that has been checked against other primary instruments around the world.

processor   An electronic device that delivers an output depending on the inputs it receives.

product   The chemical(s) formed following a reaction.

product standard   A series of tests for assessing products or materials to make sure they are of an acceptable quality.

professional development   An activity such as training that develops skills relevant to a profession or job.

proficiency tests   Tests to check the accuracy of analytical procedures.

protein   A biological polymer. The sequence of units in the polymer is determined by DNA sequences. Examples of proteins are enzymes, muscle fibres, gluten.

public health   The general health and well-being of people in the population. Foods contaminated with toxins or pathogens cause a risk to public health.

pulse   The rhythmic expansion of the arteries, coinciding with the contraction of the left ventricle of the heart.

pyrexia   Fever, a condition in which the core body temperature rises to over 37.5 °C.

qualitative   Relating to what substances are present.

quality A measure of how good a foodstuff is for its purpose. For example, milk for making butter is high quality if it has a high proportion of butterfat.

quantitative Relating to the amount of substances present.

random error An error that has a variable effect on measurements.

range The lowest and highest concentrations of a substance that can be determined by a particular analytical method.

rate of reaction A measure of how quickly a reaction is happening. It may be found by measuring the rate of disappearance of a reactant or the rate of appearance of a product. In most chemical reactions the rate changes with time.

reactant A chemical that takes part in a reaction.

reacting masses The masses of chemicals that react together, and the masses of products that are formed in a reaction. Reacting masses are calculated from the balanced symbol equation using relative atomic masses and relative formula masses.

rectum The final, straight portion of the large intestine.

red blood cells Blood cells containing haemoglobin. In humans, they are biconcave discs and have no nucleus.

referral Treatment at a hospital or other institution arranged through a GP.

reflection Sound or light is reflected when it bounces off a material, and is not absorbed or transmitted.

refraction The bending of a ray of light as it goes from one medium to another and changes speed.

refractive index A measure of how much a material changes the direction of light rays at its surface. Refractive index is the speed of light in air divided by the speed of light in the material.

regulation Regulations are the rules made by governments or other agencies to control what people do. Regulations are often designed to protect the health and safety of the public. Other regulations aim to limit harm to the environment.

rehabilitation The process of restoring activity and health to somebody who has been injured or sick.

relative atomic mass The mass of an atom of an element compared with the mass of an atom of carbon. The relative atomic mass of carbon is defined as 12. On the same scale the relative atomic mass of hydrogen is 1.

relative formula mass The combined relative atomic masses of all the atoms in a molecule. To find the relative formula mass of a molecule, you just add up the relative atomic masses of its atoms.

repeatable When the instructions for a task allow a new person to follow them exactly, a procedure is repeatable.

representative Having properties that are the same as the rest of the item or set.

reproducible When further tests give the same result, the test result is reproducible.

reproductive system All the organs of the male or female body involved in sexual reproduction.

Resazurin A chemical that changes colour (from blue to pink to white) if mixed with milk that contains active bacteria.

residue Solid substance that collects on the filter paper during filtration, or is left in the flask after distillation.

resolving power The distance between two points that can still be seen as separate under a microscope. Also known as resolution.

respiration A process in the body that generates energy from the breakdown of food.

respiratory system All the organs and tissues involved in breathing and gas exchange. The respiratory system includes the nose and nasal passages, the pharynx, the windpipe (trachea), the bronchi, and the lungs.

retardation factor (R_f) The movement of a chemical relative to the movement of the solvent front in paper or thin-layer chromatography. R_f = distance moved by a chemical/distance moved by the solvent.

rigid A rigid structure is stiff and strong.

risk The probability of an outcome that is seen as undesirable, associated with some behaviour or process.

safety Food safety ensures that foods do not make people ill.

safety curtain A fire-proof curtain that can separate the stage from the audience in a theatre.

safety margin A specification standard beyond the minimum necessary to prevent an accident, for example, ensuring the strength of a material or structure is greater than the expected maximum load.

salt A chemical formed when an acid reacts with a metal, a metal oxide, or a metal carbonate.

sample A part of something taken for analysis.

scaling up To find a way of producing a chemical on a large scale by adapting the reaction that produced it during research and development.

scanning electron microscope (SEM) An electron microscope that scans the surface of a sample.

secondary colours The secondary colours are cyan, magenta, and yellow; they are each made by combining two of the primary colours.

secondary standard A secondary standard instrument is an instrument that has been checked against a primary standard instrument.

selective breeding Improvement of stock by choosing animals or plants to breed from that show good combinations of characteristics.

semen Liquid produced by a male animal during sexual reproduction, containing sperm and nutrients.

semi-quantitative Giving an indication of the amounts of substances present.

senescence The process of ageing; in a culture of bacteria, the stage when more bacteria are dying than are being formed.

sensor An electronic device that sends a signal indicating the conditions in a controlled system.

shivering The involuntary contraction of muscles to warm the body.

shutter A mechanism in a camera that operates when a picture is taken. It opens for a fraction of a second to let light onto the film or electronic screen.

skimmed milk Milk from which most of the butterfat has been removed.

skinfold measurement Measurements of a fold of body tissue. A skinfold measurement includes skin and the fatty tissue beneath, but no muscle.

soluble Describes a chemical that dissolves in water (or another named solvent).

solute Substance (usually a solid) that is dissolved by a solvent to form a solution.

solution A liquid containing dissolved chemicals.

solvent A liquid in which chemicals dissolve to make a solution.

solvent front The level reached by the top or the moving solvent during paper or thin-layer chromatography.

sonography Another name for ultrasound scanning.

spectrophotometer A piece of laboratory equipment that measures how much light a sample absorbs over a range of different wavelengths.

speed How far something travels in a given time. Usually measured in m/s or km/h.

sperm The male gametes in animals.

sphygmomanometer An instrument used for measuring blood pressure in the arteries. It usually consists of an inflatable cuff, a stethoscope, and a pressure gauge.

spina bifida A birth defect in which the vertebrae around the spinal cord have not formed fully as the fetus has grown. It may cause physical difficulties for the individual affected.

spring wheat A variety of wheat that can be planted in spring and will germinate and grow quickly to provide a late summer harvest.

standard A standard measuring instrument is one with a known high accuracy.

standard procedure Precise instructions written so that scientists can carry out a preparation, analysis, or test in the same way every time.

standard reference materials Materials of known composition that scientists use for reference.

state symbol Letters showing the form that a reactant or product takes including (s) solid, (l) liquid, (aq) aqueous solution, and (g) gas.

stationary phase The medium through which the mobile phase passes in chromatography.

stationary phase A stage in the growth of a population of bacteria where the size of the population remains constant because the growth rate and the death rate are the same.

step test Tests of cardiovascular fitness that involve stepping on and off a bench or step for a period and then taking a pulse reading. The quicker you return to a resting heart rate, the fitter you are.

sterile Sterile materials are clean and free of any living organism, especially microorganisms.

sterilisation (of equipment) Destroying all the microorganisms present by heat or chemical treatment.

stiff   A stiff material resists being deformed when bent.

stroke   An interruption of the blood supply to the brain caused by a blood clot, head injury, or burst blood vessel in the brain.

strong   A strong material withstands a large force before it breaks.

structure   The way a material is shaped or put together, or the way its atoms or molecules are arranged.

sugars   Sugars dissolve in water and taste sweet. They belong to the family of chemicals called carbohydrates. The sugar from cane or beet is sucrose. Other sugars include glucose, fructose, and lactose.

suspension   A mixture of an insoluble solid with a liquid. The solid particles are dispersed through the liquid. You cannot see through a suspension.

sweating   The secretion of a watery fluid onto the skin. Sweating helps the body to cool down as it evaporates from the skin.

switch   A switch in an electric circuit allows components to be turned on and off.

symbol equation   A summary of a chemical reaction using the chemical symbols for the materials involved.

symptoms   The way we describe what we are feeling when we are feeling ill.

synovial fluid   Fluid found in the cavity of a joint. Synovial fluid lubricates and nourishes the joint and prevents two bones from rubbing against each other.

synovial membrane   Loose connective tissue lining the inside of a joint capsule in a synovial (free-moving) joint. It secretes synovial fluid.

synthetic routes   Methods of synthesising a chemical using different series of chemical reactions or different starting reactants.

systematic error   An error that has the same effect on all measurements.

systolic blood pressure   The blood pressure when blood is pumped from the left ventricle to the rest of the body.

tampering   Tampering with a sample means deliberately changing the contents of the sample.

technical grade   The cheapest and least pure variety of a chemical, only suitable where the presence of some impurities would not cause a problem.

temperature receptors   A group of cells that allow the body to detect changes in temperature.

tendon   Connective tissue that joins muscle to a bone.

tensile strength   The stretching force needed to break a material.

tension   A material is in tension when it is being stretched by forces pulling apart.

test kits   A kit, often of coloured chemicals bonded to strips of card or paper, that can be used to make quick semi-quantitative tests of samples.

theoretical yield   The amount of product that would be obtained in a reaction if all the reactants were converted to products exactly as described by the balanced chemical equation.

thermal conductance   The rate of heat transfer through something for a given temperature difference. Something made from a good thermal conductor such as a metal generally transfers a lot of heat.

tinnitus   A ringing sound in the ear.

tough   A tough material is not easily broken and resists cracking.

toxic   A poisonous chemical that can damage the body through skin contact, inhalation, or swallowing.

toxins   Poisonous chemicals. Toxins are produced by some bacteria and moulds.

trachea   The windpipe, which carries air to the lungs.

translucent   A translucent material scatters light as it passes through. Clear images cannot be seen through the material.

transmission electron microscope (TEM)   An electron microscope that passes a beam of electrons through the sample.

transmit   Pass through.

transparency   How much light can pass through a material.

transparent   A transparent material transmits light – it allows light to pass through it. Clear images can be seen through the material.

triage   The process of deciding what the priorities are when deciding who to treat when faced with several patients.

triage nurse   A nurse whose job in an A&E department is to decide who should be treated first.

trigger voltage   The voltage of an electrical signal in a control system that is at a level to trigger a reaction in another part of the system.

true value   The real value of anything that is measured.

turbid   Cloudy. You cannot see through a turbid liquid.

turbidity   A measure of the cloudiness of a sample of liquid, for example, water. Light does not pass through turbid samples.

UHT   Ultraheat treatment involves heating milk to a high temperature for a short time to kill pathogenic microorganisms in the milk.

ultrasound scanning   The use of ultrasonic waves (very high frequency sound waves) to examine internal organs.

ultraviolet (UV)   Electromagnetic radiation beyond the violet end of the visible spectrum. It is ionising and can cause damage to body cells.

umbilical cord   A cord made of blood vessels and connective tissue that attaches an embryo or fetus to its placenta. It carries nutrients and oxygen from mother to fetus, and removes waste products.

uncertainty   An indication of the confidence a scientist has in the reliability of a measurement. It can be expressed as a range of values within which the result must lie.

Universal indicator   A mixture of indicator chemicals that gives a specific colour at each pH across a wide range.

urine   A watery solution of salts and urea produced by the kidneys.

uterus   Another name for the womb.

vaccination   A way of protecting us from catching a disease by introducing a sample from a disease-causing organism into the body. This stimulates our immune system.

vagina   The canal that leads from the uterus to the outside.

valid   Suitable for what was being analysed or tested.

vasoconstriction   Narrowing of blood vessels, which reduces blood flow to the area of the body supplied by the blood vessels.

vasodilation   Widening of blood vessels, which increases blood flow to the area of the body supplied by the blood vessels.

vein   Blood vessels that carry blood to the heart.

ventilation   Ventilation allows fresh air to flow through a building.

ventilation   The exchange of air between the lungs and the environment. Also known as breathing.

ventricle   One of the lower chambers of the heart. The right ventricle pumps blood to the lungs. The left ventricle pumps blood to the rest of the body.

viewfinder   The device in a camera that enables the user to set up the desired image. Looking through the viewfinder, the user adjusts the camera direction and lens position.

viruses   The smallest type of microorganism. A virus consists of genetic material inside a protein coat. Viruses cannot reproduce outside another living organism.

visible light   Visible light is the part of the electromagnetic spectrum that our eyes can detect.

vitamin   A nutrient in our food that is essential for good health but needed in only small amounts.

weak   A weak material breaks with only a small force acting on it.

white blood cells   Blood cells that defend the body against disease.

white light   White light is made up of all the colours of the part of the electromagnetic spectrum that our eyes can detect.

winter wheat   A variety of wheat that can be planted in autumn and will survive the winter after germinating.

womb   The muscular organ that holds a developing baby. The uterus protects the fetus during pregnancy.

word equation   A summary in words of a chemical reaction.

yeasts   A group of single-celled fungi that, in the right conditions, can multiply quickly by budding.

yield   The mass or volume of a product produced by growing and harvesting a crop.

yield   The amount of product obtained from a chemical reaction. It may be measured as the actual yield or the percentage yield.

yoghurt   The result of fermentation of milk using lactic acid bacteria.

Index

accident and emergency (A&E) 54, 55
accreditation 96, 97
acids 276
acoustics 194, 195, 207–211
aerobic respiration 24, 26–28, 36, 45, 237, 238
aerodynamics 159
aflatoxins 229, 244, 245
agar plates 252, 253
agriculture 218–261
AI *see* artificial insemination
air monitoring 92, 93
airways (respiratory system) 26
alcohol 25, 33
alkalis 274, 277
alveoli 26
ammonia 277
amnion 70, 75
amphetamines 33
amplifiers 207
amplitudes 208
anaerobic respiration 237, 238
animals 232, 233, 271
antacids 277
antenatal care 60, 61
antioxidants 24
apertures 202
APGAR scores 66
Arctic animals 86
arteries 24, 30, 31
artificial insemination (AI) 232, 233
aseptic technique 237
atria 31
automatic control systems 240

bacteria 228, 229, 234, 235, 239, 244, 245
baking 222
balanced equations 281, 288, 289
bank cards 163
bases *see* alkalis
batch culture 238
batch processes 287
beauty products 264, 265
BHF *see* British Heart Foundation
bioinformation 142, 143
biomass 237

bioreactors 240, 241, 249
biotechnology 142, 143, 219, 238, 239, 242, 243
birth 73, 77
blankets 161
blood 24, 25, 29–32, 43, 46, 74, 75, 78
body mass index (BMI) 17, 32, 44
body temperature 36, 37, 41
bone 34, 172
botany 105
bread 222, 230, 231
breathing rates 28
breeding 232
brewing 220, 221, 228
British Heart Foundation (BHF) 22
British Standards Institution (BSI) 96, 163

calibration chains 164, 165
calibration graphs 130, 146, 147
cameras 202, 203
cancer 22, 24, 58, 59, 124, 125
Cape Farewell 86, 87
capillaries 30
carbon cycle 158
carbon dioxide 27, 87
carbonates 276
cardiac arrest 56, 57
cardiovascular system 29–31
care plans 57
careers (environment) 90, 91
cartilage 34
catalysts 283
caustic alkalis 274
CEN *see* Committee for Standardization
ceramics 171
cervix 68
chains of food production 225
CHD *see* coronary heart disease
cheese 227, 235, 254, 255
chemical fertilisers 231
chemical formulae 288, 290
chemical industry 268, 269
chemical products 262–303
chemical quantities 280, 281
Chlamydia 65
chromatography 123, 124, 132, 135, 148, 149

chymosin 242, 243
clinic visits 67
coaches 16–18, 20
coatings 198
coffee 235
collisions 284
colony counts 237, 252, 253
colourimetry 130, 131, 146, 147
colour 107, 108, 114, 130, 145, 189, 265
colourings 273
comparators 241
complementary materials 174
composites 158, 166, 167, 171
compound light microscopes 136
compounds 288
compression 168
computer control (lights/special effects) 193
concentrate 272, 273
concentrations 108, 130, 145, 283, 284, 291, 298
concrete 171
conductivity 175, 180
consumer protection 126, 127
contamination 103
continuous culture 238
continuous scaling up processes 287
contrast (images) 106
Control of Substances Hazardous to Health (COSHH) 275
converging lenses 202, 203, 212
core body temperature 36, 41
coronary arteries 31
coronary heart disease (CHD) 22, 24, 25
corrosive chemicals 274
cortisone 33
COSHH *see* Control of Substances Hazardous to Health
cosmetic creams 265
cosmetics 265
cows 232, 233
crime 122, 123, 128, 129, 142
crystallisation 278
curtains 190
cycle helmets 174

dairy industries 226, 227
damping vibration 211
data loggers 240
decibels (dB) 209, 213
degrees Celsius (°C) 41
density 178
deodorants 265
depth of field 106, 137, 138
development checks (postnatal care) 66
diabetes 24, 71
diastolic blood pressure 31, 43, 78
dimmers 204
diverging lenses 203
DNA 123, 140, 141, 143, 243
doctors 63
double-glazing glass 198, 210
drawings (record keeping) 105
Drinking Water Inspectorate (DWI) 92
drugs 25, 33, 77, 266
dry mass measurements 257
drying and storage (bread production) 231
DWI *see* Drinking Water Inspectorate (DWI)
dyes 124, 125, 247
dyestuffs 268

ECA *see* emergency care assistants
ECG *see* electrocardiograms
effects filters 189
EHOs *see* environmental health officers
Elasmax fibres 181
elasticity 169
electrocardiograms (ECG) 17, 32
electrodes 140, 141
electrolysis 285
electron microscopy 138, 139, 221, 244
electrophoresis 140, 141
elements 288
embryo 69, 70, 73
emergency care assistants (ECA) 64
emulsions 273, 292, 293, 299
energy 169, 204
Environment Agency 90, 92, 93

Environment Protection Officers 92
Environmental Health Officers (EHOs) 229
environmental monitoring/ protection 84–119
enzymes 242, 243
equipment maintenance/ operation 95, 271
Escherichia coli (*E. coli*) 223
European Committee for Standardization 163
evaporation 36
explosions 283
eyepiece lenses 136

Fallopian tubes 68
Farewell (Cape) 86–87
farming *see* agriculture
fat 24
FAU *see* Food Advertising Unit
feedback systems 241
female reproductive systems 68
fermentation 235–238, 243
fertilisation 69, 72, 73
fertilisers 231, 268
fetus 69, 70
fibres 122, 160, 167, 177, 181
filters/filtration 189, 197, 278, 286, 287
fingerprints 128, 136, 142
fitness *see* sports and fitness
Fitness Industry Association (FIA) 21
flavourings 273
fluid-filled dampers 211
fluorescent lamps 196
flutter echo 194, 195
focal lengths 203, 212
focal planes 202
food 22, 24, 124, 125, 218–261, 272
Food Advertising Unit (FAU) 22
Food Standards Agency (FSA) 229
footprints 128
forces 35, 164, 168, 169
forensics 122, 123, 128, 129, 142
formulae 100, 288, 290
formulations 266, 267, 270, 272, 273
freezing points 246
frequency (Hz) 209, 211
fruit 24, 224, 272, 273

FSA *see* Food Standards Agency
fungi 228, 229, 235, 244, 245

gas exchange 26
general practitioners (GP) 63
genetic modification (GM) 242, 243
germination rates 256
gestational diabetes 71
ghost illusions 201
glass 171, 198, 206, 210
GM *see* genetic modification
good laboratory practice 94, 95, 110, 111
GP *see* general practitioners
growth charts 67

haemoglobin 29
hair 123, 139
hard cheeses 254, 255
hard light 188
harvesting 231
hazards (chemicals) 274, 275
head injuries 174
health 20–22, 24, 32, 71, 124, 125
health-and-safety regulations 21, 94, 271, 275
healthcare 52–83
heart 30, 31
heat management 160, 161
hertz (Hz) 209
heterogeneous samples 103, 104
high blood pressure 24, 25, 43
hops 220
hormones 68, 72
hydroxides 276
hypothermia 41

IAAF *see* International Association of Athletics Federations
identity (crime) 122
implantation (embryo) 69, 73
in vitro fertilisation (IVF) 72, 73
incandescent lamps 196
indicator organisms 88
indicators 107, 276
industry 226, 227, 268, 269, 287
infrared (IR) light 197
injury treatment 18, 19
inks 268
inorganic chemicals 268, 276
insoluble salts 278, 279, 296

instrumentation systems 240
insulation 194
International Association of Athletics Federations (IAAF) 39
International Organization for Standardization (ISO) 96, 162
IR *see* infrared light
irritant chemicals 274
IVF *see* in vitro fertilisation

joints 19, 34, 35

kick sampling 113
kidneys 33
Kitemarks 163

labelling samples 103
laboratories 93–96, 110, 111, 127, 249, 270, 271, 286
labour 77
lamps 196
land monitoring 92
lasers 196
law enforcement 128–129
lenses 136, 202, 203, 212
ligaments 34
light-sensitive surfaces 202
lighting 188, 189, 196–199, 204, 205
loudness (sound) 209
loudspeakers 192, 207
luminaires 204, 205
lungs 26–28

magnification 106, 136, 138
malt 220
mass 257, 281
material classes *see* sports equipment
medicine 266, 270
menstrual cycle 68
metals 170, 198, 269, 276, 283
microorganisms 228, 229, 234–238, 244, 245, 252, 253
microphones 192, 207
microscopes 47, 136–139, 221, 244
milk 224, 226, 227, 246, 247
mirrors 200, 201
mould 228, 235
multiple birth 73
muscle 17, 19, 28, 34, 35
mycoprotein fermentation 238

nail polish 265
nanocosmetics 265
National Blood Service 65
National Health Service (NHS) 62–65
negative feedback loops 241
Newtons (N) 35
nurses 62

obesity 24
objective lenses 136
opaque surfaces 198
organic acids 235, 276
outliers 101
ovaries 68
ovulation 69, 72
oxides 276
oxygen 27

pain relief (labour) 77
paint 267, 268
palpated pulse rates 42
paper chromatography 133, 148, 149
parachutes 162
paramedics 64
pathogens 223, 226
performance arts 186–217
performance-enhancing drugs 33
perfume 265
Personal Protective Equipment (PPE) at Work 275
personal trainers 20
pesticides 231
petrochemicals 268
pH levels 88, 89, 114, 276
pharmaceuticals 269
photographs 106
physiotherapist 21
phytoplankton 87, 104
pipettes 111, 291
pitch 209
placenta 70
plants 271
plasma 29
plastics 167–169, 171, 174
platelets 29
plywood 167
poisoning (food) 223, 229, 244
pollution 92, 93, 108
polymers 161, 170, 174, 268
postnatal care 66, 67
PPE *see* Personal Protective Equipment at Work
pre-eclampsia 71
precipitation reactions 279

pregnancy 69–71, 73
preservation 103, 244, 250, 273
pressure 164
primary colours 197
producers (food) 235
proficiency tests 96
protein 238, 242, 243
public analysts 127
pulse rates 31, 32, 42
purification processes 243
purity (chemical) 282
PVC 174
pyrexia (fever) 41

qualitative analyses 107, 134
quantitative analyses 107, 108, 130, 134

random errors 100
range (colourimetry) 131
rates of reaction 283, 284, 297
reacting masses 281
recording of images 202, 203
recording information (microscopy) 137, 139
reference materials 99, 134
reflection 200, 208
refractive indices 206
reinforced concrete 171
relative mass 280
repeatable measurements 98
representative samples 103
reproducible measurements 98
reproductive systems 68
resolving power 138
respiration 26–28, 36, 237, 238
retardation factors (R_f) 134, 135, 149
revolving power 137
RICE injury treatment 18, 19
rigid structures 172, 173
risk assessments 21
room lighting 198, 199
rubber 211
running shoes 23

safety 21, 94, 174, 190, 191, 229, 244, 245, 271, 274, 275
safety curtains 190
safety margins 173
salinity 87
salts 24, 274, 277–279, 290, 291, 294–296
samples/sampling 103, 104, 110, 113, 134, 271
satellite images 106
saturated fat 24

scaling up processes 286–287
scanning electron microscopes (SEMs) 138, 139
Secchi discs 109
secondary colours 197
self-cleaning glass 198
semi-quantitative tests 107, 108
SEMs see scanning electron microscopes
senescence 239
sensitivity (colourimetry) 131
sensors 240
sharpness of focus (images) 106
shivering 36
shoes 23
shutters 202
sketches 105
skin 32, 36, 58, 59
skull (protection) 174
slides 46, 144
smoking 25
soft light 189
soil preparation (bread production) 231
soil tests 89, 98
soluble salts 278, 279, 294, 295
solutes 133, 291
solutions 291, 298
solvents 133
sonography 76
sound 76, 194, 195, 207–211, 213
sowing 231
soy sauce 235
'space' blankets 161
speciality chemicals 269
specimen preparation (microscopy) 137
speed 38
sphygmomanometers 78
sport and fitness 14–51
sports equipment 155–185
staff training (environment) 95
stage lighting 204, 205
standard pressure tests 165
standard procedures 88, 98, 176–181
standard reference materials 99
standards 126, 163
state symbols 289

step tests 45
sterilisation 226
steroids 33
stiffness measurements 176
streak agar plates 252
stressful lifestyles 25
stroke 25
Sudan dye 124, 125
sugar 24
sunlight 196, 198
surface area 283, 284
suspensions 292, 293
sweating 36
sweeteners 273
swimsuits 159
switches 204
symbols 288, 289
sympathetic vibration 211
synovial membrane/fluid 34
synthetic routes 285
systematic error 100
systolic blood pressure 31, 43, 78

tampering (samples) 104
technicians 271
temperature 36, 37, 41, 86, 160, 161, 175, 180, 241, 283, 284
temporary slides 144
TEMs see transmission electron microscopes
tendons 34
tensile strength 177, 181
tension 168
test kits (environment) 89
thin-layer chromatography 123, 133
tinnitus 210
toxic chemicals 274
Trading Standards officers 126
translucent glass 198
transmission electron microscopes (TEMs) 139
transmission (light/sound) 197, 208
transparent glass 198
treating cancer 58, 59
triple-glazing glass 210
turbidity 109, 115, 237
two-way chromatography 134, 135

UKAS see United Kingdom Accreditation Service
ultrahigh-temperature (UHT) sterilisation 226

ultrasound scanning 76
ultraviolet (UV) 197
umbilical cord 70
uncertainty 100, 131
United Kingdom Accreditation Service (UKAS) 96, 97, 127
Universal indicator solution 107
urine 33, 74, 79
uterus 68
UV see ultraviolet

vagina 68
vasoconstriction 37
vasodilation 37
vegetables 24, 224
veins 30
Velcro 139
ventilation 28, 204, 205
ventricles 31
vibration see frequency
video analyses, shoes 23
videos 106
viewfinders 202
viruses 228, 229
visible light 197
visual information 105, 106
vitamins 235
volumes 291

water 88, 89, 92, 104, 108, 220, 273
wet mass measurements 257
wheat 222, 230, 231
white light 197
windows 198
wire loops 251
womb 68
wood 166, 171
word equations 276, 288
world records 38–40
World Wide Fund (WWF) 90, 91
written records 105

X-rays 265

yeast 221, 235, 242, 243, 248, 249
yields 257, 282
yoghurt 227, 234, 235

zinc chelate 286, 287

Appendices

Useful relationships

You will need to be able to carry out calculations using these mathematical relationships:

A1 Sport and fitness

$$\text{body mass index (BMI)} = \frac{\text{body mass (kg)}}{[\text{height (m)}]^2}$$

$$\text{speed} = \frac{\text{distance}}{\text{event time}}$$

turning moment (Nm) = force (N) × distance of the line of action of force from joint (m)

A3 Monitoring and protecting the environment

$$\text{area (m}^2) = \frac{\text{length of}}{\text{longer side (m)}} \times \frac{\text{length of}}{\text{shorter side (m)}}$$

A4 Scientists protecting the public

Microscopes

$$\text{magnifying power} = \frac{\text{magnification}}{\text{(eyepiece)}} \times \frac{\text{magnification}}{\text{(objective)}}$$

Chromatography

$$\text{retardation factor } (R_f) = \frac{\text{distance travelled by substance}}{\text{distance travelled by solvent}}$$

B1 Sports equipment

$$\text{force constant, } k = \frac{\text{force, } F}{\text{extension, } x}$$

$$F = kx$$

$$\text{density (kg/m}^3) = \frac{\text{mass (kg)}}{\text{volume (m}^3)}$$

B3 Agriculture, biotechnology, and food

germination rate

$$= \frac{\text{number of seeds that germinated}}{\text{total number of seeds}} \times 100\%$$

Units that might be used in the Additional Applied Science course

Length: kilometres (km), metres (m), centimetres (cm), millimetres (mm), micrometres (μm)

Height: metres (m), centimetres (cm)

Mass: kilograms (kg), grams (g)

Time: seconds (s)

Temperature: degrees Celsius (°C)

Area: metres^2, centimetres^2, millimetres^2

Volume: millilitres (ml), litres (l), centimetres^3

Density: g/cm^3, kg/m^3

Concentration: g/dm^3, parts per million (ppm)

Speed: m/s, km/s

Force: newtons (N)

Energy: joules (J)

Frequency: hertz (Hz), kilohertz (kHz)

Loudness: decibels (dB)

Blood pressure: millimetres of mercury (mm Hg)

Prefixes for units

micro	milli	kilo	mega
one millionth	one thousandth	× thousand	× million
0.000001	0.001	1000	1000 000
10^{-6}	10^{-3}	× 10^3	× 10^6

Useful information and data

A1 Sport and fitness

Interpreting BMI results

BMI	Condition
Under 20	Underweight
20.0 – 24.9	Advisable range
25.0 – 29.9	Overweight
30.0 – 34.9	Obese
35 and over	Severely obese

A3 Monitoring and protecting the environment

Litmus is an indicator. It is red in acid and blue in alkali.

B1 Sports equipment

The area under a force–extension graph tells you how much energy is stored in the stretched material.

B3 Agriculture, biotechnology, and food

Aerobic respiration

glucose $+$ oxygen $\longrightarrow$ carbon dioxide $+$ water $+$ energy

$C_6H_{12}O_6 + 6O_2 \longrightarrow 6CO_2 + 6H_2O +$ energy

Anaerobic respiration of yeast

glucose $\longrightarrow$ alcohol $+$ carbon dioxide $+$ (less) energy

$C_6H_{12}O_6 \longrightarrow 2C_2H_5OH + 2CO_2 +$ (less) energy

Anaerobic respiration of *Lactobacilli*

glucose $\longrightarrow$ lactic acid $+$ (less) energy

$C_6H_{12}O_6 \longrightarrow 2C_3H_6O_3 +$ (less) energy

Interpreting results of the Resazurin test for milk freshness

Colour of sample after test	Quality
blue (no change)	excellent
blue to mauve	good
mauve to pink	fair
pink to very pale pink	poor
white	bad

B4 Making chemical products

1 ml $= 1$ cm^3 and 1 litre $= 1000$ ml

Reactions of acids

Metal $+$ acid $\longrightarrow$ salt $+$ hydrogen

Metal carbonate $+$ acid $\longrightarrow$ salt $+$ carbon dioxide $+$ water

Metal oxide (or hydroxide) $+$ acid $\longrightarrow$ salt $+$ water

Chemical formulae

Ammonia NH_3

Acids: sulfuric acid H_2SO_4, hydrochloric acid HCl, nitric acid HNO_3

Metals: magnesium Mg, zinc Zn, iron Fe.

Metal oxides: magnesium oxide MgO, zinc oxide ZnO, copper(II) oxide CuO

Metal hydroxides: sodium hydroxide NaOH, potassium hydroxide KOH, magnesium hydroxide $Mg(OH)_2$

Metal carbonates: magnesium carbonate, $MgCO_3$, zinc carbonate $ZnCO_3$, calcium carbonate $CaCO_3$, copper(II) carbonate $CuCO_3$, cobalt phosphate $Co_3(PO_4)_2$

Other salts: iron(II) sulfate $FeSO_4$, magnesium sulfate $MgSO_4$, zinc sulfate $ZnSO_4$, copper(II) sulfate $CuSO_4$, ammonium sulfate $(NH_4)_2SO_4$

OXFORD
UNIVERSITY PRESS

Great Clarendon Street, Oxford OX2 6DP

Oxford University Press is a department of the University of Oxford.
It furthers the University's objective of excellence in research,
scholarship, and education by publishing worldwide in

Oxford   New York

Auckland   Cape Town   Dar es Salaam   Hong Kong   Karachi
Kuala Lumpur   Madrid   Melbourne   Mexico City   Nairobi
New Delhi   Shanghai   Taipei   Toronto

With offices in

Argentina   Austria   Brazil   Chile   Czech Republic   France   Greece
Guatemala   Hungary   Italy   Japan   Poland   Portugal   Singapore
South Korea   Switzerland   Thailand   Turkey   Ukraine   Vietnam

Oxford is a registered trade mark of Oxford University Press
in the UK and in certain other countries.

© University of York and the Nuffield Foundation 2011.

The moral rights of the authors have been asserted.

Database right Oxford University Press (maker).

First published 2011.

British Library Cataloguing in Publication Data.

Data available.

ISBN 978-0-19-913827-2

10 9 8 7 6 5 4 3 2

Printed in China by Print Plus.

Paper used in the production of this book is a natural, recyclable product made
from wood grown in sustainable forests. The manufacturing process conforms to
the environmental regulations of the country of origin.

Acknowledgements
The publisher and authors would like to thank the following for their permission
to reproduce photos:
P14: Amriphoto/Istockphoto; **P16l:** Brian Bell/Science Photo Library; **P18br:** Vm/
Istockphoto; **P23tl:** Beyza Sultan Durna/Istockphoto; **P23tm:** Webphotographeer/
Istockphoto; **P23tr:** Olivier Blondeau/Istockphoto; **P24:** Empics; **P29:** Dr. Gopal
Murti/Science Photo Library; **P33:** Popperfoto/Getty Images; **P38:** Bill Kostroun/
AP Photo; **P39t:** Michael Steele/Staff/Getty Images; **P39bl:** Aaintho/Istockphoto;
P39br: Staff Sgt. Stephen Otero/DOD Media; **P44:** Mauro Fermariello/Science
Photo Library; **P49:** Pictor/Imagestate/Alamy; **P52:** 1joe/Istockphoto; **P55:**
Lester Lefkowitz/Corbis; **P56t:** Sarah Codrington/Nuffield Foundation; **P56b:** ER
Productions/Corbis; **P56-57:** Yarinca/Istockphoto; **P58t:** Carolina K. Smith,M.D./
Istockphoto; **P58b:** AJ Photo/Science Photo Library; **P59l:** Mark Thomas/
Science Photo Library; **P59r:** Life In View/Science Photo Library; **P61l:** Science
Photo Library; **P61r:** Antonia Reeve/Science Photo Library; **P60-61:** Don Bayley/
Istockphoto; **P63b:** Adam Hart-Davis/Science Photo Library; **P64:** Josh Sher/
Science Photo Library; **P65:** Family Planning Association; **P66t:** Don Bayley/
Istockphoto; **P66bl:** StockLite/Shutterstock; **P66br:** Yarinca/Istockphoto; **P71:**
Jose Luis Pelaez Inc/Blend Images/Getty Images; **P73t:** Dem10/Istockphoto; **P73b:**
Donna Coleman/Istockphoto; **P74t:** David Gold/Istockphoto; **P74bl:** Li Wa/
Shutterstock; **P74br:** Peter Raneri/Istockphoto; **P75:** FotoYakov/Shutterstock;
P76l: Michael Donne/Science Photo Library; **P76r:** OUP to provide; **P78t:**
Sven Hoppe/Shutterstock; **P78b:** Ian Hooton/Science Photo Library; **P79:** Dr P.
Marazzi/Science Photo Library; **P84:** Jim Edds/Science Photo Library; **P86t:** David
Buckland/Cape Farewell; **P86b:** John Shaw/Science Photo Library; **P87t:** Cape
Farewell; **P88t:** Simon Fraser/Science Photo Library; **P88b:** Paul Rapson/Science
Photo Library; **P89t:** Stella Stella/Photolibrary; **P90:** Environment Agency Wales;
P91: Environment Agency Wales; **P93:** Pullman; **P94:** Ria Novosti/Science Photo
Library; **P96t:** Bill Van Aken/Csiro/Science Photo Library; **P96b:** Keith/Custom

Medical Stock Photo/Science Photo Library; **P97l:** Aquaculture Pictures Library;
P97r: Geoff Kidd/Science Photo Library; **P98:** Health Protection Agency/
Science Photo Library; **P99:** Andrew Brookes/Corbis; **P103** Zooid Pictures;
P104t: Cape Farewell; **P108l:** Martyn F. Chillmaid/Science Photo Library;
P108br: Paul Rapson/Science Photo Library; **P109:** Doug Steley A/Alamy;
P113: MShieldsPhotos/Alamy; **P120:** James King-Holmes/Science Photo Library;
P123: Tek Image/Science Photo Library; **P124:** Burcin Tuncer/Istockphoto;
P125: Flashgun/Istockphoto; **P126:** Peter Titmuss/Alamy; **P127:** Photo
courtesy of LGC; **P129:** Peter Menzel/Science Photo Library; **P130:** Kryczka/
Istockphoto; **P131:** Martyn F. Chillmaid/Science Photo Library; **P132:** Analtech
Inc.; **P133b:** Analtech Inc.; **P137bl:** Power and Syred/Science Photo Library;
P137br: Pasieka/Science Photo Library; **P138:** Dr Jurgen Scriba/Science Photo
Library; **P139t:** Dee Berger/Science Photo Library; **P139b:** Steve Gschmeissner/
Science Photo Library; **P140:** J.C. Revy/Science Photo Library; **P142l:** Patrick
Landmann/Science Photo Library; **P142r:** Philippe Psaila/Science Photo Library;
P143t: Mauro Fermariello/Science Photo Library; **P143b:** TEK Image/Science
Photo Library; **P151:** Dorling_Kindersley/Istockphoto; **P154:** Oliver Furrer/
Getty Images; **P156t:** David Brodie; **P156-157:** David Cannon/Getty Images
Sport/Getty Images; **P158t:** Hannah Johnston/Getty Images Sport/Getty Images;
P158b: David Cannon/Getty Images Sport/Getty Images; **P159:** Al Bello/Getty
Images Sport/Getty Images; **P158-159:** Shane White/Shutterstock; **P160t:**
Lance Bellers/Istockphoto; **P160b:** Shariff Che\'Lah/Istockphoto; **P161l:** Scott
Cramer/Istockphoto; **P161r:** NASA; **P160-161:** Patrik Mezirka/Shutterstock;
P162: Oliver Furrer/Getty Images; **P164t:** VisualField/Istockphoto; **P164b:**
Sergiy Zavgorodny/Istockphoto; **P166bl:** Lise Gagne/shutterstock; **P166br:**
AdShooter/Istockphoto; **P167t:** Michael Flippo/Fotolia; **P167m:** Vladimir
Kondrachov/Istockphoto; **P167b:** Darren K. Fisher/Shutterstock; **P170t:** Scott
Hailstone/Istockphoto; **P170b:** Ronen/Istockphoto; **P171t:** Julio Etchart/Alamy;
P171b: Shane White/Shutterstock; **P172l:** Lester V. Bergman/Corbis; **P172m:**
Pascal Goetgheluck/Science Photo Library; **P172r:** Dennis Kunkel/Photolibrary;
P173t: Nik0s/Shutterstock; **P173m:** Renkshot/Shutterstock; **P173b:** Robert
Hallam/Rex Features; **P174:** Sovereign,ISM/Science Photo Library; **P175t:**
Patrik Mezirka/Shutterstock; **P175b:** Dorling Kindersley/the Agency Collection/
Getty Images; **P186:** Rex Features; **P188-189:** Imagebroker/Alamy; **P190t:**
Illustrated London News Ltd/Mary Evans Picture Library; **P190b:** Rich Legg/
Istockphoto; **P191l:** MG/Istockphoto; **P191bl:** Code6d/Istockphoto; **P191br:**
Rido/Shutterstock; **P190-191:** Christopher Futcher/Istockphoto; **P192t:**
Imagebroker/Alamy; **P192bl:** Vladimir Vladimirov/Istockphoto; **P192br:**
Luminis/Alamy; **P193t:** Roberto A. Sanchez/Istockphoto; **P193bl:** Denis
Vrublevski/shutterstock; **P193br:** Naphtalina/Istockphoto; **P194t:** Page One;
P194b: Photofusion Picture Library; **P196t:** Xyno6/Istockphoto; **P196m:** Igor
Smichkov/Istock; **P196b:** Stan Conti/Istockphoto; **P197:** Kyoshino/Istockphoto;
P198b: Pilkington PLC; **P200:** Emre ogan/Istockphoto; **P202:** Olivier Blondeau/
Istockphoto; **P204:** Imagebroker/Alamy; **P206:** Leslie Garland Picture Library/
Alamy; **P207t:** Mike Bentley/Istockphoto; **P207b:** Don Nichols/Istockphoto;
P210: Rex Features; **P218:** Herve Conge, ISM/Science Photo Library; **P220t:**
Ted Spiegel/Corbis; **P220b:** Holt Studios International; **P221:** Manfred Kage/
Science Photo Library; **P222b:** Bruce Peebles/Corbis; **P223l:** David Munns/
Science Photo Library; **P223r:** B. Boissonnet/Science Photo Library; **P222-
223:** Brand X Pictures/Getty Images; **P224t:** Brand X Pictures/Getty Images;
P224m: Vicki Reid/Istockphoto; **P224b:** Wayne Hutchinson/ AgstockUSA/
Science Photo Library; **P225tl-bl:** Uzi Tzur/Istockphoto, BA LaRue/Alamy,
Charlie Riedel/AP Photo, Donald Weber/VII Network/Corbis, Susie Slatter/
Alamy; **P225tm-bm:** David Aubrey/Science Photo Library, Justin Kase z03z/
Alamy, P. Magielsen/Corbis, Maximilian Stock Ltd/Spl/Photolibrary, Lee Beel/
Alamy; **P225r:** RichardBakerWork/Alamy; **P226r:** Nigel Cattlin/Alamy; **P227:**
Rosenfeld Images Ltd/Science Photo Library; **P228:** Morgan Lane Studios/
Istockphoto; **P229t:** David Scharf/Science Photo Library; **P229b:** Shout/Alamy;
P230: Eyewave/Istockphoto; **P231tl:** Jeremy Walker/Science Photo Library;
P231ml: Jack K. Clark/AgstockUSA/Science Photo Library; **P231bl:** Jack K.
Clark/AgstockUSA/Science Photo Library; **P231tr:** Larry Fleming/ AgstockUSA/
Science Photo Library; **P231br:** Jeremy Walker/Science Photo Library; **P232t:**
David Marsden/Photolibrary; **P232m:** David Aubrey/Science Photo Library;
P232b: Wayne Hutchinson/Alamy; **P233t:** Wayne Hutchinson/Science Photo
Library; **P233b:** Wayne Hutchinson/Alamy; **P234:** Guojieyi/Istockphoto; **P235t:**
Sidney Moulds/Science Photo Library; **P235b:** Martyn F. Chillmaid/Science
Photo Library; **P237b:** Cordelia Molloy/Science Photo Library; **P238:** St. Austell
Brewery Company Limited; **P240t:** Tyler Miles/Istockphoto; **P240bl:** Dimple
to provide; **P240tr:** Tjanze/Istockphoto; **P244:** Dr. Gary Gaugler/Science Photo
Library; **P245:** Photofusion Picture Library; **P262:** Val Thoermer /Shutterstock;
P264: Picture Contact BV/Alamy; **P265:** Gustoimages/Science Photo Library;
P266l: Richard Philpott/Zooid Pictures; **P266r:** Rosal/Science Photo Library;
P267: R. Maisonneuve, Publiphoto Diffusion/Science Photo Library; **P266-
267:** Sean Locke/Istockphoto; **P268t:** William Taufic/Corbis; **P268b:** Geoff
Tompkinson/Science Photo Library; **P271t:** Andrew Brookes/Science Photo
Library; **P271m:** ED Young/AgstockUSA/Science
Photo Library; **P271b:** Philippe Benoist/Look At Sciences/Science Photo Library;
P272: Danisco; **P274:** Andrew Lambert Photography/Science Photo Library;
P275: Tek Image/Science Photo Library; **P276:** Andrew Lambert Photography/
Science Photo Library; **P277:** Zooid Pictures; **P279:** Lawrence Migdale/Science
Photo Library; **P283:** Crown Copyright/Health & Safety Laboratory/Science
Photo Library; **P292:** Zooid Pictures; **P293:** Martyn F. Chillmaid/Science
Photo Library.

315

Illustrations by IFA Design, Plymouth, UK, Clive Goodyer, and Q2A Media.

The publisher and authors are grateful for permission to reprint the following text and any artwork we have redrawn:

Although we have made every effort to trace and contact all copyright holders before publication this has not been possible in all cases. If notified, the publisher will rectify any errors or omissions at the earliest opportunity.

Project Team acknowledgements
These resources have been developed to support teacher and students undertaking the OCR specification GCSE Additional Applied Science. They have been developed from the 2006 edition of the Twenty First Century Science GCSE Additional Applied Science resources.
We would like to thank Anne Wolstenholme, Byron Dawson, and the examining team at OCR, who produced the specification for GCSE Additional Applied Science.

Authors and editors of the first edition
We thank the authors and editors of the first edition, Michael Brimicombe, David Brodie, Peter Campbell, Ken Gadd, Anna Grayson, Ginny Hales, Ruth Holmes, Andrew Hunt, Merryn Kent, Mike Kent, David Sang, Caroline Shearer, and Mike Tingle.
Many people from schools, colleges, universities, industry, and the professions contributed to the production of the first edition of these resources. We also acknowledge the invaluable contribution of the teachers and students in the pilot centres.
The first edition of Twenty First Century Science was developed with support from the Nuffield Foundation, The Salters' Institute, and the Wellcome Trust. A full list of contributors can be found in the Teacher and Technician Resources.

The continued development of *Twenty First Century Science* is made possible by generous support from:
• The Nuffield Foundation
• The Salters' Institute